ABOMINABLE SCIENCE!

DANIEL LOXTON AND DONALD R. PROTHERO

ABOMINABLE SCIENCE!

ORIGINS OF THE YETI, NESSIE, AND OTHER FAMOUS CRYPTIDS

COLUMBIA UNIVERSITY PRESS  NEW YORK

COLUMBIA UNIVERSITY PRESS
Publishers Since 1893
New York Chichester, West Sussex
cup.columbia.edu

Library of Congress Cataloging-in-Publication Data

Loxton, Daniel, 1975–
Abominable science! : origins of the Yeti, Nessie, and other famous cryptids /
Daniel Loxton and Donald R. Prothero.
pages cm
Includes bibliographical references and index.
ISBN 978-0-231-15320-1 (cloth : alk. paper)
ISBN 978-0-231-52681-4 (e-book)
1. Cryptozoology. 2. Animals, Mythical. 3. Pseudoscience. I. Prothero, Donald R. II. Title.

QL88.3.L69 2013
001.944—dc23

2013008424

Columbia University Press books are printed on permanent and durable acid-free paper.
Printed in the United States of America

C 10 9 8 7 6 5 4 3 2

COVER DESIGN: PHILIP PASCUZZO (PEPCOSTUDIO.COM)
COVER IMAGE: DANIEL LOXTON
BOOK DESIGN: VIN DANG

The foreword appeared in a modified form in *Scientific American*, and other material appeared in a modified form in *Skeptic* magazine and on Skepticblog.org.

THIS BOOK IS DEDICATED TO

Ronald Binns, David Daegling, Benjamin Radford, Joe Nickell, Blake Smith, Karen Stollznow, Darren Naish, Sharon Hill, Matt Crowley, and all the brave skeptics who have blazed the trail of scientific inquiry in cryptozoology despite the tremendous cultural pressures to follow the crowd and give in to easy answers

CONTENTS

5

6

7

FOREWORD
SHOW ME THE BODY

IN JANUARY 2003, the world lost the creators of two of its most celebrated bio-hoaxes in modern times: Douglas Herrick, father of the risibly ridiculous jackalope (half jack rabbit, half antelope), and Raymond L. Wallace, godfather of the less absurd and more widely believed Bigfoot. The jackalope enjoins laughter in response to such peripheral hokum as hunting licenses sold only to those whose IQs range between 50 and 72, bottles of the rare but rich jackalope milk, and additional evolutionary hybrids like the jackapanda. Bigfoot, though, while occasionally eliciting an acerbic snicker, enjoys greater plausibility for a simple evolutionary reason: large, hirsute apes presently roam the forests of Africa, and at least one species of a giant ape—*Gigantopithecus*—flourished several hundred thousand years ago alongside our early human ancestors. Footprints in the mud really did mean that another bipedal primate was lurking about.

Is it possible that a real Bigfoot lives despite the confession by Wallace's family, after his death, that the tracks found by one of his employees were just a practical joke by a fun-loving prankster—a guy in an ape suit? Certainly. After all, although proponents of Bigfoot do not dispute the evidence that Wallace tromped around in strap-on Shaq-o-size wooden feet, they correctly note that tales of the giant Yeti living in the Himalayas and Native American lore about Sasquatch wandering around the Pacific Northwest emerged long before Wallace pulled his prank in 1958.

Throughout much of the twentieth century, it was entirely reasonable to speculate about and search for Bigfoot, as it was to look for the monsters of Loch Ness and other lakes and to investigate the visits to Earth by extraterrestrials. Science traffics in the soluble, so while jackalopes do not warrant our limited exploratory resources, for a time these other creatures did.

The study of animals whose existence has yet to be proved is known as "cryptozoology," a term coined in the late 1950s by the Belgian zoologist Bernard Heuvelmans. Cryptids, or hidden animals, begin life as muddy footprints, blurry photographs, grainy videos, and anecdotes about strange things that go bump in the night. Cryptids come in many forms, including the aforementioned giant pongid and lake monsters, as well as sea serpents, giant octopi, snakes, birds, and even living dinosaurs (the most famous being Mokele Mbembe, purportedly slogging through the rivers and lakes of the Congo Basin in Central and West Africa).

The reason that cryptids merit our attention is that there have been enough discoveries of previously unknown animals by scientists based on local anecdotes and folklore that we cannot dismiss all claims a priori. The most famous examples include the gorilla in 1847 (and the mountain gorilla in 1902), the giant panda in 1869, the okapi (a short-necked relative of the giraffe) in 1901, the Komodo dragon in 1912, the bonobo (or pygmy chimpanzee) in 1929, the megamouth shark in 1976, the giant gecko in 1984, the beaked whale in 1991, and the spindlehorn ox from Vietnam in 1992. Cryptozoologists are especially proud of the catch in 1938 of a coelacanth, an archaic-looking fish believed by zoologists to have gone extinct in the Cretaceous period, as if to say "See, Bigfoot really is out there, and we just have to keep looking."

Although discoveries of new species of bugs and bacteria are routinely published in the annals of biology, the gorilla, beaked whale, and other examples are startling because of their recentness, size, and commonality to the famous Bigfoot, Nessie, Mokele Mbembe, and other cryptids. But all of them have one thing in common—a body! In order to name a new species, taxonomists must have a type specimen—a holotype—from which a detailed description can be made, photographs taken, models cast, and a professional scientific analysis published.

Anecdotes are a good place to begin an investigation, but anecdotes by themselves do not constitute a new species. In fact, in the words of the social scientist Frank J. Sulloway—words that should be elevated to a maxim: "Anecdotes do not make a science. Ten anecdotes are no better than one, and a hundred anecdotes are no better than ten."

I employ Sulloway's maxim every time I encounter Bigfoot hunters, Loch Ness seekers, or alien abductees. Their anecdotal tales make for gripping narratives, but they do not make for sound science. After a century of searching for these chimerical creatures, until a body is produced skepticism is the appropriate response. So whenever someone regales you with such stories, I recommend the following rejoinder: "That's nice. Show me the body."

In *Abominable Science! Origins of the Yeti, Nessie, and Other Famous Cryptids*, two of today's leading skeptical thinkers have teamed up to produce a marvelous history and science of all matters cryptozoological. Daniel Loxton—editor, writer, and illustrator of *Junior Skeptic*; regular contributor of investigative pieces to *Skeptic* and other skeptical publications; and author of books and articles on many of the elusive creatures that fascinated our ancestors and compel our attention—offers new insights that I have not encountered before in the skeptical literature. Donald Prothero—trained in paleontology, biology, and geology, and writer of crystal-clear scientific prose readable by anyone with just a modicum of curiosity—brings to the field fresh ideas on how to think about these storied creatures, which may or may not exist, and illuminates how scientists think about such matters.

Together, Loxton and Prothero have written what may well be the most important work to date on cryptozoology, taking its rightful place in the annals of skeptical literature in particular and scientific literature in general. *Abominable Science!* is the defining work on cryptozoology of our generation.

MICHAEL SHERMER

PREFACE

THIS BOOK EMERGED from our very different but complementary experiences and backgrounds in cryptozoology. For that reason, we chose to write each chapter separately and maintain our individual voices, describing our personal experiences in our respective chapters.

Daniel Loxton (chapters 2, 4, 5, and 7) brings with him a lifelong love of monster mysteries and a sense of fellowship with all those who seek to solve them. He spent his childhood obsessed with cryptids—engrossed in the school library, planning his future monster-hunting expeditions, or passionately arguing with the other kids on the playground about the Sasquatch or the Loch Ness monster. When he discovered the skeptical literature while in high school and university, Loxton learned about the other, lesser-known half of the cryptozoological story, but never lost the sense of curiosity and adventure that brought him to the topic in the first place. Today, he writes as a "professional skeptic" with a special interest in cryptozoology. In that role, he secretly hopes (or at least wishes) that some of those he critiques may turn out to be right after all.

Donald Prothero (chapters 1, 3, 6, and 7 [all with contributions by Loxton]) approached this topic in a different way. He was one of those kids who got hooked on dinosaurs at age four, but never wavered in his determination to become a paleontologist. During his thirty years as a professional paleontologist and geologist, Prothero had to meet the hard-nosed requirements of the scientific community, according to which every statement has to meet the most rigorous standards of scientific scrutiny. In particular, the problems with creationism and its effects on science and education have been a major concern during his career. In his undergraduate education, Prothero learned the methods of field biology and ecology, so he is acquainted with the more rigorous approaches to understanding animals in their environment and the techniques and assumptions that modern field zoologists employ. Prothero's background as a geologist and paleontologist in particular gives him a broader understanding of what the fossil record says and does not say, as well as how it counters the arguments of cryptozoology.

We did not attempt to make the book an encyclopedia of every known cryptid, since that would make its length prohibitive. Instead, we focused on a handful of the best-known and most-familiar examples because they require the most detailed treatment in understanding their history and debunking their myths. Most of the caveats and issues that apply to these creatures also apply to all cryptids, so discussing them all would be largely redundant.

Thus our approaches to the topic are very different, but both of us are guided by the rules of naturalistic scientific inquiry and a commitment to critical thinking and skepticism. As we point out again and again, such an approach is sorely lacking in most cryptozoological research, yet it must be applied to cryptids, as it is to the rest of science, if cryptozoologists wish to be taken seriously as scientists.

WE WOULD LIKE TO THANK the entire team at Columbia University Press who partnered with us on *Abominable Science!*: Patrick Fitzgerald, science publisher; Irene Pavitt, senior manuscript editor; Bridget Flannery-McCoy, associate editor; Jennifer Jerome, director of production and design; Vin Dang, designer; Derek Warker, publicist; and Brad Hebel, director of sales and marketing. Our deep appreciation goes to all these professionals for helping us realize our dreams for this book.

We thank Benjamin Radford, Blake Smith, Adrienne Mayor, Darren Naish, Karen Stollznow, and Sharon Hill for their reviews of the manuscript and their many helpful corrections and suggestions. Ben Radford's help in securing some of the images is much appreciated.

We thank our wives, Cheryl Hebert and Teresa LeVelle, for their love and unwavering support during the long writing process. And we thank our children (including Erik, Zachary, and Gabriel Prothero). We hope that they will inherit a world somewhat less ignorant and anti-scientific than the one in which we grew up.

Also deserving of special mention are Michael Shermer and Pat Linse of the educational, nonprofit Skeptics Society. We thank them for their guidance, encouragement, and open-minded devotion to the spirit of scientific inquiry. A tip of the hat as well to *Skeptic*'s creative team, including Webmaster William Bull and illustrator Jim W. W. Smith. (Some of Jim's work appears in this book, to its very great benefit.)

Daniel Loxton wishes as well to thank his cryptozoological counterparts John Kirk, Loren Coleman, Paul LeBlond, and Ed Bousfield for their collegiality and for their generosity in sharing sources and information.

Many people provided eyestrain and patience in hunting down key sources for this research. Special mention must be made of James Loxton's many days of scrolling through microfilm, as well as Jason Loxton, Patrick Fisher, Stefan Bourrier, Jennifer Griffith, Greg Carr, and Colin Walsh for contributions to our research. Hans-Dieter Sues, Doug Henning, Kristjan Wager, and Matthew Kowalyk are warmly thanked for important translations. Barbara Drescher is thanked for her expert assistance on the topic of the psychology of paranormal belief. Our thanks as well to Tony Harmsworth, Charles Paxton, Peter Gillman, Michael Fredericks, Donald Glut, David Goldman, Janet Bord, Max Crowther, Stephen Cosgrove, and many others for providing or helping us to secure important images and resources for our research or for reproduction or both.

Thanks are due to Lesley Kennes and Gavin Hanke at the Royal British Columbia Museum (and Jim Cosgrove, now retired) for fielding questions and providing valuable access. The RBCM has stood in the middle of cryptid central for decades and has weathered it with good humor and just the right attitude of accessibility, outreach, and public service. Similarly, we thank Edwina Burridge, Norman Newton, and Susan Skelton of the Inverness Reference Library; Caroline Cameron, Catriona Parsons, and the Gaelic Council of Nova Scotia; and Leslie K. Overstreet and Kirsten van der Veen at the Smithsonian Institution Libraries.

Thank you, everyone.

ABOMINABLE SCIENCE!

CRYPTOZOOLOGY

REAL SCIENCE OR PSEUDOSCIENCE?

EVER SINCE WE WERE YOUNGSTERS, we have been enthralled with ideas about monsters and magnificent creatures with mythic and ancient roots. Indeed, we have never really gotten over being smitten with magical beasts. That is why one of us became a writer and an illustrator of books for young readers about creatures great and small, and the other became a paleontologist who studies the history of life on Earth as revealed through fossils.

To understand the field of cryptozoology, it is necessary to appreciate the core concepts and procedures of science. We begin our investigation with a story about the sighting of one of the most famous cryptids: Bigfoot.

THE GEORGIA "BIGFOOT"

On July 9, 2008, among the usual stories of political news and celebrity scandals, the Internet and televised media were buzzing with a report of two men who supposedly had found a body of Bigfoot in the woods of northern Georgia. In this age of electronic media, the discoverers' account was posted on YouTube before it was even covered by the conventional televised news or newspapers. The images were blurry and difficult to decipher, but the insatiable twenty-four-hour news and Internet cycle demands filler with some sort of content, no matter how suspicious. Major mainstream media—including the BBC, CNN, ABC News, and Fox News—gave the story considerable coverage.[1] Even if some news announcers read this, like other Bigfoot stories, with a tone of incredulity or mocking sarcasm, the media nonetheless reported the Georgia Bigfoot story—and provided many of the Bigfoot "experts" with an opportunity to claim their Warholian "fifteen minutes of fame" by testifying to the existence of the creature.

The two discoverers of the Georgia Bigfoot, Rick Dyer and Matthew Whitton, were soon celebrities, interviewed repeatedly on television and for stories posted on the Internet.[2] Searching for Bigfoot, Inc., a cryptozoological organization dedicated to proving that Bigfoot is real, paid them $50,000 for their evidence. The head of the organization, Tom Biscardi, a long-controversial Bigfoot "hunter,"[3] examined their find and endorsed it.[4] Biscardi has his own radio show, which promotes paranormal ideas, and he interviewed Dyer and Whitton on the show. Finally, on August 15, Biscardi, Dyer, and Whitton held a press conference[5] and displayed the body of Bigfoot for examination, frozen on blocks of ice.[6] When the specimen was thawed, it turned out that the corpse was not real, but a rubber Sasquatch Halloween costume with fake hair, padded and propped up to look like a Bigfoot. Dyer and Whitton soon admitted that their discovery had been a hoax all along, telling Atlanta's WSB-TV that they had "bought a costume off the Internet and filled it with possum roadkill and slaughterhouse leftovers."[7]

Once the deception was revealed, the story died in the mainstream media, but members of the community of Bigfoot "researchers" were at one another's throats over the debacle. Loren Coleman, a leading cryptozoologist, argued that Biscardi had been either in on the hoax or so anxious to find proof of his beliefs that he ignored the obvious evidence even when it was in front of him. "He's a huckster, a circus ringmaster," Coleman said.[8] "It's all about money with him. It probably didn't matter to him whether it was real or not." Why would Dyer, a former security guard, and Whitton, a Clayton County police officer, pull the stunt in the first place? "They probably started out small, as a way to promote their Bigfoot tracking business, and got in way over their heads," Coleman said. "These are not very intelligent individuals." Of Biscardi and the "discoverers," Coleman commented, "In a way, both sides may have been trying to out-con each other." The Bigfoot Field Research Organization called for the arrest of the three principals in the fiasco. Whitton was fired from the police force for his part in the stunt.[9]

Most Bigfoot believers were angry that their work had been besmirched by a silly hoax, yet they continue to promote equally unlikely claims on a regular basis. Their gullibility should not be surprising: cryptozoology is built on openness to first-person testimony. More disturbing is the attitude of the mainstream media, which has long tended to approach "silly season" paranormal stories and monster yarns with a looser standard than that applied to other news items. That unfortunate double standard seems to be worsening, rather than improving. Would the Georgia Bigfoot—a story of questionable origin, with no independent sources to back it up—have made the evening news a generation ago, when there were only three networks and limited programming on television? Perhaps not. Those who remember the Watergate scandal and the book *All the President's Men* will recall that Bob Woodward and Carl Bernstein had to track down multiple reliable sources before publishing their allegations about the misdeeds in the White House. Today, the situation has changed: the modern television and Internet news are so hungry for anything to fill their airtime and Web space that they are much more likely to accept and report dubious stories from questionable sources without adequate (or in some cases, any) journalistic due diligence. With twenty-four-hour cable television, such a story may run many times before anyone has a chance to check it. Even more to blame is the huge number of "documentaries" of pseudoscience on cable television, including those on stations that used to broadcast actual science documentaries. These stations promote pseudoscientific ideas just to draw audiences, earn high ratings, and justify their existence.

WHAT SCIENCE IS—AND IS NOT

Lost in the media firestorm generated by the Georgia Bigfoot stunt and in the huffing and puffing by the various pro-Bigfoot organizations was any scientific, critical perspective on the entire process. Almost all the "experts" who were interviewed about the hoax were committed believers in the reality of Bigfoot; their claims to "expertise" were that they had a Web site, wrote a book, or headed an organization dedicated to finding Bigfoot. Almost no actual biologists or other real scientists with relevant training were interviewed, and the story died so quickly that there was little commentary on it among scientists, even in the blogosphere. About the only exception was Thomas Nelson, a wildlife biologist and ecologist at North Georgia College and State University. With twenty years of experience in the area, he knows the wildlife of the "Bigfoot habitat" better than most. As he told local reporters, "Science is always open to new discoveries; however, the chances that a Bigfoot species exists anywhere in the world is highly unlikely. If Bigfoot inhabited North Georgia, people would have known about them long ago. Although the southern Appalachians might seem remote to some people, there are thousands of deer hunters, hikers, mushroom hunters, and campers using these forests throughout the year."[10] Nelson pointed out that there are legends of giants and monsters worldwide, but few of these stories have proved to be real. In particular, he pointed out that there are many possible sources of Bigfoot myths, including bears and vivid imaginations. "Or perhaps," he said, "we just like to think that there are wild things in our world that are bigger and wilder than us. There is an upside to the media coverage, even if it did turn out to be a hoax. The coverage provided a 'teachable moment' to demonstrate to people how science works. We are open to new ideas, but skeptical until there is good supporting evidence—even of Bigfoot."

Nelson raised an important point about the "teachable moment": despite the gullibility of the media and the public, the Georgia Bigfoot hoax helps us think clearly about the scientific method. Science is about *testing hypotheses,*

Figure 1.1 A native black swan in Perth, Australia. (Photograph courtesy of Kylie Sturgess)

or offering ideas that may explain some facet of nature and seeing if they hold up to critical scrutiny. As philosopher Karl Popper pointed out, science is not about proving things true, but proving them false. You can make the general inductive statement "All swans are white," but all the white swans in the world do not prove that the statement is true. One black swan easily falsifies the claim (figure 1.1). Scientists are open to any and every idea that can be proposed, no matter how crazy it may sound—but ultimately, for the idea to be accepted within the scientific community, it has to withstand the critical evaluation and peer-review process of the legions of scientists who are dedicated to shooting it down. Scientific hypotheses must always be tentative and subject to examination and modification, and they never reach the status of "final truth."[11]

Scientists are not inherently negative sourpusses who want to rain on everyone else's parade. They are just cautious and skeptical about any idea that is proposed until it has survived the process of repeated testing and possible falsification and has come to be established or acceptable. They have good reason to be skeptical. Humans are capable of making all sorts of mistakes, entertaining false ideas, and practicing self-deception. Scientists cannot afford to blindly accept an unsubstantiated claim by one person or even a group of people. They are obligated as scientists to criticize and carefully evaluate and test it before it is acceptable as a scientific idea.

This cautious approach is necessary in part because scientists are human and thus subject to the same foibles that all mortals are. They love to see their ideas confirmed and to believe that they are right. Yet in all sorts of ways, scientists can misinterpret or over-interpret data to fit their biases. As the Nobel Prize–winning physicist Richard Feynman put it, "The first principle is that you must not fool yourself—and you are the easiest person to fool." That is why many scientific experiments are run by the double-blind method: not only do the subjects of the experiments not know what is in sample A or sample B, but neither do the investigators. They arrange to have the samples coded, and only after the experiment is run do they open the key to the code and learn whether the results agree with their expectations.

So if scientists are human and can make mistakes, why does science work so well? The answer is testability and peer review. Individuals may be blinded by their own biases, but once they present their ideas in a lecture or publication, their work is subject to intense scrutiny by the scientific community. If the results cannot be replicated by another group of scientists, then they have failed the test. As Feynman put it, "It doesn't matter how beautiful your theory is, it doesn't matter how smart you are. If it doesn't agree with experiment, it's wrong."

This happens routinely in scientific research, although most people never hear about it because it is so common that it never catches the attention of the media. The media love to report on flashy ideas that have just been proposed, such as a meteorite impact causing a particular mass extinction, but soon move on to other glamorous topics and never report the debunking of the initial assertion a year or two later. The infamous case of "cold fusion" claimed by Stanley Pons and Martin Fleischmann in 1989 was a rare exception, since it was so astonishing that laboratories all over the world dropped everything and tried to replicate their results. The news media saw the story go from flashy proposal to discredited hypothesis in a matter of weeks. But whether it happens in weeks or years, scientists eventually test important ideas. Over enough time, most of the good ideas in science have been thoroughly vetted by this process and have passed from hypothesis or speculation to a well-established reality. As the late great astronomer Carl Sagan said in his famous television show *Cosmos*, "There are many hypotheses in science which are wrong. That's perfectly all right; they're the aperture to finding out what's right. Science is a self-correcting process. To be accepted, new ideas must survive the most rigorous standards of evidence and scrutiny."

SCIENCE, BELIEF SYSTEMS, AND PSEUDOSCIENCE

The scientific method is often in direct conflict with many notions and beliefs that humans use to make decisions and guide their lives. Sometimes, a belief not only contradicts the discoveries of science and even common sense, but can harm the believers, as when members of snake-handling sects incorporate rattlesnakes and copperheads into their rituals in the belief that God will protect them—but nearly every member of such groups is bitten sooner or later, and more than seventy people have died over the past eighty years. Adherents of some religions and even the recent rash of vaccination deniers refuse to accept the advantages of modern medicine, even though the evidence is overwhelming that vaccines are a safe and effective means to protect themselves and their children from the risk of serious illness (or even death). As Sagan put it, "If you want to save your child from polio, you can pray or you can inoculate."[12]

Many nonscientific belief systems try to sound "scientific," even though their ideas and methods do not correspond to those of science. They try to appropriate the respectability of science—science has done so much for our civilization and is held in high regard by many people—without actually doing science. They include such decidedly nonscientific belief systems as Christian Science (which actually rejects modern medicine) and Scientology, as well as the attempts by Christian fundamentalists to frame their biblical ideas as scientific creationism or creation science. But if their ideas were examined by strict scientific methods, they would fail the test; because their tenets cannot be falsified in the minds of their followers, they have nothing to do with real science.

The creationists, in particular, are tireless in their attempts at subterfuge. They first went by the relatively honest label "biblical creationism" until the federal courts threw out their efforts to include creationism in public-school science curricula because it would be a violation of the First Amendment principle of separation of church and state. So in the late 1960s, they simply renamed themselves scientific creationists and deleted the overt references to God in their textbooks. But it is clear from their documents that scientific creationism is simply old-fashioned biblical creationism without the obvious religious terminology. Even more revealing, creationists in many organizations must swear to an oath of literal acceptance of the Bible, and any scientific ideas they propose must be bent to conform to their version of Genesis. This is hardly the practice of real scientists, whose conclusions must be tentative and always subject to rejection and falsification. A series of federal court decisions saw through this creationist sham, so since the 1980s the creationists have tried to sneak their religious ideas into public schools by increasingly subtle methods. For about a decade, they used the ruse of intelligent design until they were shot down in December 2005 by a court decision in Dover, Pennsylvania. Now intelligent design is on the decline, no longer pushed on school districts. Currently, the creationists are using even sneakier tactics: "teach the controversy," "teach the strengths and weaknesses of evolution," and the like. But they always leave a paper trail. If you track down who sponsored the idea and what they have revealed about their religious motivations, it always boils down to yet another attempt to conceal their religious motivation. Once it reaches the court system, this newest attempt to insinuate an anti-evolution agenda into public schools will be rejected.

"BALONEY DETECTION" AND PSEUDOSCIENCE

The practice of making claims that appear to be scientific, but do not actually follow the scientific method of testability and falsification of hypotheses, is usually called *pseudoscience*. There are many kinds of pseudoscience—trying to give us (or sell us) answers to questions that we want answered or attempting to appeal to our sense of wonder and mystery—but all fail the criteria of science: testability, falsifiability, peer review, and rejection of ideas when they do not pan out. Most educated people have developed a sense of skepticism when it comes to everyday claims, like the lies, half-truths, and exaggerations that we hear in sales pitches, see in commercials, and read in advertisements. Our "baloney detection kit" (as Carl Sagan put it) is good at filtering out these assertions, and in most practical aspects of our lives we follow the maxim "caveat emptor" (let the buyer beware). Yet we ignore these filters when something promises to give us comfort or help with uncertainty. When a psychic claims to be able to communicate with our dead relatives, a snake-oil salesman or faith healer offers us phony cures for our ailments, or an astrologer or a fortune-teller professes to be able to forecast our future, they are preying on our vulnerability.

Even though Americans have one of the highest standards of living in the world, and one of the best educational systems, we are as gullible as ever. Poll after poll shows that high percentages of Americans believe in demonstrably false or pseudoscientific ideas—from UFOs to extrasensory perception (ESP), astrology, tarot cards, palm reading, and so on. It does not seem to matter that these notions have been repeatedly debunked. As *Skeptic* magazine publisher Michael Shermer points out, people have a *need* to believe in such things, since they provide comfort or help cope with the uncertainties of the future.[13] We can all sympathize to some degree with how people might find reassurance in the predictions made by a fortune-teller or solace in the communiqués from the other world transmitted by a psychic. But it is puzzling that so many people believe in UFOs or give any credence to the vile and scary anti-Semitism of Holocaust deniers.

A number of general tools can be used to detect "baloney." Most of them were offered by Shermer and Sagan,[14] but this list will focus on those directly relevant to the subject of this book.

- *Extraordinary claims require extraordinary evidence*: This famous statement by Carl Sagan (based on Marcello Truzzi's maxim "An extraordinary claim requires extraordinary proof") is the most important to the subject of cryptozoology.[15] As Sagan noted, hundreds of routine discoveries and assertions are made by scientists nearly every day, but most are just small extensions of what is already known and do not require thorough testing by the scientific community. But it is typical of crackpots, fringe scientists, and pseudoscientists to make revolutionary pronouncements about the world and argue strenuously that they are right. For such claims, it is not sufficient to have just one or two suggestive pieces of evidence—such as blurry photographs, eyewitness accounts, and ambiguous footprints of, say, Bigfoot—when most of the proof goes against cherished hypotheses. Extraordinary evidence, such as the actual bones or even the corpse of the creature, is required to overcome the high probability that it does not exist.

- *Burden of proof*: In a criminal court, the prosecution must prove its case beyond a reasonable doubt, and the defense need do nothing if the prosecution fails to do so. In a civil court, the plaintiff has to prove its case based on a preponderance of the evidence, and the respondent need do nothing. In science, extraordinary claims have a higher burden of proof than do routine scientific advances because they aim to overthrow a larger body of knowledge. When evolution by means of natural selection was first proposed 150 years ago, it had the burden of proof because it sought to overturn the established body of creationist thought. Since then, so much evidence has accumulated to show that evolution has occurred that the burden of proof is back with the creationists, who seek to overthrow evolutionary biology. And the burden of proof lies with those who believe in the existence of most of the cryptids discussed in this book, since so much of what is claimed about them goes against everything we know from biology, geology, and other sciences.

- *Authority, credentials, and expertise*: One of the main strategies of pseudoscientists is to cite the credentials of the leading proponents of their claims as proof that they are credible. But a doctoral degree or advanced training is not enough; the "authority" must have advanced training in the *relevant fields*. Pointing to an advanced degree is a strategy to intimidate the members of an audience into believing that the holder of the degree is smarter than they are and an expert in everything. But those who have earned a doctorate know that it qualifies the holder to talk about only the field of her training, and during the long hard slog to get a dissertation project finished and written up, a doctoral candidate actually tends to *lose* some of his or her breadth of training in other subjects. Most scientists agree that anyone who is flaunting a doctoral degree while making his arguments is "credential mongering." A good rule of thumb: if a book says "Ph.D." on the cover, its arguments probably cannot stand on their own merits.

  Creationists do this all the time by pointing to lists of "scientists who do not accept evolution." Those on the list, though, have degrees that are irrelevant to the fields they are critiquing: they often have degrees in engineering, hydrodynamics, physics, chemistry, mathematics, veterinary medicine, dentistry, and perhaps biochemistry. But among them are vanishingly few paleontologists or geologists and almost no biologists trained at major institutions. The absurdity of this strategy, which also is used by those who deny the reality of global warming, was revealed by the National Center for Science Education (NCSE), which fights creationist attacks on public-school science curricula. Rather than try to compile a list of the 99.99 percent of qualified scientists who *do* accept the reality of evolution (compared with the creationist list of a few dozen), the NCSE organized Project Steve.[16] This list consists of just scientists whose name is Steve (or Steven, Stephen,

Stefan, or Stephanie) and who accept evolution, which would make up less than 1 percent of all scientists who accept evolution—and it is much longer than the creationists' total list, with more than 1,200 names so far!

The same can be said for proponents of cryptozoology. Few of those who promote cryptozoology or engage in cryptozoological research—including those on a "hall of fame" list of famous cryptozoologists[17]—have advanced degrees or relevant training in fields that could make them authorities in finding or documenting mysterious beasts. These fields include wildlife biology and ecology, systematic biology, paleontology, and physical anthropology. Most monster hunters are amateur enthusiasts who are not familiar with the rules of science and are not trained in the basics of ecology, paleontology, or field biology. Others boast impressive-sounding titles such as "zoologist" or "Dr." while important clarifying information is omitted. For example, the late Richard Greenwell, a co-founder and long-time secretary of the defunct International Society of Cryptozoology (ISC) who collaborated on Roy Mackal's cryptozoological expeditions, has been referred to as "Dr." by Bigfoot proponent John Bindernagel and others.[18] However, Greenwell had no formal training in relevant fields; his "doctorate" was an honorary degree from the University of Guadalajara in Mexico.[19] Similarly, William Gibbons, a champion of Mokele Mbembe, sometimes is described by cryptozoological sources as "Dr. Bill Gibbons" and boasts several degrees.[20] Gibbons's bachelor's and master's degrees in religious education were granted by a religious institution, however, and have no academic standing,[21] while his doctorate in cultural anthropology was procured from Warnborough College, Oxford—an institution not associated with Oxford University that has been determined (and fined) by the Department of Education to be unauthorized to grant degrees in Great Britain, and therefore not eligible to participate in federal student financial assistance programs in the United States.[22] (If the Oxford location makes Warnborough sound more impressive than it is, this is unlikely to be accidental. As the *New York Times* explained, "Officials at Oxford University say that Warnborough is only the latest and most shameless in a string of institutions that unfairly trade on Oxford's name and reputation abroad," with students alleging that they had been misled into the false belief that Warnborough was formally associated with Oxford University.)[23]

There are exceptions to the generalization that cryptozoologists lack advanced degrees in relevant fields (some of which we will discuss in chapter 7), but even these exceptions may be more complicated than they appear. Some, like Ben Speers-Roesch, got hooked on cryptozoology as teenagers but became critics of cryptozoological research as their scientific training advanced (and as their familiarity with cryptozoology's shortcomings deepened).[24] Others, like paleozoologist Darren Naish, are friendly to cryptozoology in principle, yet highly critical of its claims in practice.[25] Among those more wholeheartedly supportive of cryptozoology, Roy Mackal (another co-founder of the ISC) is frequently cited as a legitimate scientist with a doctorate in biology. This is true: he is retired from the University of Chicago, where from the late 1950s to the late 1960s he did highly respectable research in molecular biology and microbiology. Mackal's later research and writing related to the search for the Loch Ness monster and the alleged Congo dinosaur, Mokele Mbembe, however, displaced his mainstream scientific work—and had no close connection to his area of scientific expertise. Notably, Mackal has no credentials in field biology, paleontology, or ecology. Stronger exceptions include wildlife biologist John Bindernagel and anatomist Jeffrey Meldrum, both of whom are leading proponents of Bigfoot whose formal training is relevant to their area of cryptozoological focus. But the interest of a handful of formally trained authorities tells us little about the robustness of the evidence for Bigfoot (or any cryptid). Thousands of similarly trained anthropologists are just as qualified to examine and critique the evidence for Bigfoot, and they are virtually unanimous in considering the evidence for Bigfoot to be useless or marginal at best.[26] This is the process of peer review at work. Even if someone does have proper training, what he or she thinks is not necessarily true or authoritative. The ideas must pass muster in the process of peer review, and he or she must defend the evidence for the rest of the scientific community to take the claims seriously.

- *Special pleading and ad hoc hypotheses*: One of the marks of pseudoscientists is that when the evidence is strongly against them, they do not accept the scientific method and abandon their cherished hypotheses. Instead, they resort to special pleading to salvage their original ideas, rather than admitting that they are wrong. Such attempts are known as ad hoc (for this purpose) hypotheses and are universally regarded as signs of failure. When a psychic conducts a séance and does not contact the dead, she may plead that the skeptic "just didn't believe hard enough" or the "room wasn't dark enough" or the "spirits didn't feel like coming out this time." When a creationist is told that Noah's Ark could not have housed the tens of millions of species of animals, he may use an evasion like "only the created kinds were on board" or "fish and insects don't count" or "it was a miracle." Science, however, does not permit such creative, convenient flexibility; ideas in science must stand or fall according to the evidence. This is, as the great Victorian naturalist Thomas Henry Huxley said, "The great tragedy of Science—the slaying of a beautiful hypothesis by an ugly fact."[27]

EYEWITNESS TESTIMONY: INSUFFICIENT EVIDENCE

Humans are storytelling animals, and they are easily persuaded by the testimony of other individuals. Telemarketers and advertisers know that if they get a popular celebrity to endorse a product, it will sell briskly, even if no careful scientific studies or Food and Drug Administration approvals back up their claims. The endorsement of your next-door neighbor may be good enough to make simple decisions, but anecdotal evidence counts for very little in science. As celebrated science historian Frank Sulloway put it, "Anecdotes do not make a science. Ten anecdotes are no better than one, and a hundred anecdotes are no better than ten."[28]

Most scientific studies require dozens to hundreds of experiments or cases, and detailed statistical analyses, before scientists can tentatively accept the conclusion that event A probably caused event B. In the testing of medicines, for example, there must be a control group, which receives a placebo rather than the treatment under study, so the experimenters can rule out the possible influence of the power of suggestion as well as random effects. Only after such rigorous testing—which can eliminate the biases of the subjects and the observers, random noise, and all other uncontrolled variables—can scientists make the statement that event A *probably caused* event B. Even then, scientists do not speak in finalistic terms of "cause and effect," but only in probabilistic terms that "event A has a 95 percent probability of having caused event B." Independent replication by other scientists helps to refine the findings—but no conclusion, no matter how often or robustly confirmed, is ever wholly free from uncertainty.

If scientists are obligated to consider the outcomes of the most tightly controlled laboratory experiments to be tentative, they must naturally approach eyewitness testimony with even greater caution. Eyewitnesses may have some value in a court of law, but their anecdotal evidence is necessarily regarded as highly suspect in most scientific studies. Thousands of experiments have shown that eyewitnesses are easily fooled by distractions such as a weapon, are confused by stress, or otherwise are misled into confidently "remembering" events that did not happen.[29] The extreme fallibility of eyewitness testimony was vividly demonstrated in a famous psychology experiment, conducted in 1999, in which subjects were instructed to watch a video and count the number of times young men and women dressed in white pass a basketball.[30] During the video, a person in a gorilla suit barges into the center of the scene, looks straight into the camera, lingers to pound his chest, and then strolls out of the scene. However, roughly half of first-time viewers are shocked to learn that they overlooked the gorilla entirely! With their attention focused on counting the passes, a majority of people are unable to correctly perceive what happens right in front of their eyes.

The "inattentional blindness" effect, which renders the gorilla invisible, is just one of many ways in which eyewitness perception may critically fail. As A. Leo Levin and Harold Kramer put it, "Eyewitness testimony is, at best, evidence of what the witness believes to have occurred. It may or may not tell what actually happened. The familiar problems of perception, of gauging time, speed, height, weight, of accurate identification of persons accused of crime all contribute to making honest testimony something less than completely credible."[31] Consequently, court systems around the world are undergoing reform as DNA evidence has revealed case after case of eyewitness testimony that resulted in a wrongful conviction. As psychologist Elizabeth Loftus has shown, eyewitness accounts of events and their memory of them are notoriously unreliable:

> Memory is imperfect. This is because we often do not see things accurately in the first place. But even if we take in a reasonably accurate picture of some experience, it does not necessarily stay perfectly intact in memory. Another force is at work. The memory traces can actually undergo distortion. With the passage of time, with proper motivation, with the introduction of special kinds of interfering facts, the memory traces seem sometimes to change or become transformed. These distortions can be quite frightening, for they can cause us to have memories of things that never happened. Even in the most intelligent among us is memory thus malleable.[32]

Skeptical investigators Benjamin Radford and Joe Nickell recount several examples that make this point vividly.[33] In 2004, Dennis Plucknett and his fourteen-year-old son, Alex, were hunting in northern Florida. Alex was in a ditch some 225 yards away from his father when someone yelled "Hog!" Dennis grabbed his gun, pointed it at a distant moving object that looked like a wild hog to him, and fired. Instead, he killed his son with a single shot to the head. Alex had been wearing a black toboggan cap, not a hog costume or anything else that would have made him look remotely hog-like. Yet at that distance, and with the suggestion that a wild hog was nearby, Dennis mistook a toboggan cap for a boar, and a tragic result occurred. Indeed, hunting accidents like this are common, since many hunters shoot first and ask questions later, often confused by distant objects that are moving and by the suggestibility of their own imaginations to "see" what they are looking for—not what is really there.

Or take the Washington, D.C., sniper panic of 2002. Early eyewitnesses told the police to look for a white or light-colored box truck or van, with a roof rack. Police officers wasted weeks of time and caused huge traffic jams stopping every vehicle that remotely matched the description. Finally, the suspects, John Lee Malvo and John Allen Muhammad, were caught—and the white van turned out to be a dark blue sedan. Police chief Charles H. Ramsey said, "We were looking for a white van with white people, and we ended up with a blue car with black people."[34] Newspapers reported that the suspects had actually been stopped several times during the period of the panic because they were near the site of one of the attacks, but each time they were released because their car did not match the descriptions.[35]

Some "memories" of strange experiences or eyewitness accounts of bizarre beasts can be attributed to sleep deprivation, dreams, and hallucinations. For example, Michael Shermer has described how he had an experience of being abducted by aliens in 1993.[36] At the time, he was undergoing the stress of competing in an ultra-marathon cross-country bicycle race; badly exhausted, his brain misperceived his own support crew as space invaders and their motor home as an alien spacecraft. As he showed, some famous accounts of alien abductions and out-of-body experiences may be attributable to dreams or hallucinations caused by stress.[37] Harvard psychologist Susan Clancy's research on alien abductees expanded on this, revealing that the seed for abduction beliefs need not be as literal as Shermer's hallucination. As she explained, "Coming to believe you've been abducted by aliens is part of an attribution process. Alien-abduction beliefs reflect attempts to explain odd, unusual, and perplexing experiences."[38] These anomalous experiences include common sleep disruptions, such as sleep paralysis (a viscerally compelling experience that occurs at the transition between wakefulness and sleep, often combining physical paralysis with hallucinations). Seeking an explanation, people find that alien abduction is one of the culturally available scripts for mysterious happenings—and once that seed is planted, it tends to grow. "Once abductees have embraced the abduction theory, everything else tends to fall into place," cautioned Clancy. "Alien abduction easily accommodates a great variety of unpleasant symptoms and experiences." Memories are malleable; over time, they may change to better match the abduction narrative. Similar processes distort the cryptozoological literature—and indeed, distort all literature that involves testimonials. Whether "eyewitness" accounts describe cryptids, ghosts, aliens, or muggers, it is an inconvenient fact of human nature that individuals quite commonly imagine things that are not really there, misinterpret known phenomena (such as sleep paralysis, bears, or the planet Venus), and, in an attempt to understand their experience, rely on memories that may include false details.

Scientific investigators must also tread carefully with indigenous, developing societies. Unlike highly industrialized culture, which has very specific ideas of what is reality and what is myth, native cultures can be much less rigid and literal. Thus when investigators go to a remote region and inquire about a legendary animal, such as the Yeti, they cannot determine whether the animal is literally or mythically real. Even worse are attempts to elicit responses by showing photographs or drawing sketches of the animal being sought or, in especially egregious instances, first providing a story and then demanding that local informants confirm it.[39] In a court of law, that is considered leading the witness and it is not allowed, since this practice often biases the testimony or plants a suggestion in the mind of the witness. Thus "witnesses" may claim to have seen any animal that remotely resembles that being sought.

What is the proper scientific approach to eyewitness testimony? As we have seen, most scientists give it very little weight unless there is strong physical evidence to support it. Even the most reliable eyewitness account does not meet the standard of "extraordinary evidence" that is necessary to substantiate an "extraordinary claim."

WHAT IS A CRYPTID? PERSPECTIVES FROM REAL SCIENCE

The term "cryptozoology" comes from the Greek words *kryptos* (hidden) and *zoos* (life) and literally means "the study of hidden animals." The origins of this and related terms (such as "cryptid" and "cryptozoological") are somewhat unclear and thus a matter for much discussion.[40] The word "cryptozoology" is often traced to Lucien Blancou, who in 1959 dedicated a book to Belgian zoologist Bernard Heuvelmans as the "master of cryptozoology," in honor of Heuvelmans's book *On the Track of Unknown Animals*, published in French four years earlier.[41] Because of his reputation as an early major figure in the field, Heuvelmans is sometimes thought to have coined the term "cryptozoology" himself. But Heuvelmans gave the credit to Scottish explorer Ivan T. Sanderson: "When he was still a student he invented the word 'cryptozoology,' or the science of hidden animals, which I was to coin much later, quite unaware that he had already done so."[42] As early as 1941, a reviewer of Willy Ley's *The Lungfish and the Unicorn* described it as covering "not only lungfish and unicorns but an array of other marvels, zoological and cryptozoological, from the *mushrush* of the Ishtar Gate to the basilisk, the tatzelwurm, the sea serpent, and the dodo."[43] In any event, the discipline that we now call cryptozoology goes back much further—certainly to Anthonie Cornelis Oudemans and his book *The Great Sea-serpent* (and arguably back to Pliny the Elder and other natural historians of classical antiquity, as we will see in chapter 5).[44]

By most definitions, any purposeful search for unconfirmed animals (or unconfirmed populations of animals) can be classified as cryptozoology, although conventionally the focus is entirely on hypothetical animals of spectacular size, such as Bigfoot, Nessie, and Yeti. The term "cryptid" was coined by John Wall to refer to the animals that are sought by cryptozoologists.[45] As Darren Naish has pointed out,[46] the definition of cryptids is highly fluid. Heuvelmans wrote that in order to be considered a cryptid, a creature must be "truly singular, unexpected, paradoxical, striking, or emotionally upsetting." He argued that these characteristics allow such an animal to become the subject of myths and thence *become* a cryptid.[47] Yet Heuvelmans's list of animals that he thought might soon be discovered includes such cryptids as a marmot-size mammal from Ethiopia, a small wildcat from the Mediterranean, and a flightless rail from the South Pacific—creatures that would not astound most zoologists if they were indeed found.[48] Naish redefined "cryptozoology" to refer to the study of animals known only from *indirect evidence*: eyewitness accounts and anecdotal clues, such as sightings, photographs, stories, casts of footprints, and questionable hair or tissue samples. This definition distinguishes cryptids from animals whose existence is based on the *direct evidence* of actual specimens that have been captured or bones that have been collected.

The distinction between cryptozoology and conventional biology is not necessarily very big. As Naish points out, several hundred previously

Figure 1.2
A mountain gorilla in Bwindi Impenetrable Forest, Uganda. (Photograph by Julie Roberts)

unknown species are discovered and described every year,[49] although most are insects and other invertebrates that do not capture the imagination of nonscientists. Nor it is unusual for a creature to be known from folk legend before scientists officially "discover" it. A number of these "former cryptids" are suggested by cryptozoology advocates as evidence that many animals are yet to be found. The "former cryptids" include the lowland gorilla, known from legend since the seventeenth century but not described until 1840, and the mountain gorilla, not described until 1901 (figure 1.2). The okapi, the only close living relative of the giraffe, was well hidden in the dense jungles of Central Africa and known to the local peoples before it was officially "discovered" in 1901 (figure 1.3). The Komodo dragon was called the "land crocodile" by Indonesians and frequently reported since 1840, but was not officially named or described until 1912. The kouprey, a large ox native to the jungles of Cambodia, was first mentioned in 1860, but not officially named until 1937.

Although the rate of discovery of large animals has dropped off considerably in the past century, less spectacular species are found every year. Naish summarizes a considerable catalogue of them.[50] They include a number of species of small brocket deer, from Central and South America, and muntjacs, from Southeast Asia, described in the past decade, as well as tree kangaroos and numerous primates, rodents, and bats. Among animals known from

Figure 1.3
The okapi, an animal native to the Democratic Republic of the Congo, was long rumored on the basis of testimony from local informants before it was formally identified for science in 1901. (Illustration by Daniel Loxton)

native accounts is the kipunji, a Tanzanian monkey, which was described in 2005. The odedi, a bush warbler from Bougainville Island in the South Pacific, was formally described in 2006. In 2008, zoologist Marc van Roosmalen claimed to have discovered a dwarf species of manatee in the Amazon River, although the DNA evidence has so far failed to show that it is really distinct from the large Amazonian manatee, and this "species" may be based on immature specimens.

Some "new" species were first known from fossils and then discovered alive. The classic example is the Chacoan peccary, a relative of the peccaries or javelinas, which are widespread across the southwestern United States, Central America, and South America. First described from a tooth fossil by the famous Argentine paleontologist Florentino Ameghino in 1904, it was known to natives of the Gran Chaco in Paraguay for decades before scientists "discovered" it and named it.[51] Van Roosmalen recently claimed to have found a fourth species, the "giant peccary" of the Brazilian Amazon basin, although it has not yet been accepted as distinct by most other zoologists.

The most famous example of a creature known first from fossils is the coelacanth, a lobe-finned fish related to lungfishes and amphibians (figure 1.4). Before 1938, it was known primarily from rare fossil specimens older than 65 million years and was thought to be extinct. (There are now fossils as young

Figure 1.4 The West Indian Ocean coelacanth (*Latimeria chalumnae*), a "living fossil," was known originally from fossils older than 65 million years old until a live fish was caught off South Africa in 1938. It has now been found in several places around the Indian Ocean, including off the coasts of Indonesia and the Comoros Islands. (Reproduced by permission of Fortean Picture Library, Ruthin, Wales)

as 15 to 5 million years old, so the gap is no longer as great as it once was.)[52] Then in 1938, a living coelacanth was found in a deep-water trawl off the coast of South Africa and was immediately recognized as a relict of the early evolution of fishes. Its great significance led to a widespread hunt for more specimens and a sizable cash reward, and eventually more coelacanths were discovered off the Comoros Islands north of Madagascar and, more recently, off the coast of Indonesia and again off South Africa. The oceans continue to reveal many new and spectacular fishes—such as the megamouth shark discovered in 1976—squids, and other creatures, so the pace of discovery is no less impressive than that for land animals. The remotest and deepest parts of the seas still may hold secrets as yet undreamed of.

REALITY CHECKS

Biology and Ecology

If so many animals are still being found every year, why is the study of Bigfoot and Nessie considered "fringe science"? There are important distinctions to make between animals newly known to science and cryptids, even when the folkloric elements are set aside. Nearly all the "new" species just mentioned are relatively small in body size and are closely related to other living animals—so it was easy to mistake them for familiar species until more detailed studies could be undertaken—and none are outside the normal range of zoological diversity. The last large land animals to be found and named by biologists were the Chacoan peccary, lowland and mountain gorilla, okapi, Komodo dragon, and kouprey, and all were described about 50 to 150 years ago. In addition, nearly all these discoveries were made in remote areas of Africa, Indonesia, and Southeast Asia, *not* in well-populated places like Scotland and the Pacific Northwest (where Nessie and Bigfoot allegedly reside).

Biologists have dealt rigorously with the question of how many species remain to be found, using a variety of sophisticated methods to model "discovery curves" (figure 1.5). Applying these techniques in 2008 to a question of interest to cryptozoologists, Michael Woodley, Darren Naish, and Hugh Shanahan estimated that as many as fifteen species of seals and sea lions (pinnipeds) conceivably remain undiscovered, although they warned that this figure was "probably a significant over-estimate of the true number likely to exist." Turning to a slightly different estimation technique, they suggested that few or no species of pinniped are yet to be found.[53] Naish expanded,

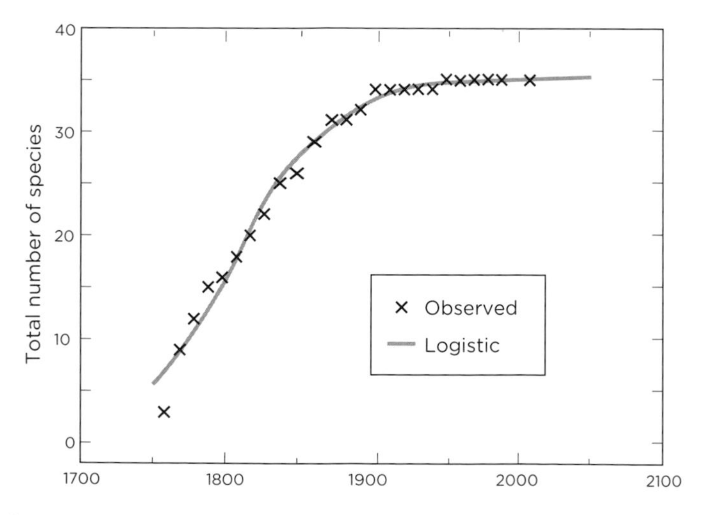

Figure 1.5 A curve showing the number of new species of pinniped discovered over the past 300 years, with observed data shown by the × and a logistic curve fit in the solid line. The curve indicates that discoveries of unknown species of marine animals such as pinnipeds have reached a point of diminishing returns, so there are few or no species remaining to be found. (Modified from Michael A. Woodley, Darren Naish, and Hugh P. Shanahan, "How Many Extant Pinniped Species Remain to Be Described?" *Historical Biology* 20 [2008]: 225–235)

"We definitely favoured the idea that there are hardly any new pinnipeds to find—that is, there are about zero new ones out there, with one or two being conceivable based on the data. This could be taken to mean that cryptozoological ideas of new pinnipeds are unlikely, since there probably aren't many (or any) new pinnipeds to find. *Or*, it could be taken as being consistent with cryptozoological ideas about crypto-pinnipeds"—provided that only a very small number of cryptozoological pinnipeds are proposed.[54] When we add to this reality check the constraining factors that seals are noisy and return often to land in order to molt and breed (where they would presumably be seen and identified), we may conclude that reports of "sea monster" cryptids are highly unlikely to be of genuine sightings of as-yet-undiscovered species of seals (despite the popularity of the "pinniped hypothesis" in the cryptozoological literature) but more likely are cases of mistaken identity. Andrew Solow and Woollcott Smith took an even broader approach and showed that the discovery rate of large marine animals has dropped dramatically in the past few decades. From this method, they estimated that, at best, fewer than ten species of large marine animals remain to be found.[55]

Thus the discoveries of the okapi and the kouprey do not really help the cause of cryptozoologists because these animals are within the realm of conventional zoology. These were zoological discoveries of the type that we would expect once qualified field zoologists began to work in remote and sparsely populated areas like tropical Africa and Southeast Asia. Similarly, the discoveries of the coelacanth and the megamouth shark may be remarkable, but these fish dwell in very deep, seldom fished waters in only a few places on Earth. It is easier for animals to remain hidden for a longer period in poorly accessible locations—and yet, notably, animals such as the okapi and coelacanth eventually were located and described.

By contrast, hundreds of zoologists and tens of thousands of hikers, campers, and loggers wander the woods of the Pacific Northwest, yet not one has managed to find any hard evidence to support the existence of Bigfoot. Nearly all the "evidence" consists of anecdotes and eyewitness accounts, dubious film footage, and a spectrum of suspicious footprint casts, plus hair of undetermined origin. The forests of the Pacific Northwest are neither remote nor underexplored; indeed, they have been decimated by clear-cut logging (and more recently, by the mountain pine beetle epidemic).[56] Every animal that lives in these forests leaves plenty of hard evidence of their existence: nests and dens, carcasses, skeletons and partial skeletons, teeth, and isolated individual bones. Yet not once has the elusive Bigfoot left so much as a single toe bone, and explanations from advocates that Bigfoots somehow decay faster than other animals or are buried by their kin are ad hoc and simply do not cut it.

This raises an even more important issue: animals the size of Bigfoot, Nessie, and Yeti cannot simply be singletons or the last surviving lonely members of their species. If they have persisted through thousands of years and been sighted hundreds of times, as their advocates claim, they must have significant populations. If they are living as viable populations, there should be hundreds of reliable sightings, many good examples of film footage, and plenty of hard evidence in the form of carcasses and bones—not only of isolated individuals, but of groups, including females and their young. Ironically, the best proof that these cryptids do not exist comes *not* from the abundance of questionable evidence that cryptozoologists try to promote, but from the dearth of *more* quality evidence, especially conclusive, concrete proof in the form of carcasses and bones.

Even more interesting is that as more and more people have been looking for these creatures in the woods of the Pacific Northwest, the snows of the Himalayas, the jungles of the Congo, and the waters of Loch Ness, the evidence for the existence of Bigfoot, Yeti, Mokele Mbembe, and Nessie has *diminished*, not increased. In the modern world, where cell-phone cameras are ubiquitous and space satellites that can read newspapers on Earth are constantly watching us, the absence of reliable images does not bode well for the existence of cryptids.

The next important consideration is that populations of cryptids as large in body size as Bigfoot, Yeti, and Nessie would have to occupy relatively extensive areas to obtain enough food and other resources to sustain themselves. Although the relationship between body size and home range is complicated, in general home range expands as body size increases, as is confirmed by the large and complex literature on the relationship of home range and body size (figure 1.6).[57] As Johan du Toit has shown,[58] for large-bodied African mammals, the home range (A_{hr}) scales by body mass (M) in the formula

$$A_{hr} = 0.024\,M^{1.38}$$

Lowland gorillas have home ranges of roughly 40 square miles, and they are much less active and mobile than, say, giraffes (which require individual ranges of about 110 square miles). Animals the size of the presumedly wide-ranging bipedal Bigfoot and Yeti would have to occupy home ranges of at least 100 square miles. But because so much of this habitat has been degraded or broken up by human habitation, they would be forced onto

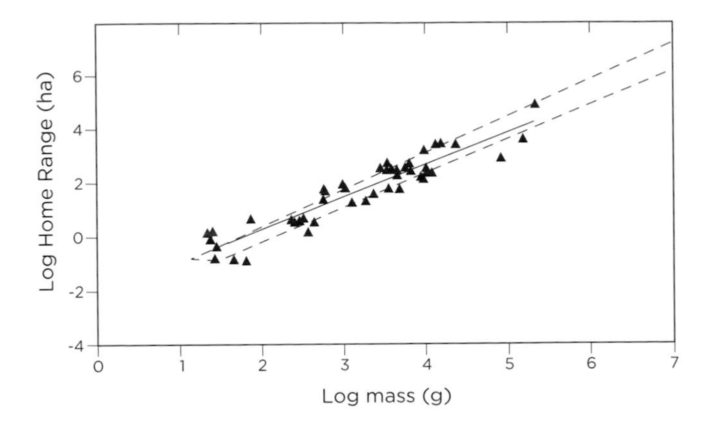

Figure 1.6 A plot of body mass versus home-range area for omnivorous land animals, showing that animals of very large body size (such as Bigfoot, the Yeti, and Mokele Mbembe) require huge home ranges in order to sustain a viable population. A similar plot applies to herbivores and carnivores as well. (Data replotted from Douglas A. Kelt and Dirk H. Van Vuren, "The Ecology and Macroecology of Mammalian Home Range Area," *American Naturalist* 157 [2001]: 637–645)

remnant patches, enhancing the likelihood of their discovery or, conversely, decreasing the likelihood of their existing.

This is the problem that large animals face in places like the tropical rain forest. In a famous monograph, ecologists Robert MacArthur and Edward O. Wilson showed that the number of species in a given region is well predicted by the total area that they occupy.[59] Using this famous relationship, rainforest ecologists have found that breaking the forest into numerous small patches separated by logged or farmed areas leads to the disappearance of large animals, since jaguars, tapirs, rhinos, and tigers need huge home ranges in order to survive. This problem would be magnified in the case of Bigfoot, whose temperate-forest habitat has been chopped into hundreds of small preserves and parks, separated by roads and towns and farms. Even within the larger national forests, most of the pristine regions are extremely reduced or have been destroyed because of clear-cut logging.

Limited range is an even bigger problem for Nessie and the wide assortment of air-breathing "lake monsters" that have been promoted as cryptids. Loch Ness and other lakes said to be the home to monsters are relatively small bodies of water that are generally not rich enough in fish and other food resources to sustain a population of the enormous creatures purported to live in them.

Geology

If the problems enumerated by ecology were not enough to reduce the plausibility of the existence of large cryptids, there are even bigger hindrances apparent to a geologist. Let's start with a very simple constraint: the Ice Age of the Pleistocene epoch (2.5 million–11,700 years ago). As recently as 20,000 years ago, the northern part of the Northern Hemisphere was covered by a sheet of ice more than 1 mile thick, similar to the ice sheets that now cover Greenland and Antarctica (figure 1.7). Thus Loch Ness (and much of the British Isles, except for the southern margin during interglacial warmings) was under a glacier for most of the past 3 million years. Either the Nessie population was frozen in the ice and then miraculously was resuscitated after thawing (a biological impossibility), or the creatures lived elsewhere and swam

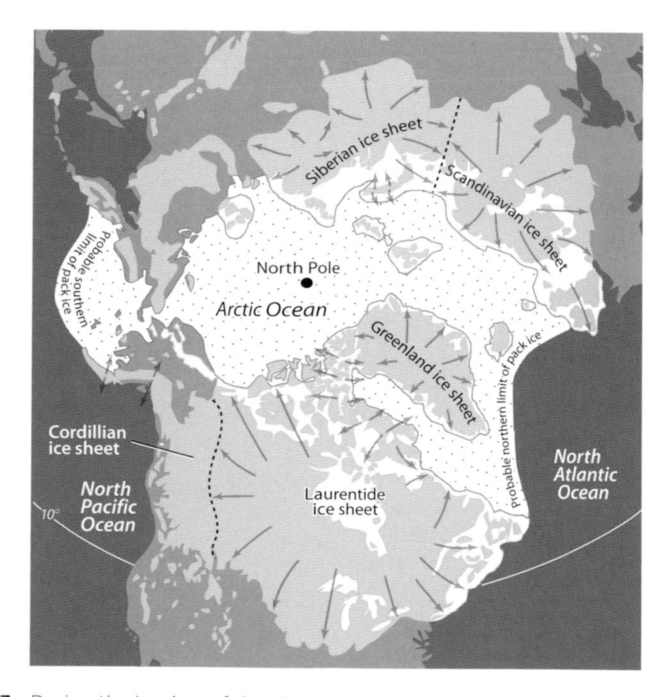

Figure 1.7 During the Ice Age of the Pleistocene, sheets of ice covered not only the polar regions, but most of northern Eurasia and North America, including all the lakes (such as Loch Ness and Lake Champlain) that allegedly are home to monsters. Thus if such creatures existed at that time, they either were frozen under a mile of ice or somehow swam into these landlocked lakes from the distant North Sea and Atlantic Ocean. (From Donald R. Prothero and Robert H. Dott Jr., *Evolution of the Earth*, 8th ed. [Dubuque, Iowa: McGraw-Hill, 2010])

into Loch Ness. Yet Loch Ness is a freshwater lake 52 feet above sea level. If Nessie were indeed an ocean-going plesiosaur, why have plesiosaurs not colonized other freshwater rivers and lakes—and where are the other saltwater plesiosaurs today? Glaciation is an even more serious problem for inland cryptids like Champ, which supposedly lives in a lake that is far inland, a long way from the nearest outlet to the Atlantic Ocean—and Lake Champlain, the home of Champ, was buried under even more ice than was Loch Ness just 20,000 years ago. Indeed, most of the lakes that allegedly support cryptids are in northern regions and were buried under ice during the last glaciation.

Paleontology

Even more problematic is the evidence—or lack of evidence—from the fossil record. No one claims that the fossil record is perfect, but paleontologists have an excellent understanding of which organisms tend to fossilize well and which ones do not.[60] Small animals with delicate bones (like birds) may be underrepresented in the fossil record, but large animals with robust bones (like large nonavian dinosaurs) have a much better chance of being preserved. An excellent fossil record exists for such large animals as mammoths, mastodonts, rhinos, and dinosaurs. For this reason, paleontologists can be fairly confident that if fossils of certain animals are not found in the appropriately aged beds along with the abundant bones of many other animals of the same body size, the former creatures were almost certainly absent from the region. Even small bone fragments of large-boned animals are extremely durable, so paleontologists can determine that the animals were present, no matter how small the fossilized fragment.

These considerations have important implications for nearly all the large-bodied cryptids, such as Bigfoot, Nessie, and Mokele Mbembe. If they have lived in their respective homelands for as long as cryptozoologists assert, they would have left fossil records. The skeletons of Bigfoot and Nessie would be exposed in the Ice Age deposits of the Pacific Northwest and Scotland, as are those of other large animals once native to North America and Great Britain, like mammoths and mastodonts. Instead, there is no evidence of their existence anywhere in the fossil record.

From the excellent worldwide fossil record, paleontologists can determine exactly which regions had which dinosaurs and marine reptiles in the Mesozoic era, or Age of Dinosaurs (250–65 million years ago). More important, the complete absence of *any* dinosaur fossils (other than those of birds, which are surviving dinosaurs) in beds younger than 65 million years is strong evidence that no nonavian dinosaurs or large marine reptiles survived the mass extinction at the end of the Cretaceous period (144–65 million years ago). This argument applies to the plesiosaur-like Nessie and the sauropod-like Mokele Mbembe.

Plesiosaur fossils are common in marine beds deposited during the later part of the Mesozoic era (Jurassic [206–144 million years ago] and Cretaceous periods), and even plesiosaur vertebrae and teeth are distinctive and identifiable. Extensive marine deposits laid down after 65 million years ago are full of fossil sharks and other fish, whales, seals and sea lions, and other marine animals from all over the world. The collections of these fossils, especially in rich beds like Calvert Cliffs in Maryland and Sharktooth Hill Bone Bed in California, are enormous. Indeed, there are so many fossils of so many different marine animals that it is extremely unlikely that a creature as large as a plesiosaur swam in Chesapeake Bay, the Temblor Sea (present-day San Joaquin Valley), and Loch Ness and yet has never shown up in any fossil beds anywhere in the world that date to the Cenozoic era (65 million years ago–present). This record is good enough that the absence of evidence *is* evidence of absence.

The excellent fossil record for mammals and other animals that have lived in Africa over the past 65 million years not only is proof that no dinosaurs—such as Mokele Mbembe—survived on that continent, but also shows what animals (including some very large ones, such as early mastodonts and the horned, rhino-like arsinoitheres) have lived in Africa since the beginning of the Cenozoic.[61] Once again, the fossil record is good enough that the absence of evidence of dinosaurs in Africa is enough evidence of their absence.

Now that we have an understanding of how scientists work and how they might investigate claims about the reality of major cryptids, we can begin our exploration of these wondrous creatures.

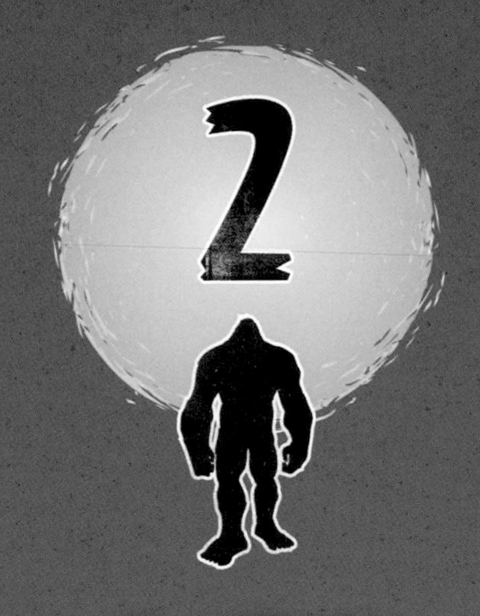

BIGFOOT

THE SASQUATCH

AS A SKEPTIC WHO INVESTIGATES cryptozoological mysteries, I am sometimes asked by exasperated proponents, "Why do you even bother? If you don't think cryptids are real, why waste your time picking on cryptozoology?"

I explain that cryptozoology is my first love. It was my youthful obsession with monster mysteries that led me to the skeptical literature in the first place, and I've never lost my fondness for the topic. I don't think of myself as picking on cryptids, but as making an all-too-rare effort to solve cryptozoological mysteries.

You may be surprised by one of the reasons I find myself drawn to such mysteries: I once personally found a Bigfoot track. My parents were silvicultural contractors in British Columbia. Thus I spent most of the summers of my childhood in tree-planting camps in the remote wilderness (a wilderness I later returned to for a ten-year career as a professional shepherd). In those days, tree-planting camps were tipi-filled refuges for the last first-generation hippies. In the daytime, the crew was trucked out to plant trees in the hills while the cooks worked in the cook trailer—and we feral hippie children (I remember young friends named Raven and Blue Sky) had the run of the camp. It was a nice way to grow up, catching frogs and fishing and playing in the mud.

One day, not far from camp, my brother and I came across a giant footprint stamped deeply into the mud at the side of a dirt road. We surmised that the other tracks were lost in the debris of the forest floor—evidence of the giant stride of the Sasquatch.

Was it a genuine Bigfoot track? I doubt it—now. The tree planters loved to play practical jokes, and my brother believes that later we may have been tipped off that this was a prank. But it doesn't matter. The point is that I know, deep in my bones, how the mystery of Bigfoot can take root in the imagination. For my entire life, I've wondered, "Could Sasquatch really exist?" Answering that question is one of the reasons I became a skeptical investigator. Here is some of what I have learned.

THE HISTORICAL VIEW

When discussing the possibility of the existence of Sasquatch, most books, articles, and news reports start in the same place. First, they describe the pop-culture notion of Bigfoot and ask if the animal is real. Then, they point to an accumulated "mountain of evidence" for the creature and look back to historical accounts that appear to resemble modern Sasquatch lore.

This approach (common in the coverage of all cryptozoological claims) starts at the wrong end of history. Talking about the myth in its current form is beside the point. We know that a legend now exists, and, as anthropologist John Napier warned, "Few would deny that today the tales of the Sasquatch are subject to intense cultural reinforcement."[1]

What we want to discover is: Where the heck did the Bigfoot idea come from in the first place? Has the legend changed over time? Does the story feel compelling only because we are so steeped in decades of Bigfoot folklore and popular culture? Even before considering the scientific plausibility of Sasquatch, the historian's view may reveal that the "mountain of evidence" is an edifice built on sand. Where does Bigfoot come from? The answers may lie as far back as the origin of storytelling itself.

Ogres

The Field Guide to Bigfoot and Other Mystery Primates is an odd but intriguing pro-Sasquatch book that attempts to describe the many varieties of human-like creatures reported in eyewitness testimony around the world. It is certainly surprising to learn that modern witnesses have routinely reported Bigfoot-like creatures as large as 12, 15, or even 20 feet tall,[2] but many readers may be most taken aback by the decision by the authors of the guide to include the monster Grendel (from the Anglo-Saxon epic poem *Beowulf*) as a "mystery primate." This seems like an outrageous stretch, but it does help to spotlight an important truth: fantastic tales of humanoid monsters, ogres, and wild men were common in almost all cultures for thousands of years before the emergence of a canonical portrait of "Bigfoot." It is natural that giant lore should be a universal feature of human storytelling. After all, giants are the easiest monsters to imagine: bigger, stronger, wilder versions of ourselves.

Like other peoples worldwide, Native Americans told stories of ogres and wild men. Today, many Bigfoot enthusiasts ("Bigfooters") retroactively link these ancient stories to the modern conception of Bigfoot: "The first Americans acknowledged these hairy races, and their tales come down to us in the records that ethnographers, folklorists, and anthropologists have preserved in overlooked essays on hairy-giant legends and myths. Examining these closely, a pattern begins to emerge of Bigfoot revealed."[3] Making this connection has the effect of conflating a rich variety of diverse cultural traditions into a single, modern composite. At best, the projection of modern monsters into Native American stories flattens context and nuance; at worst, it may become a naive and even ugly sort of paternalism in which whites tell Native Americans what their tales "really" mean. As a practical matter, many of these appropriated stories are also a very bad fit. For example, Bigfooters routinely cite tales from eastern North America that feature man-eating "giants" whose bodies are literally covered with stone.[4] These stories are detailed right down to the creatures' use of magical talismans to locate human prey and their Achilles-heel weaknesses that allow warriors to escape (such as their inability to swim, look up, or gaze on menstruating women).[5] These stone-covered cannibal-wizards are clearly not close analogues to the fuzzy, gentle, largely herbivorous Bigfoot.

To their credit, some monster advocates are uneasy about this wholesale co-optation of Native American lore as evidence for Bigfoot. Pro-Sasquatch anthropologist Grover Krantz thought that "Native stories that can confidently be related to the sasquatch occur throughout the Pacific Northwest" but complained that "it is only with some difficulty that a sasquatch image can be read into" tales from elsewhere on the continent.[6] (In particular, Krantz doubted that legends of stone giants "who strike with lightning from their fingers" had "any physical referent at all.") Yet even in the Pacific Northwest, Native stories include a wide variety of monsters with human-like forms. Are the underground dwarves or the underwater people described by the Twana of the Skokomish River drainage "really" Sasquatches? What of the soul-stealing "wet-cedar-tree ogre"?[7] Should we project our expectations of Bigfoot onto the giants described by the Quinault of the Olympic Peninsula who look almost the same as humans—except for a 6-foot-long quartz spike growing out of the big toe of their right foot? (Recording this story in the 1920s, anthropologist Ronald Olson deadpanned, "If a human is kicked with this he will likely die.")[8] And what are we to make of the cannibal-ogress tales widespread throughout the region? In her depiction by the Kwakwaka'wakw, called Dzunuk'wa (figure 2.1),[9] the ogress is frequently presented by Bigfoot proponents as a distorted, mythologized representation of their modern monster.[10] However, there is little to justify this cultural appropriation. The Native peoples of the Pacific Northwest typically drew a sharp distinction between the cannibal ogress, on the one hand, and any of several more Sasquatch-like races of giants or feral humans, on the other.[11] The cannibal ogress was usually regarded not as a race but as a terrifying singular character, similar to Hansel and Gretel's witch or eastern Europe's Baba Yaga. Luring or snatching

Figure 2.1
The cannibal ogress depicted as Dzunuk'wa on a Kwakwaka'wakw heraldic pole. Carved in 1953 by Mungo Martin, David Martin, and Mildred Hunt, it is in Thunderbird Park at the Royal British Columbia Museum, Victoria. (Photograph by Daniel Loxton)

away unwary children, the cannibal ogress carried them off in the basket on her back, roasted them over a fire, and ate them. She is not Bigfoot.

Bigfooters sometimes resolve the discrepancy between stories of cannibal giants and of gentle Bigfoot by simply inventing additional species of mystery primates as needed. "In eastern North America," writes Loren Coleman, "a specific subvariety of manlike hairy hominid allegedly exists. It exhibits aggressive behavior. . . . A few hominologists have labeled these unknown primates Taller or Marked Hominids, and others have written of them as Eastern Bigfoot. What may be in evidence is actually an Eastern geographic race, perhaps a subspecies, of the more classic Bigfoot."[12]

Whether this retrofitting forces Native traditions into one or several Bigfoot-shaped boxes, it is an exercise in confirmation bias. Bigfoot enthusiasts look back over Native lore with an expectation of finding Bigfoot. They seize on any tales about a fabulous creature that resembles the Bigfoot they expected to find, while ignoring or reinterpreting the stories that do not. Then, having projected a modern Bigfoot into disparate Native legends, enthusiasts make the circular argument that Native American traditions confirm the existence of Bigfoot.

The Origin of the Sasquatch

It is inappropriate to link all Native tales of ogres or wild men to Bigfoot, but it is true that Bigfoot mythology has its roots in *specific* Native stories.

There is a ground zero for Bigfoot, a place and time when the legend can be said to have originated. In the 1920s, in British Columbia's lush Fraser Valley, a man named John W. Burns collected the original eyewitness reports of encounters with "Sasquatch." (This was a term that Burns apparently coined as an anglicization of a word in the mainland dialects of the Halkomelem language of the Coast Salish people.)[13] All of modern Bigfoot mythology grew from this regional seed: spreading across the North American continent and beyond, mutating and hybridizing with later reports, hoaxes, and popular fiction.

The earliest, purest accounts are as important for historians of Bigfoot folklore as they are problematic for those who advocate the existence of a living mystery primate. Unfortunately, the descriptions of the original Fraser Valley Sasquatch are almost completely different from those of the modern "Bigfoot."

Burns was a schoolteacher and bureaucrat ("Indian agent") who worked at the Chehalis Indian Reserve, near the town of Harrison Hot Springs. His friends told him of local legends about a race of giant wild people who lived in the mountains. "Persistent rumors led the writer to make diligent enquiries among old Indians," Burns recalled in a magazine article. His questions led, "after three years of plodding," to eyewitnesses: "men who claim they had actual contact with these hairy giants."[14]

Burns quoted these encounter stories at length. To those used to the modern Bigfoot, the original Sasquatch stories sound very strange. The witnesses repeatedly describe Sasquatches as "men," and for good reason. As pioneering Bigfooter John Green explained, "The Sasquatch with which Mr. Burns' readers were familiar were basically giant Indians. Although avoiding civilization, they had clothes, fire, weapons and the like, and lived in villages. They were called hairy giants it is true, but this was taken to mean they had long hair on their heads, something along the lines of today's hippies."[15] For example, a man named Charley Victor reported having accidentally shot a young Sasquatch. As he examined the injured boy, an older wild woman came out of the woods. Her "long straight hair fell to her waist." She conversed with Victor and performed a magic ritual to assist the boy. Victor concluded, "It is my own opinion . . . because she spoke the Douglas tongue these creatures must be related to the Indian."[16] It was common for the original witnesses to describe such conversations with the mountain giants. Burns quoted a speech that had been given before an approving crowd of 2,000 Chehalis Natives: "To all who now hear, I Chief Flying Eagle say: Some white men have seen Sasquatch. Many Indians have seen Sasquatch and spoke to them. Sasquatch still live all around here."[17] Indeed, a Chehalis woman told Burns that she not only had spoken to Sasquatches, but had lived among them—and had given birth to a baby fathered by one of the wild men![18]

Were Burns's informants pulling his leg about these personal interactions with Sasquatches? These stories certainly sound like tall tales. (Victor claimed in an aside that "I have in more than one emergency strangled bear with my hands.") It could be that the locals were teasing the white schoolteacher. Or it could be that Burns was in on the joke. (It is hard to know whether to take this comment as naive racism or as a knowing wink to his Native friends: "The Chehalis Indians are intelligent, but unimaginative, folk. Inventing so many factually detailed stories concerning their adventures among the giants would be quite beyond their powers.")[19] In any event, it is clear that the original Sasquatch lore did not describe the modern Bigfoot. How did the one legend give birth to the other? It all began with a publicity stunt . . .

HARRISON, BRITISH COLUMBIA

In 1957, the provincial government made funding available for projects celebrating the centennial of British Columbia. The town of Harrison Hot Springs would qualify for $600 of that funding if it could propose a suitable project. The village council brainstormed and settled on a scheme to dust off and exploit a largely forgotten local legend: Why not fund a Sasquatch hunt? The council took the proposal to the British Columbia Centennial Committee.

Bigfoot researcher John Green (who was then a newspaper owner in tiny, nearby Agassiz) put this in perspective. "It was, of course, a bid for publicity," he recalled, "and it was tremendously successful. Papers all over Canada played the story on the front page."[20] The Sasquatch hunt proposal was rejected (on the grounds that it was not a "permanent project"), but that hardly mattered. The Sasquatch was a hit as a marketing ploy. As Green explains, "Perhaps never before has a tourist resort achieved such publicity without actually doing anything. . . . Newspaper and radio reporters flocked around, and a tide of delighted stories rolled around the world touching Sweden, India, New Zealand and points between."[21] One of the news stories underlined the incentives involved in the rejuvenation of the Sasquatch legend. "The government spends large sums annually for tourist trade advertising and promotion," raved the *Vancouver Sun*, "without getting anything like the publicity Harrison has received before one red cent has been spent." Noting that this publicity was directly boosting the restaurant and resort industry, the newspaper urged locals not to express skepticism about Sasquatches to visitors.[22]

Meanwhile, the Sasquatch was back in a big way. Although it had rejected the up-front funding request, the British Columbia Centennial Committee offered a reward of $5,000 to anyone who could "bring in the hairy man alive."[23] (It was wisely stipulated that conveying Bigfoot to the ceremony could not involve kidnapping.) Small-scale expeditions were launched—one led by a very driven Swiss fellow named René Dahinden, who (like Green) would remain a leading figure in the search for Bigfoot for the rest of his life. (Dahinden's expedition investigated caves in the area, since original Sasquatch lore held that the wild people built rocky shelters or lived in caves, unlike the modern Bigfoot.)

Hoaxing was already a factor. In a pattern to be repeated often throughout Bigfoot history, some investigators were themselves hoaxers. When a different expedition set out from Lillooet, a newspaper happily reported that the men "hope to return here May 20 with 'definite proof' the Sasquatch live in the area."[24] The team's confidence was understandable, although ultimately foiled by a rockslide and a raft accident: when the men returned from their botched wilderness adventure, they admitted that they had intended to create Sasquatch tracks using plywood feet.[25]

Despite the silliness of much of the press coverage, this was the watershed moment for the Sasquatch. As Dahinden put it, "The widespread publicity appears to have been the genesis of the serious consideration that the Sasquatch might indeed be a creature of considerably more substance than myth."[26] In the midst of the hubbub, John Burns, the original recorder of Sasquatch lore, returned to affirm that Sasquatches were large human beings. As the *Vancouver Sun* reported, "Burns believes the sasquatch originated in British Columbia and is of Salish descent. Various Indians he has talked to say sasquatches they encountered speak the same language."[27]

But all that was about to change.

A SERIES OF SIGHTINGS

William Roe: The Most Important Sasquatch Case in History?

As the Harrison Hot Springs media storm wore on, a new witness named William Roe went to the press with a dramatic story that is now recognized as the first fully modern Sasquatch sighting. Roe's close-encounter tale effectively created the modern Sasquatch, giving the cryptid its now canonical appearance and behavior.[28] As pioneering pro-Bigfoot researcher John Green explained, Roe "was the very first to describe a Sasquatch as an ape-like creature rather than a giant Indian."[29] Before Roe's sighting, eyewitnesses in the Fraser Valley had described Sasquatches as fundamentally human in appearance and behavior: using fire, speaking fluently, living in villages, and so on.[30] With Roe, the Sasquatch transformed into a mystery primate—a primate that white people could see (figure 2.2).

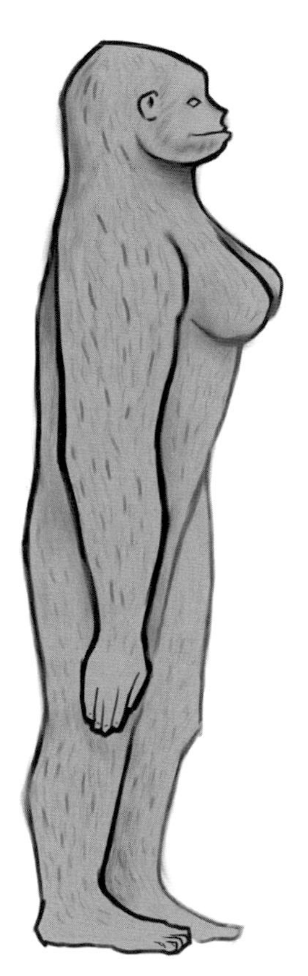

Figure 2.2
The Sasquatch as described by William Roe. (Illustration by Roe's daughter for John Green; redrawn by Daniel Loxton from John Green, *On the Track of the Sasquatch* [Agassiz, B.C.: Cheam, 1968])

In a sworn statement, Roe claimed that he had been hiking one afternoon in 1955 when he came across a large animal at the edge of a clearing. As he sat down to watch, "It came to the edge of the bush I was hiding in, within twenty feet of me . . . close enough to see that its teeth were white and even." The creature squatted down in the open to eat leaves. Roe's description is so precise, and so influential, that it is worth quoting here at length:

> My first impression was of a huge man, about six feet tall, almost three feet wide, and probably weighing somewhere near three hundred pounds. It was covered from head to foot with dark brown silver-tipped hair. But as it came closer I saw by its breasts that it was female. . . . Its arms were much thicker than a man's arms, and longer, reaching almost to its knees. Its feet were broader proportionately than a man's, about five inches wide at the front and tapering to much thinner heels. When it walked it placed the heel of its foot down first, and I could see the grey-brown skin or hide on the soles of its feet. . . . The head was higher at the back than at the front. The nose was broad and flat. The lips and chin protruded farther than its nose. But the hair that covered it, leaving bare only the parts of its face around the mouth, nose and ears, made it resemble an animal as much as a human. None of this hair, even on the back of its head, was longer than an inch. . . . And its neck also was unhuman. Thicker and shorter than any man's I had ever seen.[31]

In this account, we may see the seeds of every subsequent Sasquatch tale—and the inspiration for one famous film in particular. "Finally," wrote Roe, "the wild thing must have got my scent, for it . . . straightened up to its full height and started to walk rapidly back the way it had come. For a moment it watched me over its shoulder as it went, not exactly afraid, but as though it wanted no contact with anything strange."[32]

To a profound extent, Bigfoot lore stands or falls on Roe. If true, Roe's story remains among the most detailed and informative close-range sightings of an undiscovered primate. If Roe invented his creature, however, then a very dark shadow falls over all the reports that parrot his description—which is to say, most of the cases in the Bigfoot database. In particular, the famous footage shot by Roger Patterson and Bob Gimlin, in which an apparent Sasquatch walks across a dry riverbed, depends utterly on Roe's veracity. Patterson wrote about Roe's story in 1966 and the next year filmed a virtually identical encounter with a virtually identical creature—right down to the heels-first gait, hairy breasts, and unhurried look over the shoulder while walking away from the filmmaker. If Roe was a hoaxer, it follows that the Patterson–Gimlin film must also be a hoax.

With so much riding on Roe's sighting, I assumed that a great deal would be known about Roe. After all, Bigfoot writers have promoted this encounter for more than fifty years. Looking into the case for a magazine feature in 2004,[33] I soon noticed that all the sources seem to rely exclusively on the statement that Roe sent to Green (or, worse, on secondhand sources who relied on Green).

It began to dawn on me: no cryptozoologist had ever met Roe.

Could that really be right? I contacted Green, the Bigfooter who made Roe's famous "sworn statement" part of history, and asked him directly: Did any researcher ever personally speak to William Roe or look him in the eye? Did anyone make an attempt to meet with him? Is anything substantial known about his character or background?

Nope. According to Green, "By the time I wanted to talk to him he had moved to Alberta. I don't know who may have spoken to him in person."[34] Nor did any researcher visit the site of Roe's alleged sighting. "At a later stage we would certainly have wanted to question Roe in person, and to go to the site with him," Green told me, "but by the time I was writing a book he was dead."[35] I asked cryptozoology author John Kirk (head of the British Columbia Scientific Cryptozoology Club) for confirmation. Kirk answered, "I am unaware of anyone who met him. . . . Green did not ever meet Roe nor did any researcher I know."

It is hardly the fault of Bigfoot proponents that Roe passed away. And yet it remains bizarre that one of history's most-cited Sasquatch witnesses was never once questioned by anyone. Worse, virtually nothing is known about the circumstances of Roe's story (was he even in the area at the time the story takes place?) or how it first came to be told. As Green conceded (discussing Roe's story and another from the same period), "There was no one else involved with whom they could be checked, and no records from the time and place where they were supposed to have happened."[36] Worse, Green was unable to find anyone who had heard Roe's story before the media circus surrounding the Harrison Hot Springs "Sasquatch hunt" in 1957.[37] Roe alleged that his sighting had occurred two years earlier, but that is nothing more than a bald assertion. We are asked to believe that he saw a Sasquatch, said nothing to anyone for two years, and then took his improbable tale to the press—after the publicity started.

Hearing or reading this original press version, Green was inspired to write to Roe. Roe then provided Green with his "sworn statement"—the only version of his story that history preserves. Does Roe's later "sworn" version match his original telling? No one knows. Green was unable to tell me where he first encountered Roe's tale (or what Roe said at that time): "Sorry, I wasn't keeping records in those days."[38] Kirk wrote to me that Roe's story or stories are "believed to have been in the *Province* and *Sun*" newspapers "in 1957 prior to August"—or possibly on the radio.[39] Unfortunately, despite long searches through the microfilm archives for both newspapers, I have been unable to locate Roe's original lost account.

For all its influence, the Roe case is ultimately a story told by an unknown figure, for unknown reasons, under unknown circumstances. Because there is absolutely nothing to corroborate this story, a great deal of ink has been spilled in arguing that Roe's "sworn statement" renders the account reliable. As a Canadian, I am amused by early cryptozoology author Ivan Sanderson's goofy pitch that "while 'sworn statements' may not cut too much ice in this country [United States], they mean a very great deal in Canada and other parts of the British Empire. Canadians have an intense respect for the law."[40] (I am reminded of filmmaker Michael Moore's hilarious suggestion in *Bowling for Columbine* that Canadians do not lock their doors.) Of course, this argument is just silly. People lie in all countries—sometimes for no very good reason, and sometimes under oath. Criminal justice would be much simpler if "swearing" guaranteed truthfulness, but sadly it does not. That's why lawyers cross-examine witnesses in courtrooms, try to gauge their reactions, and compare their testimony with other evidence. No one did that with Roe. We have no idea whether his story was consistent from telling to telling. In fact, we do not even know what William Roe looked like.[41]

Raymond Wallace: The Bluff Creek Tracks

As the Canadian film and music industries know all too well, it is one thing to top the charts in Canada and another to achieve cross-border success. For all the press that the Harrison Hot Springs "Sasquatch hunt" received in 1957, the legend retained a quaint Canadian flavor until the following year. In 1958, in the region surrounding Bluff Creek, California, the Sasquatch—re-branded Bigfoot—became a break-out American hit.

It started slowly. In the early spring, some giant footprints showed up at a work site of a road construction contractor named Raymond Wallace.[42] These tracks were fake. As summer wore on into autumn, more giant footprints appeared, again at a Wallace work site. They were cast in plaster (by

Wallace's employee Jerry Crew), becoming the most famous "real" Bigfoot tracks of all time. Then Wallace personally faked further Bigfoot tracks late in the year—and remained a Bigfoot prankster for the rest of his life.

If that sequence brings you up short, I can sympathize fully. And yet, the tracks from Wallace's Bluff Creek work site are fiercely and universally defended by Bigfoot authors. John Green insists, "The tracks that were observed in the Bluff Creek drainage in northern California in the 1950's are not just another set of tracks that can easily be set aside as something tainted by claims of fakery while other tracks are still presumed to be genuine. They are the base layer of the bedrock on which the whole investigation is founded."[43]

The story unfolded like this. When bulldozer operator Jerry Crew arrived at the Bluff Creek work site on August 27, 1958, he was annoyed to find giant footprints stamped around his machine. Crew took them for a prank. After that, Ivan Sanderson relates, "nothing further happened for almost a month; then once again these monstrous Bigfeet appeared again overnight around the equipment. . . . About that time, Mr. Ray Wallace, the contractor, returned from a business trip."[44] New tracks were appearing regularly at the beginning of October. "Every morning we find his footprints in the fresh earth we've moved the day before," Crew said.[45] On October 3, Crew made a plaster cast of one of the footprints. Then, taking this cast to the local newspaper, he made history. Throughout California, headlines trumpeted the mystery of "Bigfoot."

Within days, suspicion fell on Wallace. The local newspaper reported on October 14 that a deputy had spoken to Wallace and that "the sheriff's office had sent word to Ray Wallace . . . to come in and explain the 'joke.'"[46] Wallace denied having made the prints, but the sheriff's suspicion was well placed. After Wallace's death in 2002, his family revealed one of the sets of strap-on wooden feet that Wallace had used for his track hoaxing. This particular set does not match the tracks that Crew cast in plaster, but it does closely match other tracks found in the area that year. Loren Coleman affirms forthrightly, "Yes, Wallace appears to have placed prank footprints near some of his work sites from 1958 through the 1960s."[47] Nor was that the beginning or the end of Wallace's Bigfoot-related mischief. "Ray Wallace lived a life of hoaxing, stretching the truth, and pranks," Coleman explains. "It was happening before Bluff Creek, 1958, and after."[48]

The Wallace family's announcement in 2002 set the cat among the pigeons. It exposed a deep weakness in the case for Bigfoot: a known Bigfoot prankster had his fingerprints all over a truly central Bigfoot case, and no one was talking about it. Worse, Wallace's hoaxing had been known from the earliest days. He was suspected immediately as the creator of the tracks found at his work sites; his status as a hoaxer was confirmed two years later when he claimed to have captured a live Sasquatch. Offering to sell the creature to Bigfoot researchers for $1 million, he dragged out the negotiations for weeks—only to "release" the creature. (Incidentally, Wallace claimed that the captured Sasquatch would eat nothing except 100-pound bags of Kellogg's Frosted Flakes.)[49] For decades afterward, Wallace continued to produce fake tracks, make headlines, and pester Bigfooters with outlandish tall tales. "Bigfoot used to be very tame," claimed Wallace in one typical yarn, "as I have seen him almost every morning on the way to work. I would sit in my pickup and toss apples out of the window to him. He never did catch an apple but he sure tried."[50]

Bigfoot researchers knew not only that Wallace was a prankster and that the tracks cast by Crew were found on Wallace's work site, but also that the major source of testimony about the monster allegedly throwing around heavy equipment at the site was Wallace's brother. Crew himself was Wallace's employee—as were, coincidentally, the two men who claimed to have seen a 10-foot-tall Bigfoot just days later! As the local newspaper noted, "Tracks in the dusty road were identical with those seen in the construction area," except in one strategic respect: the new incident allegedly took place in a neighboring county. Just one day after reporting that the sheriff of Humboldt County wanted to question Wallace, the front-page headline read, "Eye-Witnesses See Bigfoot: Humboldt Sheriff's Office Has No Jurisdiction in Footprint Case."[51]

Wallace was powerfully implicated in the Bluff Creek tracks all along, and yet somehow books discussed the case for decades without shining the spotlight on him. As one critical journalist put it, "Bigfooters dodge the subject of Ray Wallace—it's their dirty little secret, and the fewer questions raised about his involvement at Bluff Creek, the better."[52] Indeed, as David Daegling pointed out, pivotal books by John Green and Grover Krantz "fail to even acknowledge that Ray Wallace existed." According to Daegling, "Green and Krantz are generally meticulous in keeping track of details of events and persons, so it seems very unlikely that the omission of Wallace is a mere oversight."[53] If so, Daegling's view of this "whitewash" is scathing: for investigators to "simply opt not to disclose the contextual information of evidence when it is crucial for deciding the legitimacy of particular data" is "lying by omission."[54]

While Wallace's silly antics must have infuriated serious proponents, their silence was destined to backfire. In the 1990s, *Strange Magazine* editor Mark

Chorvinsky began to pointedly ask why Wallace had been "excluded from, or downplayed in, most of the official Bigfoot histories?"[55] When Wallace's family came forward with a set of his wooden feet for making tracks, they ended the long silence once and for all. Headlines worldwide proclaimed "the death of Bigfoot."

Forced to confront the issue, Bigfoot investigators hastened to argue that Wallace's connection to the footprints is a red herring. Many scornfully pointed out that the tracks cast by Crew do not match the wooden feet that the Wallace family had shown to the press. Calling this objection "insipid," Daegling noted that "Wallace's family maintains he had not one, but several sets of bogus feet."[56] Besides, the wooden feet revealed by the family are a *perfect* match to many of the *other* early Bluff Creek tracks.[57] "Wallace did not hoax the Jerry Crew footprint finds of 1958," in Loren Coleman's opinion, "but Wallace certainly left other prints that were not authentic, around Bluff Creek, which have ended up in Bigfoot books."[58] This contaminated database is a significant practical problem, as Coleman has continued to emphasize: "The databases of various Sasquatch researchers continue to contain examples of Ray Wallace's fakery. Because of this, the collections throw off the summary statistics and general analyses of Bigfoot track information."[59]

Nonetheless, few Bigfoot proponents seem concerned. Some have denied the problem altogether: "To even consider that Wallace, or anyone else for that matter, could produce convincing sasquatch footprints with a piece of carved wood is absurd."[60] Others argue that skeptics have the story backward. According to Green, the wooden feet revealed by the Wallace family are forgeries based on genuine Bluff Creek Sasquatch tracks. Noting that Wallace's wooden feet do, in fact, match some of the tracks found at Bluff Creek in 1958, Green concludes that "presumably they were made in imitation of those casts."[61] Readers will be forgiven if they find their patience strained by this degree of special pleading.

All of this is damning, not only to the case for Bigfoot, but also to the integrity of the search. If cryptozoology is to achieve its scientific aspirations, it cannot afford to be precious about even its own "bedrock" cases or to be unduly deferential to figures respected in the cryptozoological subculture. Coleman has ventured the opinion that the failure of Bigfooters to acknowledge the depth of the problems posed by the fake Wallace tracks is "due to the godlike status given to John Green and Jeff Meldrum," with people "afraid to speak up" in challenge to the positions taken by those prominent figures. Whatever the social forces at play, the result is as Coleman describes: "There remain bad apples in the Bigfoot baskets and they are contaminating the entire analytic apple pie. That is scientifically significant."[62] It certainly is. But how do we put it right? At the very least, we must acknowledge that the footprints cast by Jerry Crew and the other Bluff Creek tracks are contaminated by their close association with a habitual hoaxer and must be quarantined from the database of Bigfoot evidence. Indeed, I would say that they are radioactive—an almost certain hoax. (The clincher, in my view, is the level of coincidence that we are asked to accept: giant footprints found first at one Wallace work site and then at another Wallace work site, and Wallace employees later spot Bigfoot at a *third* location.)

As Wallace's son put it to reporters, "Ray L. Wallace was Bigfoot. The reality is, Bigfoot just died." Given the pivotal importance of the Bluff Creek tracks, I cannot help but think that he was right.

Roger Patterson and Bob Gimlin: The Bigfoot Film

On October 20, 1967, cowboys named Roger Patterson and Bob Gimlin rode into the California woods to film Bigfoot. And then they promptly did.

Shot on a bright autumn day at Bluff Creek in northern California—the same region where giant tracks at Ray Wallace's construction site had launched the legend of "Bigfoot" a few years earlier—the shaky, hand-held 16-mm film shows a furry, bipedal figure striding confidently across a gravel sand bar. Alleged to be a female Sasquatch (inferred from what appear to be heavy breasts on the figure's torso), this bulky-looking creature passes behind horizontal logs, low bushes, trees, and debris. As it walks away, it looks back, hauntingly, toward the filmmaker (figure 2.3).

When enthusiasts, skeptics, or mainstream journalists turn their attention to the Sasquatch, they almost invariably refer to the Patterson–Gimlin film as the "most important" evidence for Bigfoot. In social terms, this is true. The film is an enduring icon that keeps the question of the reality of Bigfoot alive in the public imagination and unifies the subculture of Bigfoot advocates. For many, it is an emotional powder keg: visceral, saw-it-with-my-own-eyes *proof* that Bigfoot exists.

As practical evidence for an undiscovered ape, however, the Patterson–Gimlin film is a dead end. I am going to say something that will be unpopular on both sides of the debate: *no one knows whether the film depicts a real Sasquatch or a man in a gorilla suit*. Moreover, after decades of argument and

Figure 2.3 The famous still from the Roger Patterson-Bob Gimlin film of Bigfoot. (Reproduced by permission of Fortean Picture Library, Ruthin, Wales)

analysis, there is no sign that the issue can be resolved unless one of three things emerges: a live or dead Sasquatch, powerful new documentary or physical evidence that exposes the film as a hoax (such as the suit itself),[63] or a confession from co-witness Bob Gimlin.

This is not an entirely traditional view of the case. Through decades of deadlocked argument, passionate voices on both sides have insisted that conclusive evidence is available, if only the other side would really *look* at the film. Indeed, many have claimed that the film itself is the only relevant line of evidence to pursue when considering the film's authenticity. "Ignore the human element," urged original investigator René Dahinden.[64] According to John Green, a friend of Patterson's and another of the first people to view the film, "The character of the person holding the camera has little bearing on whether a film is genuine, you have to study the film itself."[65]

This sounds fair, but it does not help to determine the truth. The film is a Rorschach inkblot. For skeptics, the creature obviously is a man in a gorilla suit. For believers, the creature plainly appears nonhuman. And while one might hope that analyses by qualified anthropologists would resolve the impasse, this has not turned out to be the case.

Consider the contrasting opinions of anthropologists John Napier and Grover Krantz. Both thought that the Sasquatch was a real animal, but their anthropological expertise brought them to diametrically opposed conclusions regarding the Patterson–Gimlin film. According to Napier, "There is little doubt that the scientific evidence taken collectively points to a hoax of some kind. The creature shown in the film does not stand up well to functional analysis."[66] Krantz argued the exact opposite: "No matter how the Patterson film is analyzed, its legitimacy has been repeatedly supported. The size and shape cannot be duplicated by a man, its weight and movements correspond with each other and equally rule out a human subject; its anatomical details are just too good."[67]

According to Jeff Meldrum (an anatomist and a high-profile Sasquatch proponent), "This film remains among the most compelling evidence for the existence of sasquatch, detractors and skeptics notwithstanding."[68] For their part, those detractors and skeptics pull few punches. Napier was "puzzled by the extraordinary exaggeration of the walk," asking, "Why ruin a good hoax by ordering my actor to walk in this artificial way?"[69] Daniel Schmitt, an evolutionary anthropologist and a specialist in primate locomotion, echoed Napier's conclusion, joking that "either this is a person trying to walk funny, or

Bigfoot walks in a manner that is more or less identical to people walking this way."[70]

Over forty years, an astonishing amount of ink has been spilled in analyzing the film. Re-creations have been staged and filmed, computer-animated scenes have been reconstructed, and countless diagrams have been drawn. At the end of this incredible effort, investigators still cannot agree on the basic facts about Patterson's creature, such as how tall and heavy it was.

In the absence of either a type specimen for Bigfoot or smoking-gun evidence of a hoax, the film is unable, ultimately, to speak for itself. Luckily, circumstantial factors may cast light on the plausibility of the case: Roger Patterson's character, the serendipitous circumstances of the filming, hearsay testimony that the film was a hoax, the apparently derivative structure of Patterson's narrative, and the money that Patterson earned from the film.

- Proponents traditionally defend ambiguous cases of alleged cryptid encounters with the argument that the eyewitnesses are well-respected people of good character (ideally, professionals like doctors or police officers). For example, early cryptozoology author Ivan Sanderson described Jerry Crew, who preserved the Bluff Creek tracks, as "an active member of the Baptist Church, a teetotaler and a man with a reputation in his community that can only be described as heroic."[71] That option does not exist for Roger Patterson, a man characterized by even his defenders as "no angel . . . a slapdash kinda guy that you really shouldn't do business with"[72] and a "used-car salesman type of personality."[73]

  Greg Long made the first significant progress in the case in decades. For a book published in 2004, he adopted a novel strategy: ignore the film and examine the filmmaker. What emerged was not pretty. Interviews with members of Patterson's family, friends, and colleagues (conducted over several years) paint a picture of a highly artistic, small-town hustler with dreams of the big score. A stage acrobat, carny, inventor, illustrator, Bigfoot sculptor, self-published Bigfoot author, and semi-pro rodeo rider, Patterson lived and breathed show-business schemes. He aggressively courted Hollywood long before filming Bigfoot.[74] He also seems to have ripped off everyone he met, from family[75] to friends to total strangers. (Bigfooter Dahinden accused him of "mail fraud, pure and simple.")[76] In the context of his Bigfoot film, that's a bad combination indeed. As Long concludes, "Roger Patterson's character fails the smell test. Sum up all the information about

Roger Patterson, and it comes down to two simple points. One, he had the ability to conceive of and create a Bigfoot suit, and two, he was a crook."[77]

- Whether Patterson was a saint or a sinner, the basic circumstances of the filming strain credibility. Patterson, a Bigfoot author, set out on a camping trip to film Bigfoot—and then he promptly did. As Benjamin Radford notes, this is a remarkable happenstance: "Patterson told people he was going out with the express purpose of capturing a Bigfoot on camera. In the intervening thirty-five years (and despite dramatic advances in technology and wide distribution of handheld camcorders), thousands of people have gone in search of Bigfoot and come back empty-handed."[78] Such extraordinary good fortune is not, however, unique in the annals of Bigfoot research. Others—including Ray Wallace, Ivan Marx, and Paul Freeman—have exhibited a remarkable ability to find tracks, sight Bigfoot, or film the creature more or less on demand. The explanation for such good luck has often turned out to be fraud. Could that be the case with the Patterson–Gimlin film?

- Patterson insisted that Bigfoot was "no farce, hoax or idiotic scheme. On the contrary, this could be the biggest scientific breakthrough in the study of the development of man since the beginning of time."[79] But as Daegling notes, "I have yet to correspond with anyone acquainted with Roger Patterson who thought he was above faking a film."[80] Indeed, we know that Patterson sometimes did film staged Bigfoot evidence. Explaining that fake Sasquatch footprints can be "dug into the ground with the fingers and or a hand tool," Krantz rather stunningly recalled, "Roger Patterson told me he did this once in order to get a movie of himself pouring a plaster cast for a documentary he was making."[81] Krantz added, without apparent concern, "A few days later he filmed the actual sasquatch."

Long presents extensive testimony from Bob Heironimus, who claims that he played the creature in the Patterson–Gimlin film. The "Bob H hypothesis," as it is known, is of course fiercely opposed by the Bigfoot community. Critics point out, correctly, that the truth of Heironimus's story cannot be confirmed. The case currently boils down to conflicting testimony from two close neighbors: Heironimus's confession that he took part in the hoax staged by Patterson and Gimlin, versus Gimlin's rebuttal that there was no hoax. Circumstantial evidence, though, is supportive of Heironimus. Living in the same town, Heironimus did know Patterson and Gimlin very well at the time. He even appeared in a different Bigfoot-related film project by Patterson, produced before the famous Patterson-Gimlin film.[82] Bigfooter John Green, a strong critic of Heironimus's tale, notes further corroborating evidence while also weighing in with an objection: "There is testimony that Heironimus had a fur suit, which he presumably used to play tricks on someone, but it is Heironimus' changeable word alone that connects the suit in any way with Roger Patterson."[83] Testimony from several witnesses puts a fur suit in Heironimus's trunk in the correct period,[84] and it is clear that local rumor has linked Heironimus to the Patterson–Gimlin film since the late 1960s. Whether or not his story can ever be confirmed, it is certainly plausible.

- It is also suspicious that the Patterson–Gimlin film echoes William Roe's alleged encounter with Bigfoot so exactly. From the settings (clearings) to the descriptions (the creatures are identical down to the fine details, including their "silver-tipped" hair and heavy, hair-covered breasts) to the behavior (the creatures unhurriedly walking away, each with a haunting look "over its shoulder as it went") to the reactions (the witnesses watching over rifle barrels, unwilling to shoot such human-like creatures), the two accounts are practically clones. This is an "amazing set of similarities," as Long put it. "In fact, so amazing I'd say this is where Patterson came up with the 'script' for his film."[85] We know that Patterson was willing to copy Bigfoot scenes from others, at least for some purposes: a book that he published in 1966 contains illustrations that he copied without credit from other artists.[86] We also know that Patterson was familiar with Roe's sighting because the book includes Patterson's illustration of Roe's encounter—a drawing that looks for all the world like a storyboard panel for the film that Patterson shot a year later. This suggests to many skeptics that the Patterson–Gimlin film is a hoax based on Roe's story. Proponents counter that the similarities are equally consistent with accurate independent sightings of females from the same unknown primate species. This point is debatable, but it assumes that Roe's account is genuine. If Roe's report is a hoax, we would be compelled to conclude that the Patterson–Gimlin film is also a hoax.

- Finally, Patterson made a lot of money from his Bigfoot film. Whether he earned enough money to justify a hoax is a subjective question, but the film was clearly a windfall by Patterson's perpetually broke standards. "One day he held out a $100,000 check that he got for the movie," Patterson's brother recalled. "At that time, that was quite a bit."[87]

  To promote and distribute the film, Patterson partnered with his brother-in-law Al DeAtley. Putting up the money to produce a feature-length movie from Patterson's footage, DeAtley took the show on the road. The business model was to roll from town to town, renting movie screens for private showings. With aggressive promotion of each screening as an event, they packed theaters with paying ticket holders. DeAtley recalled celebrating in their hotel room after the opening-night screening in a school auditorium: "So everything was cash—ones, fives, tens, and twenties. And we had a trash can full of money, and we were throwing it on each other on the bed and stuff!"[88] (Hearing this, I can't help thinking of the cartoon character Scrooge McDuck.)

Ivan Marx: The Bossburg "Cripple Foot" Fiasco

If the tracks at Bluff Creek, California, made Bigfoot a household name and the Patterson-Gimlin film gave the creature its enduring public face, it was the "Cripple Foot" case in Bossburg, Washington, that seduced Bigfoot's highest-credentialed scientific defenders.[89]

In 1969, Bossburg (then a dying mining town, and now a ghost town) was the new home of a Bigfoot hunter named Ivan Marx.[90] In an extraordinary coincidence, Bigfoot tracks promptly turned up at Marx's local garbage dump in late November—even more coincidentally, discovered by Marx himself.[91] These tracks, left in soft soil,[92] had a distinctive characteristic: the right foot appeared to be severely malformed (figure 2.4). For Bigfoot researchers, this was an electrifying find. Unexpected and persuasive, the malformed footprints seemed to scream, "I was made by a real biological creature." René Dahinden arrived within days; other researchers were close behind.

Further evidence (and coincidence) was not long in coming. On December 13, Marx and Dahinden went to check on some meat they had left out as bait. Getting out of the car to inspect one spot, Marx was gone for "only seconds before he came racing back," shouting that he had discovered Bigfoot tracks in the snow.[93] These 1,089 tracks, again showing the "crippled" foot, meandered

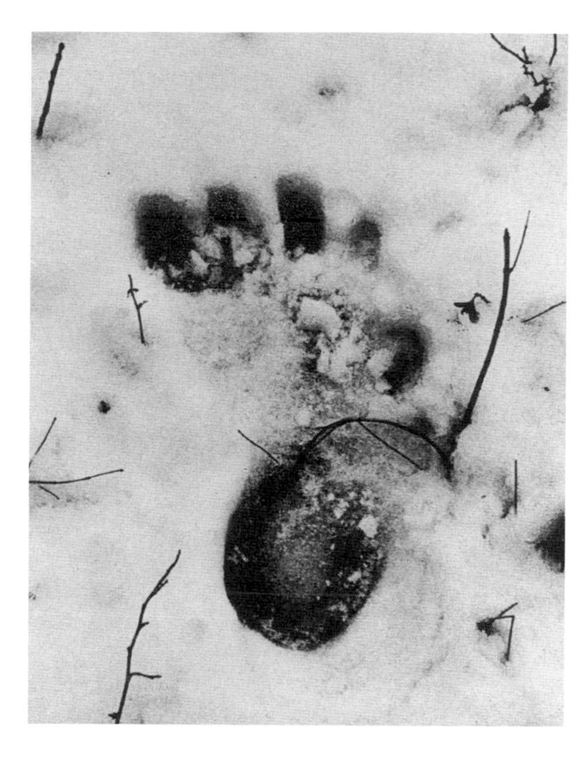

Figure 2.4
One of the "Cripple Foot" tracks, showing the unusual shape of the right foot that gave the case its name. (Reproduced by permission of Fortean Picture Library, Ruthin, Wales)

aimlessly. Conveniently starting at a river, where footprints would not be recorded,[94] the trail turned back on itself—crossing and recrossing a railway track, and crossing the same road and fence "several times"—before ending at the river, where it began.[95] This route seemed a bit artificial. As Dahinden asked himself, "Why did the tracks happen to be just there, where [Dahinden] would be sure to go every day, where he checked all the time, within a few miles of the garbage dump where the thing had been reported seen all the time? It was the obvious place for a hoaxer to plant his work."[96]

In any event, the find led to a kind of Sasquatch gold rush. Everyone involved in the search for Bigfoot descended on Bossburg, armed with everything from tranquilizer guns to aircraft. This stampede reached a fever pitch when a hoaxer named Joe Metlow came forward with the claim to have a Sasquatch for sale. This bait triggered first a bidding war; then stakeouts and even a madcap chase on January 30, 1970, as competing "groups of hunters left Colville by plane, helicopter, in various vehicles and on snowshoes in an attempt to find the beast on their own";[97] and, eventually, the embarrassment of all involved.[98]

The hunters left, but the now-infamous Bossburg fiasco was far from over. As Dahinden relates, Marx continued to find Bigfoot evidence with unbelievable regularity: "It seemed that every time he called, Marx had found something; a handprint here, a footprint there, signs of an unusually heavy creature bedding down in the bush; always something to keep the trail warm."[99]

And then, the biggest news yet: Marx announced that he not only had seen the creature with his own eyes—but also had captured full-color movie footage![100] Monster hunters rushed back to Bossburg to bid on Marx's footage, offering the kind of money that would buy a new house.[101] Then well-known Bigfoot hunter Peter Byrne swooped in with an odd arrangement: Marx agreed to put the film in a safe-deposit box, and Byrne put Marx on retainer (with a very hefty monthly salary, new camping gear, a snowmobile, and a new truck).[102] Not bad for simply having some unverified film.

Meanwhile, new Bigfoot tracks turned up in the neighboring village of Arden—a staggering 5,000 footprints stomped across Arden's fields and town dump as well as around the grocery store. Thousands of tourists flocked to see these tracks. Marx, on Byrne's payroll, went to the scene and declared his plan to tranquilize Arden's Sasquatch as soon as his dogs picked up the scent.[103]

But Byrne was less gullible than his deal with Marx seems to imply. He was literally buying time to dig into the circumstances of Marx's too-lucky film. Sure enough, he soon discovered the truth: Marx had faked his movie. Local children tipped off Byrne about where the film had been shot, leading Byrne to conclude that Marx "had misled us about the site of the footage."[104] As he told the newspapers, the film had been made "about 10 miles from where Marx claims it was."[105] With the site identified, landmarks revealed that the creature was also much smaller than Marx had claimed. Worse, the direction of the shadows in Marx's movie differed from those in his still photographs. This debunked his claim that they had been taken seconds apart; the "Sasquatch" had clearly worked with the photographer for some time. Finally, it was discovered that Marx had been spotted buying fur pieces in a nearby town.[106]

While Byrne was piecing together the true story, the Arden tracks were also revealed as a hoax. It turned out that a local bricklayer named Ray Pickens had created them, inspired by the news of Marx's nearby "Cripple Foot" tracks. "I just wanted to show that anybody could fake them," Pickens explained. His bogus feet were carved from 2- × 10-inch planks, and then nailed to boots. Notably, these confessed fakes had an inhumanly long stride, a characteristic typically cited by Bigfoot hunters as a sign of legitimacy. "The stride I made was about 54 inches—not at all hard to make," Pickens explained. "You can do the same—just trot."[107]

As news of Pickens's hoax circulated, Byrne moved to confront Marx about "the fact that the footage was obviously a total fabrication." He was too late. Marx had literally fled in the night, leaving his front door flapping in the wind and his belongings strewn across his front yard. When the safe-deposit box was opened, the only thing found were clips from old black-and-white Mickey Mouse cartoons.[108] (In subsequent years, Marx promoted several other Bigfoot films that are universally acknowledged as hoaxes.)

Where does this leave us? The bottom line is that the famous "Cripple Foot" trackways were "discovered" by a repeat Bigfoot hoaxer. This is a red flag so huge and glaring that the evidence from Bossburg should long ago have been set aside by all serious Bigfoot researchers. Bizarrely, the exact opposite has happened. Even knowing that Marx hoaxed "Cripple Foot"–related evidence at the time that he "found" the tracks, some Bigfooters find it impossible to set aside what they view as the inherent believability of the "Cripple Foot" prints. As one Bigfoot book put it, "In the case of Ivan Marx, the Bossburg 'cripple foot' prints were just too 'good' and perhaps too numerous for him, or anyone else for that matter, to have fabricated."[109] We can dispense with the statement that the footprints were "too numerous" to have been hoaxed—after all, Pickens created a much larger number of tracks in the next town—but what of the argument that the "Cripple Foot" tracks were too "good" to have been faked?

Anthropologist Grover Krantz was seduced into the Bigfoot search by these tracks. Attempting to reconstruct the underlying anatomy of the foot, Krantz became convinced that the prints had to be genuine. If someone did hoax them, Krantz asserted, "with all the subtle hints of anatomy design, he had to be a real genius, an expert at anatomy, very inventive, an original thinker. He had to outclass me in those areas, and I don't think anyone outclasses me in those areas, at least not since Leonardo da Vinci. So I say such a person is impossible, therefore the tracks are real."[110]

However, Krantz's fellow anthropologist David Daegling has very little confidence in Krantz's speculative reconstruction of the underlying anatomy of "Cripple Foot": "The problem with Krantz's argument is simply this: there was no demonstration at the time, nor has there been since, that a foot skeleton can be recreated out of a footprint with any degree of certainty. Investigation that has been done on this question suggests that footprints are simply not good indicators of underlying anatomy."[111]

For his part, John Napier felt that the Bossburg tracks were "biologically convincing,"[112] but his defense of the tracks boiled down, ultimately, to an argument from personal incredulity. "It is very difficult to conceive of a hoaxer so subtle, so knowledgeable—and so sick—who would deliberately fake a footprint of this nature," wrote Napier. "I suppose it is possible, but it is so unlikely that I am prepared to discount it."[113] Napier may have been prepared to discount the glaring probability that the "Cripple Foot" tracks were faked, but we cannot. Again, the source of the evidence is Marx, a known Bigfoot hoaxer. What can we do with evidence found by a man who (as one Bigfoot researcher who knew Marx put it) "would lie to you just to [screw] with you"?[114]

Nothing. We can't do anything with Ivan Marx's evidence. The "Cripple Foot" case is radioactive.

EVIDENCE

Filtering

The case for Bigfoot rests on two primary lines of evidence: the reports of alleged eyewitnesses and footprints. We'll consider both in a moment, but I want to first talk about a deep conceptual problem that runs through all Bigfoot evidence and, indeed, all of cryptozoology: how to filter good data from bad.

Everyone agrees that some eyewitness reports and footprints are hoaxes or mistakes. When pressed on this, cryptozoologists often concede the likelihood that *most* accounts may be inaccurate. On the face of it, this seems to be a small concession: after all, it would only take one confirmed case to prove that Bigfoot exists. But even if we assume for the sake of argument that there is a signal hidden in the noise (and there is no compelling reason to make this assumption), how could we detect it? In other words, would we know a Sasquatch if we saw one?

Eyewitness descriptions are wildly, outrageously variable; without type specimens, there is no way to determine which (if any) descriptions are accurate. For example, many Sasquatch witnesses have reported seeing creatures of heights that strain belief. Bigfooters reject such cases, based on little more than personal incredulity. As John Green noted, "From time to time someone will claim to have seen something like a Sasquatch that was 10 feet tall, or 12, or 14, but everyone just assumes they are mistaken."[115] Indeed, some witnesses report even taller creatures, but the assumption that they are mistaken seems hard to justify within a cryptozoological framework. Isn't a priori rejection of eyewitness testimony the sin of the close-minded "scoftics"?[116] Why reject firsthand accounts of 15-foot monsters while accepting 7-foot monsters on the strength of the same type of eyewitness evidence? Green reflected uneasily on this tendency: "Eight feet tall seems to be about right for a monster. It's big enough to be impressive, but not so big as to cause much controversy. . . . It would be comfortable, therefore, if one could assume that a full-grown Sasquatch is eight feet tall and leave it at that. For years, that's about what has been done."[117]

Cryptozoologists are really backed into a corner here. On the one hand, such reports seem self-evidently silly. On the other, *the entire point of cryptozoology is that eyewitness testimony about unlikely creatures should not be rejected on the basis of a priori implausibility*. How do cryptozoologists deal with this problem? Badly. Most just take it for granted that common sense can eliminate most bad data. Green notes that "trying to sort out the less-obvious fakes from the genuine information is a major task,"[118] but the situation is much grimmer. Without hard evidence (such as confessions), Bigfooters have no means to confirm that bad data is bad. What if Bigfoot really is made of titanium, as has been suggested?[119] No one can prove otherwise.

Sounding weird is not in itself evidence that a report is inaccurate; if it were, why bother with cryptozoology in the first place? And if eyewitness accounts are filtered through the preconceptions of the pro-Sasquatch subculture, how can we extract a composite picture of Bigfoot from the artificially selected database that remains?

Eyewitness Accounts

In any event, the fact remains: thousands of people say that they have seen Bigfoot. Everyone agrees that many of those witnesses are sincere. What *did* they see in the woods?

Many of these cases will never be solved. I have gone along with the convention of referring to the totality of eyewitness reports as a "database," but this conceals the deep chaos of the eyewitness record. Collected by amateurs over decades, encounter anecdotes are extraordinarily variable. This frustrates attempts to construct a composite Bigfoot or extract useful statistical information.

To begin with, many reports are fragmentary. Musing in 1970 about the possibility of doing statistical analyses of eyewitness reports, John Green was all too aware that the information in his physical card file, "cannot be considered very accurate, since many cards say no more than 'John Doe reported to have seen Sasquatch,' with no definite time or place or description and considerable room to question whether anyone actually saw anything."[120] Grover Krantz noted further obstacles to such analyses, since available reports "include vague descriptions, uncertain locations, erroneous observations, and some hoaxes—and we generally do not know which of the reports are tainted in these ways."[121]

Worse, as we have discussed, eyewitness reports are variable and diverge often from the popular, canonical Bigfoot. Some accounts feature Sasquatches with huge pointed ears, complex markings, or heights over 12 feet tall. Bigfoot is reported in many colors, at many sizes, with many diverging anatomies. In some reports, Bigfoot can speak human languages. And, although this is systematically downplayed in the mainstream Bigfoot literature, it is very common for witnesses to claim that Bigfoot has paranormal features and abilities, such as eyes that literally glow, psychic powers, or flying saucer–type vehicles.[122]

John Napier was hopeful that composite Sasquatch data could be useful under certain restricted circumstances, but he was not naive. Noting that "the reports of solid citizens can be as false as the ramblings of the town drunk,"[123] Napier gave a warning that should echo throughout Bigfoot research and throughout cryptozoology: "Eyewitness accounts offer considerable problems of interpretation. Individually they can probably be ignored; who knows under what conditions of exhaustion, mental stress, alcoholism or drug addiction, etc., the sighting was made? Was the eyewitness . . . hallucinated, fooled or simply lying through his teeth? There is no certain way of telling what factors were influencing his judgment."[124]

Misidentification Errors

One fundamental problem with eyewitness tales of Bigfoot encounters is that people make up stories. Another, even larger problem is that people make *mistakes*. As Grover Krantz noted, "With enough imagination almost any object of about the right size and shape can be seen as a sasquatch."[125]

Krantz was right about that. I know because I have seen it happen. I was a shepherd for about ten years, working in the remote wilderness along the

British Columbia side of the Alaska panhandle. We used crews of three people to herd flocks of 1,500 sheep on tree plantations (as a brush-removal tool for silviculture). One day, I returned to the flock to find shepherds Jill Carrier and Jolene Shepherd (a much remarked-on coincidence, that surname) talking excitedly. "While you were gone," they began, laughing a little nervously, "we, uh, saw a Sasquatch." They hastened to add that they were, well, *pretty* sure that it was a tall stump. After all, it stood still for a very long time, and half-burned stumps and broken logs are everywhere on overgrown clearcuts in British Columbia. Such logging debris (slash) can look like anything. A Sasquatch is a perfectly predictable occasional illusion, although this time it was especially compelling. The women laughed, confessing that they could have sworn that the Sasquatch turned its head to follow them.

I will not keep you in suspense. It was indeed a stump. (We were able to examine it later.) But the illusion from the original vantage point was very powerful. Even staring right at it, we could not be sure that it was not a Sasquatch. If we had not had the chance to check it out, this sighting could well have become a strong case for the existence of Bigfoot. Think of it: an extended, broad-daylight sighting by multiple witnesses, all of them experienced outdoor professionals. And memory is collaborative and malleable. If we had left the plantation at the end of the summer believing that we had seen a Sasquatch, our memories of the event would have "improved" over time.

Nor are stumps the only source of misidentification errors. All skeptical discussions of Bigfoot note that the creature looks very much like a person and even more like a bear (figure 2.5). Moreover, the bulk of Sasquatch sightings come from bear habitat (and all Sasquatch sightings come, by definition, from human habitat). In 2009, a paper in the *Journal of Biogeography* found that Sasquatch distribution is essentially identical to black bear distribution. "Although it is possible that Sasquatch and *U[rsus] americanus* share such remarkably similar bioclimatic requirements," the authors note wryly, "we nonetheless suspect that many Bigfoot sightings are, in fact, of black bears."[126]

A typical response from Bigfoot enthusiasts is to simply deny that there is a compelling resemblance. "While we cannot discount these possibilities in some cases," wrote Christopher Murphy, "the difference between sasquatch, bears, and hikers are very obvious."[127] John Bindernagel similarly asserted, "Although it has been suggested that both black bears and grizzly bears are regularly misidentified as sasquatches, I would suggest that this occurs only rarely and is most unlikely in the case of sightings or encounters lasting more

Figure 2.5
Two brown bears spar in Alaska. (Photograph by Dave Menke; U.S. Fish and Wildlife Service)

than one or two seconds."[128] Often, an illustration compares these "usual suspects" with an artist's conception of a Sasquatch—under the brightly lit, unobstructed, close-up conditions of a police lineup.

Some Bigfooters argue that bears are a poor fit on the grounds that it is anatomically impossible for bears to walk more than a few steps on their hind legs. I'll happily concede that extended upright walks are unusual for bears, but such walking behavior can occur in nature. I have seen video footage of an adult black bear striding along confidently on her hind legs for periods of twelve seconds (an eternity during a high-adrenaline Bigfoot sighting), and the effect is heart-stoppingly Sasquatch-like.[129]

But this is somewhat beside the point. Bigfoot sightings happen out in the world, where viewing conditions are never perfect and direct comparisons are not possible. Even if we were to grant that humans, bears, and Sasquatch look dissimilar, the simple fact remains: people make whoppingly huge misidentification errors all the time.

This was certainly the case during my shepherding career. It's not that my colleagues and I were poor observers. Quite the opposite. From dawn to dusk, a shepherd's primary job is to see everything: if 1 sheep out of 1,500 has a sore foot or is about to jump over a dangerous log, it's the shepherd's job to notice it. We were highly experienced with the terrain, wildlife, and observing conditions. But we still made misidentification errors. The reason is that we saw a tremendous number of things, all of them under variable real-world conditions.

We had no end of close encounters with wildlife, from wolves to grouse to moose to grizzlies. Added to those were thousands of sheep sightings per minute, plus our wide-ranging guardian dogs, plus our co-workers and herding dogs, plus a vast unknown number of trees, bushes, stumps, and logs. As the number of seeing events continued to increase, the relatively rare errors began to accumulate. Some of them were doozies. For example, several times I watched "grizzly bears," only to have them morph into sheep before my eyes. The opposite was also true: several times the object I initially identified as a sheep or a group of sheep turned out to be a grizzly or black bear. Likewise, seeing stumps as bears was so common that we gave this illusion a (not very imaginative) name. "Nah, just a stump-bear," we would say. (And sometimes we were wrong about *that*, too.)

In one memorable instance, Shepherd was napping in the bushes beside the sheep as they snoozed and ruminated. She was awakened by a small commotion (such catnaps are a one-ear-open sort of affair) and looked around to see a strange dog on the far side of the resting flock. Not stopping to wonder what a domestic dog would be doing in the remote wilderness, she charged through the bushes toward it, yelling and waving a stick. She burst through the brush . . . and stopped. Towering over her, almost close enough to touch, three grizzly bears (a mother and two grown cubs) stood up in surprise.

Mistakes are predictable—for anyone who goes into the wilderness. Viewing conditions vary and are never ideal. The time of day, the season, the surrounding foliage, the weather, and human variables (sleepiness, experience, expectation, fear, eyesight, and so on) all conspire to reduce people's reliability as witnesses. As long as humans are fallible to even a tiny degree, this is just a numbers game. Over sufficient time, a large enough number of observations will lead to dramatic errors. Millions of people see animals in North America every year: deer flashing across a highway, bears scavenging at a dump, an unidentified shape moving among the trees. Given that everyone in North American is exposed to the idea that Bigfoot might exist, the vast total number of animal sightings virtually guarantees that tales about encounters with Sasquatches will emerge—even in a world without Sasquatches.

Footprints

This brings us to the evidence that led to the very word "Bigfoot" (figure 2.6). Giant footprints are found throughout North America (and, indeed, throughout the world). Without that persuasive trace evidence, John Green mused, the Sasquatch could be chalked up to legends, lies, and hallucinations: "But none of these easy explanations can be applied to holes in the ground. Something has to make them."[130]

For cartoonishly clear giant footprints, there are only two explanations: human hoaxers and genuine Sasquatches. But many vaguer tracks amount to wishful thinking. Green conceded early on that many supposed tracks are purely in the eye of the beholder: "I have learned that people can see clear prints that to me are completely shapeless. They can see toes that I can't see at all."[131] Based on this, he surmised that the record of footprint finds was already contaminated with "a lot" of unverified, mistaken tracks.

But there are also a lot of the bold, unmistakable cookie cutter–type Bigfoot tracks. All sources agree on two key points regarding such footprints: they can be hoaxed, and they routinely are. The hopes of Bigfoot proponents are thus

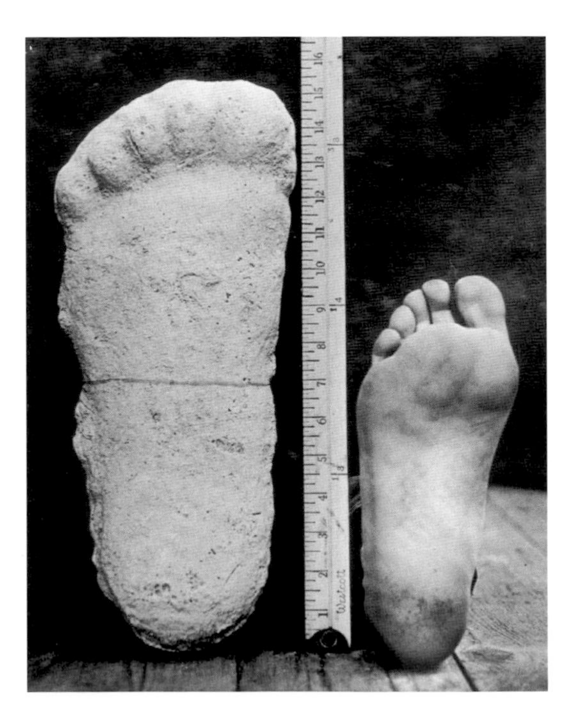

Figure 2.6
A normal-size man's foot compared with a cast of an alleged Bigfoot footprint found in Bluff Creek, California, in 1967, after the shooting of the Patterson-Gimlin film. (Reproduced by permission of Fortean Picture Library, Ruthin, Wales)

pinned on a "where there's smoke, there's fire" argument. "It doesn't matter if 10 per cent of these reports are mistaken, or 50 per cent of them, or 90 per cent of them," Green argued. "If Sasquatches are to be wished back into the books on mythology, every last one of these reports has to be wrong."[132]

But is there good reason to suppose that any such tracks are authentic? Bigfooters cite the consistency of footprints in support of their authenticity, but this argument does not follow at all. Consistency need not imply authenticity when independent hoaxers can draw on the same readily available pop culture. We should expect some consistency whether or not there are Sasquatches. Worse, Bigfoot tracks are *not* very consistent. As Benjamin Radford has pointed out, "Most alleged Bigfoot tracks have five toes, but some casts show creatures with two, three, four, or even six toes."[133] Footprints can be anything from 4 to 27 inches long (and even greater whoppers are sometimes reported). Some are as narrow as 3 inches; some as wide as 13 inches.[134] With such staggering discrepancies, we can infer with confidence that many tracks must necessarily be hoaxes. But which ones? And is there some genuine residue left over, as proponents believe?

Again, we confront the chronic problem: the lack of a type specimen. To evaluate Sasquatch tracks, we need to look an actual Sasquatch foot. Who is to say that real Sasquatch feet do not have three toes? Perhaps five-toed footprints are always a sure sign of a hoax. As it stands, we have no way to know. The authenticity of new maybe-bogus tracks cannot be determined by comparing them with old maybe-bogus tracks.

To his credit, pro-Sasquatch anthropologist Grover Krantz boldly underlined this need for a standard: "There is no point in trying to weigh the pros and cons of the less-than-obvious specimens unless and until both ends of the scale, fake and real, are established beyond a reasonable doubt."[135] And yet, this was a smaller admission than it appears because Krantz believed that he had a reliable standard. He claimed that he could "recognize real sasquatch tracks" using two diagnostic clues. This would be remarkable if true—certainly, he would be the only scientist on Earth known to have such a technique. Unfortunately, his "useful method for spotting fakes" is difficult to evaluate. According to Krantz, "I have told these traits to no one and have never written them down." It is easy to understand why Krantz would have wanted to keep his sure-fire forgery-detection method out of the hands of hoaxers, but as his fellow anthropologist David Daegling observed, "In so doing, Krantz could make no legitimate scientific claims. His proposed criteria for pronouncing tracks genuine could not be tested or evaluated by his peers or anyone else."[136] And by the one criterion on which Krantz's secret authentication system can be assessed—the ability to detect fraudulent tracks—it can be judged a failure. When a bogus track was sent to Krantz to test his system, he falsely concluded that it was authentic.[137]

Given both the absence of a type specimen and the demonstrable fact that footprints are often faked, the scientific value of Bigfoot tracks is very limited.

Hair and DNA

For many considering Bigfoot for the first time, the solution seems obvious: Why not test Bigfoot's hair or DNA? Samples of possible Sasquatch hair, scat, and fluids have been collected fairly often. As it turns out, many promising samples have been analyzed by experts or even sequenced for DNA.

To date, they have all been a bust. This is disappointing, but it is not the end of the bad news. It seems that testing hair would be unable to solve the mystery *even if Bigfooters were to discover samples of genuine Sasquatch hair*. It is the same type-specimen problem that plagues footprint evidence and eyewitness reports: no one has any idea what Sasquatch hair would look like. When hair experts examine an unknown sample of hair, the procedure is simplicity itself: they first look at it under a microscope and then painstakingly compare the characteristics of the unknown sample with those of *known* samples and try to find a match. However, as John Green succinctly put it, "when no collection contains known Sasquatch hairs there is no way to prove that the hair you find comes from a Sasquatch. At best you will only be unable to prove that it didn't."[138]

Some Bigfoot enthusiasts are encouraged when a sample cannot be identified, as though this failure implies that the hair comes from an unknown animal. This conclusion is premature and improbable. As Grover Krantz explained, "Hair characteristics vary on different parts of the same animal, and no comparative collection exists of all types of hair of all mammals. A hair that is unlike anything in a North American collection might be from the armpit of a bear or from an escaped llama."[139] So even in theory, the prospects for hair analysis are grim. Until "hair is pulled directly from a sasquatch body by a qualified analyst," Jeff Meldrum agrees, the best-case scenario is a dead end: "any hair truly originating from a sasquatch would necessarily languish in the indeterminate category."[140]

As a practical matter, the situation is worse. Many purported hair samples have been successfully identified, and they definitely do not come from Bigfoot. A large number of them have turned out to be artificial fibers. (David Daegling cites one geneticist's "tongue-in-cheek report that the likely origin of the sample was from the interior of a couch.")[141] Other samples come from known creatures, including humans and domestic animals. In one widely publicized case in 2005, DNA extracted from hair from an alleged Sasquatch heard in the Yukon woods proved to match that of the local bison.[142] In another prominent DNA case, hair taken from a plaster cast of a supposed Bigfoot body print (the "Skookum cast") turned out to be human.[143] Such results are typical.

It is unclear whether DNA alone could settle the Sasquatch controversy under any circumstances. As a working scientist, Meldrum offered these words of caution:

> The conventions of zoological taxonomy require a type specimen, traditionally in the form of a body or a sufficiently diagnostic physical body part, to decisively establish the existence of a new species. Whether DNA alone will ultimately satisfy that standard remains to be seen. I am doubtful. I am not aware of a precedent for determining a new species on the basis of DNA evidence, in the absence of a physical specimen.[144]

Not everyone is so restrained. The Internet had been buzzing about the long-anticipated release[145] of a paper purporting to present DNA evidence that "conclusively proves that the Sasquatch exist as an extant hominin and are a direct maternal descendent of modern humans."[146] With DNA sourced, according to the report, from among 111 "samples of blood, tissue, hair, and other types of specimens," this is the most prominent Sasquatch DNA case to date. Full expert review of the team's data and methods is not yet available; however, science writers identified several serious red flags within hours of the paper's release. To begin with, it seems that it was roundly rejected by mainstream science journals. "We were even mocked by one reviewer in his peer review," complained lead author Melba Ketchum.[147] So how did the paper get published? Although Ketchum has insisted that this circumstance did not influence the editorial process, it seems that she bought the publication.[148] Indeed, her study is the only one in the inaugural "special issue" of the *DeNovo Scientific Journal*. Benjamin Radford pointed out that no libraries or universities subscribe to the newly minted *DeNovo*, "and the journal and its website apparently did not exist three weeks ago. There's no indication that the study was peer-reviewed by other knowledgeable scientists to assure quality. It is not an existing, known, or respected journal in any sense of the word."[149] Invertebrate neuroethologist Zen Faulkes noted further that *DeNovo* lists no editor, no editorial board, no physical address—not even a telephone number: "This whole thing looks completely dodgy, with the lack of any identifiable names being the one screaming warning to stay away from this journal. Far, far away."[150]

Beyond these irregularities, there are also signs of serious problems with the study's data, methods, and conclusions. Ketchum and her colleagues found, for example, that all the mitochondrial DNA recovered from their samples tested as "uniformly consistent with modern humans," but argued despite this finding that anomalies in their nuclear DNA analyses "clearly support that these hominins exist as a novel species of primate. The data further suggests that they are human hybrids originating from human females."[151] This scenario, in which "Sasquatch is a human relative that arose approximately 15,000 years ago as a hybrid cross of modern *Homo sapiens* with an unknown primate species" (as publicized in an earlier press release about the unpublished paper) is not especially plausible.[152] As skeptic Steven Novella explained, "It is highly doubtful that the offspring of a creature that looks like bigfoot and a human would be fertile. They would almost certainly be as sterile as mules. Humans could not breed with our closest living relatives, the chimpanzees, or any living ape." He added, "The bottom line is this—human DNA plus some anomalies or unknowns does not equal an impossible human-ape hybrid. It equals human DNA plus some anomalies."[153] These problems have only multiplied with the release of Ketchum's paper and data. John Timmer, the science editor of the Web site Ars Technica and an experienced genetic researcher, offers the preliminary opinion that "the best explanation here is contamination":

> As far as the nuclear genome is concerned, the results are a mess. Sometimes the tests picked up human DNA. Other times, they didn't. Sometimes the tests failed entirely. The products of the DNA amplifications performed on the samples look about like what you'd expect when the reaction didn't amplify the intended sequence. And electron micrographs of the DNA isolated from these samples show patches of double- and single-stranded DNA intermixed. This is what you might expect if two distantly related species had their DNA mixed—the protein-coding sequences would hybridize, and the intervening sections wouldn't. All of this sug-

> gests . . . that the sasquatch hunters are working on a mix of human DNA intermingled with that of some other (or several other) mammals.[154]

HOAXING

The great problem for Bigfoot research is that hoaxing definitely occurs. Worse, it is common and has been so throughout the evolution of Bigfoot mythology. Most of the foundational cases probably were fabrications, and hoaxes definitely comprise a large percentage of overall footprint cases and eyewitness accounts.

In the absence of a type specimen, the Sasquatch can be supported by (and defined by) only two pillars: star cases and everything else. If either level of the database is contaminated by hoaxes, this is a catastrophic problem for Bigfoot proponents. They face stark choices: give up the Sasquatch hypothesis, rely on outright guessing or faith, or make a serious effort to purge hoaxes from the database.

The third option, sadly, gets little traction. Partly, this is because the practical problem is so daunting. As John Green noted forty years ago, it is one thing to know that hoaxes occur and another to identify them: "Things have reached a point where it is to be expected that some people will make up such stories in hope of fame or even financial gain, and I have no doubt that there are people doing so. It is hard to weed them out with any certainty, because there is ample information in print now to give anyone the details for a convincing story."[155]

But members of the Bigfoot subculture defend reports that they should reject. Indeed, it is a given for cryptozoologists to promote famous cases long after a possibility of deliberate fraud emerges. Evidence "discovered" by long-known hoaxers continues to feature in even the newest and best-respected pro-Bigfoot books. For example, both the "Cripple Foot" tracks found by film hoaxer Ivan Marx[156] and the handprints discovered by admitted footprint hoaxer Paul Freeman[157] are presented as strong evidence by Jeff Meldrum.[158] In such circumstances, the cryptozoological project is revealed as partisan advocacy rather than objective inquiry. In any scientific or rigorous discipline, a serious suggestion of fraud places a claim in a kind of intellectual quarantine. Bigfooters are very reluctant to set aside questionable evidence.

Cryptozoologist Loren Coleman agrees that some cases persist beyond their expiration date, although he disagrees with my assessment of the health of cryptozoology. For example, he writes that the "story of Jacko—that of a small, apelike, young Sasquatch said to have been captured alive in the 1800s—is a piece of folklore that refuses to die," despite the discovery in 1975 that news accounts busted the yarn in 1884.[159] I grew up regarding Jacko as a major discovery and not surprisingly: it is prominently featured in virtually all Bigfoot books, without disclosure of the evidence that it was a hoax. Nor am I alone in having considered this case to be stronger than it is known to be. As Coleman writes, "Unfortunately a whole new generation of hominologists, Sasquatch searchers, and Bigfoot researchers are growing up thinking that the Jacko story is an ironclad cornerstone of the field, a foundation piece of history proving that Sasquatch are real. But in reality Jacko seems to be a local rumor brought to the level of a news story that eventually evolved into a modern fable."[160] This state of affairs strikes me as a very bad sign for the methods and integrity of cryptozoology, but Coleman is more upbeat. He told me that the persistence of the tale "says more about the transmission of information and the lack of available outlets to publish our findings." According to Coleman, "Self-correcting occurs all the time in cryptozoology, although it may take years to transmit the information."[161]

A Taxonomy of Mischief

Why would anyone go to the trouble of concocting a Bigfoot hoax? The Bigfoot world seems to approach this question from two contradictory positions. On the one hand, proponents often argue that most Bigfoot evidence must be real because perpetrating a hoax is a prohibitive hassle. On the other, anyone who spends time on Bigfoot's trail learns early, and through bitter experience, that hoaxing is extremely common. A wide range of people independently fabricate tracks and sightings for a variety of idiosyncratic reasons. Some of these hoaxers are Bigfoot investigators.

The sheer quantity of Bigfoot evidence is surely impressive. To account for that "mountain of evidence" by hoaxing, proponents argue, we would have to posit a conspiracy of global reach and Machiavellian cunning. According to a rather fishy back-of-the-envelope calculation by Grover Krantz, faking all the Bigfoot tracks would require "at least 1,000 paid professionals," whose efforts would not come cheap: "During the last forty years well over a billion dollars must have been expended on this project." Either that, he argued, or we must imagine an unpaid horde of "something like 100,000 casual hoaxers who go out for a weekend of track making one or two times a year." Or, finally, we could accept what Krantz proposed as the most parsimonious explanation: an undiscovered species of giant primate stomping all over North America.[162]

Krantz's math—which was based not on the number of tracks that have actually been discovered, but on his own estimate that 100 million mostly unknown track-making events must have occurred—can be safely ignored. But what of his basic arguments? Would independent hoaxers have the wherewithal to create persuasive tracks? And what would they have to gain? Luckily, we do not have to guess on either question. Many hoaxers have run rings around Bigfoot investigators, and in many cases their motivations are also a matter of record.

Noting "a tendency among Bigfoot pundits to assume that any potential hoaxer is, by virtue of their poor character, also laughably inept," skeptic David Daegling pointed out that dishonesty does nothing to diminish the intelligence or creativity of hoaxers.[163] For a stunning demonstration, consider a case shared by Krantz himself, in which Bigfoot tracks ran up a steep slope in inhumanly long 8-foot strides: "It was later found that a high school athlete had made the tracks; he wore fake feet that were put on backwards, and he ran down the slope. Whenever a new account is recorded of incredible feats of footwork, I try to remember this case and wonder how the new one might have been faked."[164]

From the most dazzling feats to the most pedestrian fibs, hoaxing comes in many forms and is undertaken for many reasons. Some deceptions are intended to increase the stature of the hoaxer (as when Bigfoot investigators regularly discover evidence that is too good to be true). Others are simply gags. Some are meant to make money (a proposed motivation for the probable film hoax by Roger Patterson and Bob Gimlin). Another not uncommon class of fabrications might be likened to guerrilla gardening, flash mobs, or certain types of graffiti: anonymous works of art defined by their audacity and their ability to baffle onlookers.

Many hoaxers deliberately attempt (destructively or constructively, depending on your point of view) either to make paranormal researchers appear foolish or to expose flaws in the methodology of researchers or media. A widely cited example of this genre is the "Carlos" hoax of 1988, in which Australian news media were presented with the arrival of a supposed New Age channeler from America—actually a performance artist who was recruited for the role by skeptical author James Randi.[165] Although media coverage of Carlos was skeptical in tone, none of the major Australian news media that featured Carlos undertook any sort of due-diligence background check on the psychic performer—which easily would have revealed that he was a completely fictitious recent invention.[166] This proved very embarrassing when Randi's collaborators on the Australian version of *60 Minutes* revealed the hoax on national television—to the outrage of many.[167]

This testing type of hoax (what we might think of as a "secret shopper" or "believer-baiting" hoax) is very common in the Bigfoot world. It does not require the resources that Randi brought to his "Carlos" stunt. Recall that bricklayer Ray Pickens laid fake Bigfoot tracks on several occasions in 1971, using nothing more than wooden feet attached to boots.[168] (This was minor revenge. When a Bigfoot researcher described Pickens and his friend as "hicks," Pickens was provoked to create his first fake tracks.)[169] In a more recent instance, a young filmmaker named John Rael produced a fake Bigfoot video and seeded it to YouTube in 2010—only to reveal the hoax in an intentionally juvenile sequel in which Bigfoot literally gets kicked in the groin. When I asked Rael about this, he told me that he hoped to "trick the 'true believer' into believing something that can be shown, unequivocally, to be fake" in order to "inspire the true believer to begin questioning some of their other beliefs."[170]

THE PARADOX OF BIGFOOT

At the heart of the mystery of Bigfoot is a paradox: many people see Sasquatches, but no one can find one. This is an extremely uncomfortable dilemma for Bigfoot advocates. Either Sasquatches are too rare to locate, in which case they should also be too rare to see; or the common and widespread sighting reports are by and large accurate, in which case science should long ago have located a specimen.

Proponents are aware that the absence of a carcass is the central problem for their field—a problem that grows more severe with every passing day. At the dawn of the Sasquatch era, it was sensible to ask skeptics to be patient. As John Green reflected forty years ago, "The most likely way for the Sasquatch question to be settled is for some deer hunter to down one—and I believe that becomes more probable each year as more and more people learn that the things are real, that they are not human, and that the first man to bring one in is sure of fame and perhaps fortune."[171]

But after decades of fruitless patience (and unclaimed rewards of hundreds of thousands of dollars),[172] Bigfooters have to explain why no one has yet located a species of animal that is supposed to be as large as a Kodiak bear. "The mantra among career Bigfoot pursuers is that absence of evidence is not evidence of absence," explains anthropologist David Daegling, "which is a point well taken, but the skeptic can still reasonably ask why the evidence that should be there isn't."[173] The most common response is an argument of necessity: Bigfoot hunters cannot find the creature, but they should not, after all, be expected to find it. Sasquatches are rare and clever, and after they die their remains are quickly erased by posthumous predation and weather. According to Bigfooters, that's just how nature works. This was the argument of cryptozoology pioneer Ivan Sanderson: "Ask any game warden, real woodsman or professional animal collector if he has ever found the dead body of any wild animal—except along roads of course, or if killed by man. I never have, in 40 years and 5 continents! Nature takes care of its own, and damned fast, too."[174] This idea was taken up by Grover Krantz, a pro-Sasquatch anthropologist, who asked, "If bears are real, why don't we find their bones?"[175] In Krantz's opinion, "We should expect to find a dead sasquatch about as easily as we find a dead bear (assuming they occurred in the same numbers and had similar life spans) . . . and dead bears are almost never found."[176]

Skeptics are in the business of soberly considering strange claims—such as pyramid power, alien abduction, or the psychic ability to cause someone to urinate against his or her will[177]—as though they were perfectly plausible possibilities. But even by that standard, and with all the goodwill in the world, I cannot consider the "no dead bears" argument to be anything but foolish. The reason is simple: dead bears *are* found (figure 2.7). Often.

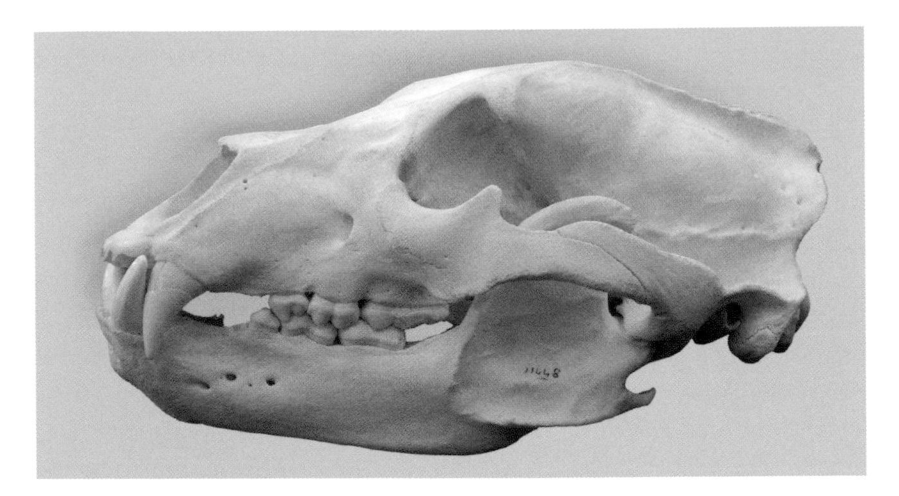

Figure 2.7 The skull of a grizzly bear in the collection of the Royal British Columbia Museum, Victoria. (Photograph by Daniel Loxton)

Indeed, many members of the tiny fraternity of Bigfoot researchers and skeptics have *themselves* come across dead bears! Pro-Sasquatch wildlife biologist John Bindernagel told me that he has found bear skulls—twice.[178] Bigfoot skeptic David Daegling came upon a "cranium of a black bear, beautifully preserved," from which he wryly concludes that "perhaps you can appreciate that, based on my personal experience, I find Sanderson's and Krantz's declarations to be rather hollow."[179] I agree that the "no dead bears" argument is hollow and for the same reason: I found a bear carcass myself, during my career as a shepherd near the Alaska panhandle. Similarly, my brother Jason stumbled across a bear carcass while doing geochemical sampling in the far Yukon. The skull of that bear now sits on his bookcase.[180]

The "no dead bears" argument is special pleading and worse: it is wildly, laughably wrong. But there is no amount of special pleading that cannot be rescued by still further special pleading. Bigfooters hasten to emphasize the "except by human actions" clause of their argument. Krantz insisted, "I've talked to hunters, many game guides, conservation people, ecology students, and asked them how many remains of dead bears have you found that died a natural death? Over twenty years of inquiry my grand total of naturally dead bear bones found is zero!"[181]

There are two problems with this "natural death" stipulation. To begin with, cause of death usually cannot be determined from a skull or old bones. As Bindernagel told me, "I really don't know how the two bears whose skulls I encountered died—'natural' death or otherwise."[182] Nor can my brother account for the death of the bear whose skull graces his bookcase, although he found the carcass in an extremely remote and uninhabited, boat-access-only area. But just for the sake of argument, let's grant the highly dubious assumption that all locatable remains are those of bears that were killed by humans. Why would Bigfoot be immune to the same cars, trucks, and guns that kill bears?

If we are to accept eyewitness testimony (and, without any bodies, Bigfooters are stuck with it), we must grant that Sasquatches are routinely exposed to the same mortal risks as bears. "There are a number of people who claim not only to have seen a sasquatch on the road but to have run into one," explained Green, presenting as an example the testimony of a logging truck driver who allegedly hit a 7-foot Bigfoot in Oregon in 1973.[183] Many others—including pivotal witnesses like William Roe and Bob Gimlin—have described watching Sasquatches over the barrel of a rifle. A considerable number of people have even claimed to have shot Bigfoot.[184] Even if we grant that Sasquatches are wily, tough, and rare, their luck must sometimes run out. Common sense dictates that hunters and truckers should occasionally bring one down. In the early days, that's just what Bigfooters expected to see. And yet, that has not happened. Let's imagine that Bigfoot is as rare as a wolverine, as stealthy as a cougar, and as powerful as a grizzly. What of it? People see wolverines, photograph cougars, and shoot grizzlies. As Daegling notes, "Wolverines are scarce animals, but we do know what their skeletons look like for the simple reason that we have them."[185]

TOO MUCH SMOKE

The two great clichés ruling the Bigfoot literature are the "mountain of evidence" and the connected argument: "where there's smoke, there's fire." This is a non sequitur: where there's smoke, there's smoke.[186] And in this case, that "smoke" is so widespread that it actually argues against the existence of Bigfoot.

It is one thing to suppose that some species of large animals may exist undetected in remote parts of the world (indeed, surveys of deep ocean life have confirmed this for marine animals). It is a different thing altogether to suppose that a species of large animal could thrive in all regions of the globe without anyone ever finding one. Sasquatches are thought to live not only in the temperate rain forest of the Pacific Northwest, but across North America and beyond. They are spotted in every state in the United States, from New Mexico to New Jersey. They lurk in Florida swamps, Texas thickets, and Ohio farmland. Bigfoot even turns up in Hawaii.[187] (Meditate on that one first, and it seems almost natural to hear that Sasquatches have been reported on Staten Island.)[188] Globally, Bigfoot stomps not only through North, Central, and South America, but across Europe, Africa, Asia, and even Australia (where it is known as the Yowie).

Grover Krantz staked his scientific career on the belief that Sasquatches exist, but the planetary extent of reports of Bigfoot-like creatures was a problem that gave him pause. He warned that "when it is suggested that a wild primate is found native to all continents, including Australia, then credibility drops sharply." Even if we were to leave aside the plausibility of a global primate remaining undiscovered, Krantz knew that very few animal species have anywhere near as wide a distribution: "Beyond a certain point, it can be argued that the more widespread a cryptozoological species is reported to be, the less likely it is that the creature exists at all."[189]

3

THE YETI

THE ABOMINABLE SNOWMAN

THE YETI IN POPULAR CULTURE

DURING A TRIP AROUND THE WORLD in the summer of 1973—through Africa, India, Thailand, China, and Japan—my college roommate and I stopped in Kathmandu, Nepal. We visited many of the usual tourist sites, including Buddhist temples and stupas and Lamaist monasteries. We attended a religious ceremony during which villagers slashed the throat of a sacrificial goat right before our eyes, and we drove to a spot where (when the clouds finally cleared) we could see the summit of Mount Everest. But as we walked through the open-air markets in several parts of the Nepalese capital, what was most surprising was the huge number of items for sale that were related to the Yeti, the mysterious ape-like creature that is native to the Himalayas. We saw casts of alleged Yeti footprints, supposed Yeti fur, and images of the Yeti on almost every item imaginable. Just like Nessie and Champ, the Yeti is big business, attracting many free-spending tourists to Nepal and bringing expeditions of cryptozoologists determined to find it. There is a five-star hotel in Kathmandu called the Yak & Yeti, as well as the Yeti Bar, and the official airline of Nepal is known as Yeti Airlines. Simon Welfare and John Fairley write that the Yeti "is tangled in a web of fantasy, religion, legend, chicanery, and commercialism. The Yeti is a highly commercial legend, perhaps even Nepal's principal foreign currency earner."[1]

The Yeti has become one of the most enduring figures of legend. It has been mentioned by a wide variety of mountaineers who have scaled the highest mountains of the world. Because of its alleged existence in an exotic world largely inaccessible to most Westerners, it has been tied to the myth of Shangri-la as well. As Brian Regal wrote, "Big and hairy, the Yeti lived in a remote part of the world, the locals feared it, and it reportedly behaved like, well, a monster. It conveniently occupied a space, both temporally and physically, at the crossroads of the worlds of East and West, and so it took place alongside other oriental 'mysteries.'"[2]

The Yeti (often misnamed the Abominable Snowman) has been the star of numerous movies, starting with *The Snow Creature* (1954), which was filmed and released only a year or two after the first widespread reports of the Yeti in popular culture, and continuing with *The Abominable Snowman* (1957) and *Snowbeast* (1977); makes a cameo appearance, voiced by John Ratzenberger, in *Monsters, Inc.* (2001); and fights Chinese soldiers in *The Mummy: Tomb of the Dragon Emperor* (2008). It has appeared in some form in dozens of television shows, including Bumble, the Abominable Snow Monster, in the stop-motion animated Christmas classic *Rudolph, the Red-Nosed Reindeer* (1964); in episodes of *Looney Tunes*, *Spiderman*, *Electric Company*, *Scooby-Doo*, *Duck Tales*, *Jonny Quest*, *Doctor Who*, *Power Rangers*, *Tintin*, and even the preschool cartoon *Backyardigans*; and in numerous "monster documentaries" on cable channels. In literature, the Yeti appears not only in *Tintin in Tibet* (1960) by Hergé (Georges Remi), but also in *The Abominable Snowman of Pasadena* (1995), the thirty-eighth book in R. L. Stine's Goosebumps series, and in Marvel comics, among others. There are even audio-animatronic Yetis in the Matterhorn Bobsled ride in Disneyland in California, and in the Expedition Everest ride at Walt Disney World in Florida. Several rock songs refer to it, and a rock band is named Yeti.

The legend is now so widespread in popular culture that most portrayals of the creature bear no resemblance to those in the original accounts reported by explorers in the Himalayas. Thanks to the "Abominable Snowman" label, most pop-culture depictions envision the Yeti as a tall bipedal white-furred beast, not the dark-furred, gorilla-like or bear-like creature that inhabits the legends of the cultures of the Himalayas.

THE MONSTER OF THE MOUNTAINS

What do we know of this mysterious creature that allegedly haunts the world's highest mountains? There is the confusion due to the numerous names that have become associated with the Yeti legend, with multiple names not only among languages and cultures, but also within each language (and, indeed, multiple or contested meanings for many of those words—not to mention a variety of spellings). Author and mountaineer Reinhold Messner has listed and attempted to define many of these terms: *Yeti* is the mispronunciation of the Sherpa name for the creature, *Yeh-teh* (animal of rocky places), or, possibly, a derivation of *Meh-teh* (man-bear). Tibetans also call it *Dzu-teh* (cattle bear), also a name for the Himalayan brown bear (*Ursus arctos isabellinus*), which can walk on its hind legs, or *Mi-go* (wild man). Messner found that the Tibetan word *chemo* or *dremo* is most commonly used for both the Yeti and the Himalayan brown bear. In the different Chinese dialects, it is *beshong* or *bai xiong* (white bear), probably the giant panda; *mashiung* or *ma xiong* (brown bear); and *renshung* or *ren xiong* (man-bear).[3] The wide spectrum of names for the creature reflects the facts that the cryptozoological legend is an amalgam of many different cultural traditions and legends, and that (as Messner discovered) there is much confusion between the legendary *chemo* and the rare Himalayan brown bear.

The odd and inappropriate name Abominable Snowman originally derived from the Everest Reconnaissance Expedition, sponsored by the Alpine Club and the Royal Geographical Society and led by Lieutenant Colonel Charles Howard-Bury in 1921. This was one of the first groups to scout a route for an attempt to climb Mount Everest. Ascending past 20,000 feet of elevation, Howard-Bury and his team were surprised to find tracks in the snow. "We were able to pick out tracks of hares and foxes," he wrote, "but one that at first looked like a human foot puzzled us considerably."[4] He soon realized that "these tracks, which caused so much comment, were probably caused by a large 'loping' grey wolf, which in the soft snow formed double tracks rather like those of a barefooted man"—that is, a wolf that had placed its back paws in the depressions that had been made by its front paws.[5] However, his Sherpa guides "at once jumped to the conclusion"[6] that this "was the track of a wild, hairy man, and that these men were occasionally to be found in the wildest and most inaccessible mountains."[7] Howard-Bury was scornful of this interpretation. "Tibet, however, is not the only country where there exists a 'bogey man,'" he wrote. "In Tibet he takes the form of a hairy man who lives in the snows, and little Tibetan children who are naughty and disobedient are frightened by wonderful fairy tales that are told about him. To escape from him they must run down the hill, as then his long hair falls over his eyes and he is unable to see them. Many other such tales have they with which to strike terror into the hearts of bad boys and girls."[8]

Be that as it may, the "wild man" idea was newspaper gold. Journalist Henry Newman interviewed both the expedition members and the porters when they returned to Darjeeling, India. The Europeans rather boringly agreed that the tracks were "unquestionably made in the snow by some four-legged creature about the size of a wolf," so Newman delightedly recounted the porters' more colorful tales of what they called *Metoh Kangmi*—a term, he wrote, that "as far as I could make out, means 'abominable snow men.'"[9] Looking back on this fateful translation, Newman described himself as "the showman responsible for the introduction of the 'Abominable Snowman' to the world of literature and art" and described how the name came to be:

> I fell into conversation with some of the porters, and to my surprise and delight another Tibetan, who was present, gave me a full description of the wild men—how

> their feet were turned backward to enable them to climb easily and how their hair was so long and matted that, when going downhill, it fell over their eyes. . . . When I asked him what name was applied to these men, he said "Metoh Kangmi." *Kangmi* means "snow-men" and the word *metoh* I translated as "abominable." The whole story seemed a joyous creation, so I sent it to one or two newspapers. It was seized upon. . . . Later, I was told by a Tibetan expert that I had not quite got the force of the word *metoh*. It did not mean "abominable" quite so much as "filthy" and "disgusting," somebody dressed in rags. There is an Urdu word, *dalkaposh*, meaning somebody wearing filthy and tattered clothing. The Tibetan word means something like that, but is much more emphatic.[10]

Based on that meaning, Newman believed that the wild-men myth was based on actual human beings: bandits and ascetics who live in the wilderness regions. It is important to note once again, however, that stories about Yeti-like creatures extend throughout the Himalayas and beyond, leading to a very large number of terms in different languages, most with multiple or contested meanings.[11] The interpretation of *Metoh Kangmi* is no exception. In 1956, zoologist William L. Straus Jr. described an alternative interpretation of *Metoh Kangmi* that had recently been put forward by Indian religious leader Srimat Swami Pranavananda: "*Mi-te*, which has been translated by some Himalayan expeditionists as 'abominable, filthy, disgusting to a repulsive degree, dirty,' actually means 'man-bear.' *Kangmi*, or 'snowman,' is merely an alternate word for the same animal. Hence the term *miteh-kangmi*, from whence 'abominable snowman' represents an incorrect combination, owing to mistranslation, of two terms that are essentially synonymous."[12] Based on this translation, Straus argued that the creature was "no other than the Himalayan red bear."

Whatever the origin of the name, Newman's newspaper account christened a cryptid, with Abominable Snowman becoming the most common appellation for the creature in the English language. As Regal puts it, "This striking name caught people's attention, and, more than any of the sightings or stories to this point, made the creature a topic of interest outside the Himalayas. With this catchy but inaccurate label, the creature went from being a quaint local legend to an international phenomenon. The same type of makeover and jump to celebrity status occurred later in the century, when Sasquatch became Bigfoot."[13]

THE LEGEND GROWS

The first known sighting of a Yeti-like creature to be recorded in the region by a Westerner appeared more than 180 years ago in a paper by Brian Hodgson, an Englishman living in Nepal. A classic gentleman-explorer typical of the British Empire at that time, Hodgson was not only a diplomat, a biologist, and an administrator, but also an adventurer. He was the first Englishman permitted to visit the Nepalese royal court and the first Westerner to record in detail the traditions of Nepalese Buddhism and other cultural traditions. He wrote more than 300 scientific papers about his discoveries in the Himalayas. In one of them, published in the *Journal of the Asiatic Society of Bengal* in 1832, Hodgson wrote, "My shooters were once alarmed in the Kachár by the apparition of a 'wild man,' possibly an ourang, but I doubt their accuracy. They mistook the creature for a cacodemon or rakshas [mythological humanoid beings], and fled from it instead of shooting it. It moved, they said, erectly: was covered with long dark hair, and had no tail."[14] As anthropologist John Napier reports, "Hodgson was somewhat scornful that his hunters had run away in terror from the creature instead of standing their ground and shooting it dead like a true British sportsman. Had they done so perhaps this book would never have needed to be written!"[15]

In 1889, the first report of Yeti footprints (high in the Himalayas in what is now Sikkim) appeared in what Ivan Sanderson described (amusingly, as it is pretty clear that Sanderson had not read it)[16] as "a somewhat uninspired and uninspiring book about it, uninspiringly named *Among the Himalayas*" by Laurence A. Waddell, a soldier, lawyer, and historian:

> Some large footprints in the snow led across our track, and away up to higher peaks. These are alleged to be the trail of the hairy wild men who are believed to live amongst the eternal snows, along with the mythical white lions, whose roar is reputed to be heard during storms. The belief in these creatures is universal among Tibetans. None, however, of the many Tibetans I have interrogated on this subject could ever give me an authentic case. On the most superficial investigation it always resolved itself into something that somebody heard tell of. These so-called hairy wild men are evidently the great yellow snow-bear (*Ursus isabellinus*), which is highly carnivorous, and often kills yaks. Yet, although most of the Tibetans know this bear sufficiently to give it a wide berth, they live in such an atmosphere of superstition that they are always ready to find extraordinary and supernatural explanations of uncommon events.[17]

As Napier points out, "The willingness of the Sherpas to assure travelers that tracks of *any* description discovered in the snow are those of the Yeti is a recurrent theme in the literature on the subject. Such conning appears to be quite without malice aforethought and probably reflects the extreme politeness of the Sherpas, who regard it as excessively bad-mannered to disappoint anyone. In addition, as many authorities have reported, the Sherpas possess an element of mischievousness and love a good story."[18]

Mischievousness aside, the specific identification of bears as "wild men" was documented by other early authors. United States ambassador William Woodville Rockhill described this encounter in 1891:

> One evening, a Mongol told me of a journey he had once made to the lakes in company of a Chinese trader who wished to buy rhubarb from the Tibetans that annually visit their shores. They had seen innumerable herds of wild yak, wild asses, antelopes, and *gérésun bamburshé*. This expression means literally "wild men"; and the speaker insisted that such they were, covered with long hair, standing erect, and making tracks like men's, but he did not believe they could speak. Then, taking a ball of tsamba he modeled a *gérésun bamburshé*, which was a very good likeness of a bear. To make the identification perfect, he said that the Chinaman cried out, when he saw one, "*Hsiung, hsiung*," "Bear, bear"; in Tibetan, he added, it is called *dré-mon*. The Mongols do not class the bear among ordinary animals; he is to them "the missing link," partaking of man in his appearance, but of beasts in his appetites. The bear takes the place of the "king of beasts" among them and the Tibetans, for they hold him the most terrible of animals when attacked, and a confirmed man-eater![19]

In the early twentieth century, numerous expeditions to the Himalayas were mounted by explorers and big-game hunters, including Major C. H. Shockley, Colonel Gerald Burrard, and Colonel A. E. Ward.[20] These three men wrote books about their explorations, in which they detailed reports of animal sightings. They described (and shot) many known and previously unknown Himalayan animals, but

> never once did they give a hint of a Yeti—a suspicious track or an ominous footprint. This negative evidence is important, for these men were expert big-game hunters and were accompanied by experienced *shikaris* and would assuredly have been keeping a sharp lookout for strange animals or for an unusual spoor. Perhaps they were traveling in the wrong area; perhaps their peregrinations were at too low an altitude; perhaps, being soldiers and government officials and anxious to preserve

the respect of their colleagues down on the plains, they saw but said nothing. Or, again, perhaps there was nothing to see.[21]

After the Everest Reconnaissance Expedition, which led to the misnomer Abominable Snowman, the next significant mention of the Yeti was in 1925 by the British-Greek explorer and photographer N. A. Tombazi. Hiking at about 15,000 feet, on an expedition for the Royal Geographical Society, he saw a creature more than 200 to 300 yards away near Zemu Glacier:

> Unquestionably, the figure in outline was exactly like a human being, walking upright and stopping occasionally to pull at some dwarf rhododendron bushes. It showed up dark against the snow, and as far as I could make out, wore no clothes. Within the next minute or so it had moved into some thick scrub and was lost to view.
>
> Such a fleeting glimpse, unfortunately, did not allow me to set up the telephoto camera, or even to fix the object carefully with the binoculars; but a couple of hours later, during the descent, I purposely made a detour so as to pass the place where the "man" or "beast" had been seen. I examined the footprints, which were clearly visible on the surface of the snow. They were similar in shape to those of a man, but only six to seven inches long by four inches wide at the broadest part of the foot. The marks of five distinct toes and the instep were perfectly clear, but the trace of the heel was indistinct, and the little that could be seen of it appeared to narrow down to a point. I counted fifteen such footprints at regular intervals ranging from one-and-a-half to two feet. The prints were undoubtedly of a biped, the order of the spoor having no characteristics whatever of any imaginable quadruped. Dense rhododendron scrub prevented any further investigations as to the direction of the footprints.[22]

Tombazi later said that he did not believe it was a Yeti after all, but probably a Hindu hermit who lived in isolation in the local mountain caves.[23] Although it seems to be possible for at least some acclimatized people to go barefoot in the snow at high elevations for long periods,[24] the measurement of the footprints as only 6 to 7 inches long but 4 inches wide is, as Napier points out, not typical of that of a human footprint. It is, however, the right proportions for a bear paw print. The prints had narrow heels, which is also a bear-like feature. Human footprints are broad in the back because the heel supports the body weight each time a human foot strides and steps. As Napier concludes, Tombazi's description is inconsistent: the creature's bipedal motion is human-like, but it left bear-like prints.[25] Either it was an unknown bipedal creature with bear-like feet, or it was just a walking bear and Tombazi's "fleeting" glimpse of the creature is distorted (as memories often are).

Additional accounts from the different Himalayan climbing and surveying expeditions occurred through the 1920s, 1930s, and 1940s, although none made much of an impact on Western consciousness until the name Abominable Snowman emerged as a grotesque label for the creature.

THE NAZI AND THE SNOWMAN

Brooke Dolan II, the son of a wealthy Philadelphia industrialist, led two expeditions to China and the Himalayas in the 1930s. In 1938, the German biologist Ernst Schäfer wrote a popular account of his second (1934–1936) expedition under Dolan.[26] Schäfer was attached to these American-funded surveys as a scientist, and he made numerous observations on the natural history of Tibet, including some of the first records of the giant panda (he was apparently "the second white man to shoot a panda"),[27] the Himalayan brown bear, and other wildlife. He also reported contemptuously on lore and traces of the Yeti in the region. "For the rest of his life Schäfer would poke fun at believers in the apemen of the Himalayas," explained author and documentarian Christopher Hale. "In 1938 he became exasperated as day and night his men fearfully discussed Migyuds [ape-man god of Buddhist mythology, associated with the Yeti], and he took to playing tricks on them by faking its footprints in the snow."[28] Rather than being some sort of mythological beast, Schäfer concluded, the Yeti was actually the Himalayan brown bear (figure. 3.1). He described an adventure that cinched it for him and shared the opinion of his assistant Wang:

> On the morning of the second day, a wild-looking Wata with a rascally face comes to me and tells the fantastic story of a snowman that haunts the tall mountains. This is the same mythical creature about which Himalaya explorers always like to write because it envelops the unconquered peaks of the mountain chains with the nimbus of mystery. It is supposed to be as tall as a yak, hairy like a bear, and walk on two legs like a man, but its soles are said to point backward so that one can never track its trail. At night it is supposed to roam, descend deep into the valleys, devastate the livestock of the native people, and tear apart men whom it then carries up to its mountain home near the glaciers. After I listen calmly to this bloody tale, I convey to the Wata that he does not have to make up such a tall tale; however, if he

Figure 3.1 Himalayan brown bear mother and cub, photographed in the district of Kargil in the Ladakh region of Jammu and Kashmir, India. (© 2013 Aishwarya Maheshwari/ WWF–India)

> could bring me to the cave of such a "snowman," and if the monster is actually in its lair, then the empty tin can in my tent, which appears to be the object of his great pleasure, would be his reward. But should he have lied to his lord, added Wang, he could expect a beating with the riding crop. Smiling, with many bows, the lad bids his leave with the promise to return early the next morning and report to me. Wang is also of the opinion that there are snowmen and draws for me the face of the mystery animal in the darkest colors, just like he has heard about it from the elders of his native tribe countless times: Devils and evil spirits wreak havoc up there day and night in order to kill men.[29]

"But Wang," Schäfer scoffed, "how can you as my senior companion believe in such fairy tales?" Wang explained that these forces were manifest all around them. After all, "the same evil demons already tried to menace us many times as we traverse the wild steppes. They also sent us the violent snowstorms that fell on our weak little group like supernatural forces and at night wanted to rip apart our tents with crude fists as if they had rotten canvas before them." Schäfer insisted "that this snowman is nothing other than a bear, perhaps a 'Mashinng,' a really large one; but with our 'big gun,' I will easily shoot him dead before he even leaves the cave!"

The Wata returned within the day with a witness who had, while searching for lost sheep, found a cave in which "he beheld for the first time with his own eyes the yellow head of a snowman." Following his guides to the den of the Yeti, Schäfer shot it at point-blank range when it emerged, roaring angrily, from its nap—and it was indeed a Himalayan brown bear. He later described the Abominable Snowman idea as a "sham," saying that he "established the yeti's real identity with the pictures and pelts of my Tibetan bears,"[30] some of which he shot and sent back to American and European museums.

Schäfer even alleged that in 1938, mountaineers and important Yeti track-finders "[Frank] Smythe and [Eric] Shipton came to me on their knees, begging me not to publish my findings in the English-speaking press. The secret had to be kept at all costs—'Or else the press won't give us the money for our next Everest expedition.'"[31] If true, this claim, made in 1991, would be extremely damning. If Shipton had been engaged in hoaxing or playing along with a popular legend in the 1930s, that would reflect badly on his famous footprint find in 1951—a print that remains the Yeti's most powerful icon, but that has (as we will see) recently been implicated as a hoax. It happens that Shipton did indeed report other Yeti tracks in 1936—footprints that predate the publication of Schäfer's book.[32] In 1937, likewise, Smythe found "the imprints of a huge foot apparently of a biped and in stride closely resembling my own tracks." Smythe was able to follow these tracks all the way to the creature's cave lair. The problem for Schäfer's claim that Smythe and Shipton begged him to stay silent (apart from being uncorroborated and reported decades after the fact) is that in 1938 Smythe had no need to fear that Schäfer would reveal that the Yeti is really a bear: Smythe already had publicly debunked his own well-documented Yeti tracks as those of a bear, declaring that "a superstition of the Himalayas is now explained" as being based on bears.[33] Still, Schäfer's allegation should not be ruled out in Shipton's case. In 1937, Shipton's climbing partner H. W. Tilman responded to an article on the Yeti written by Smythe with a tongue-in-cheek letter of denunciation (or, rather, two letters, one prankishly penned under his nickname, Balu [Bear]):[34] "Mr. Smythe's article, if it was an attempt to abolish that venerable institution the 'Yeti,' was hardly worth the paper on which it was written."[35] Tilman's comments about the Yeti were, throughout his career, filled with playful, cheeky jokes—with "unbecoming levity," as one author put it.[36] "Admittedly difficult though it is, the confounding of scientific skeptics is always desirable, and I commend [track witness] Wing Commander [E. B.] Beauman's suggestion that an expedition should be sent out," becomes a telling comment in that humorous context—as does, perhaps, Tilman's claim to have found mysterious tracks himself in 1937: "I was crossing the Bireh Ganga glacier when we came upon tracks made in crisp snow which resembled nothing so much as those of an elephant. I have followed elephant spoor often and could have sworn we were following one then, but for the comparative scarcity of those beasts in the Central Himalaya."[37] It is tempting to speculate that Shipton and Tilman (rather than Smythe) truly could have encouraged Schäfer to play along with all the Yeti fun. After all, as Ralph Izzard found when he interviewed Tilman, "It was obvious that he also belongs to the school which considers that the mystery of the Yeti should be left uninvestigated; that once the unknown becomes known and the glamour is dispelled, the interest evaporated."[38]

Be that as it may, the important zoological work that Schäfer had done on his Himalayan expeditions was besmirched by a much more serious matter: his joining Adolf Hitler's Schutzstaffel (SS; Protective Squadron) in 1933.[39] Asserting that "science only grows on a racial basis," Schäfer described the values of science and the values of the SS as one.[40] Heinrich Himmler, the head of the SS, used Schäfer's expertise in Tibet to plan Operation Tibet, a secret Nazi mission to incite the local Tibetans and Nepalese to rise up against the British troops in the region. In addition, Hitler and Himmler had strange notions that the ancestral Aryan race had originated in the mountains of Tibet, and they sent Schäfer there to find evidence in support of their bizarre genealogy. He was not successful, and his association with Hitler and Himmler completely overshadowed his scientific reputation after the war, so almost all of his solid prewar scientific work on the natural history of Tibet was forgotten. So, too, were his sober descriptions of the Yeti's legends and traces, and his evidence that the footprints had been made by Himalayan brown bears was discounted or forgotten until Reinhold Messner translated them and revived the idea in connection with his own quest for the Yeti.[41]

THE STAMPEDE TO THE HIMALAYAS

World War II slowed down the exploration of the Himalayas for a while because most men were engaged in the war effort, and Central Asia was part of the battleground between the Allies and the Japanese. Not until the late 1940s and early 1950s did the pace of exploration and the search for the Yeti pick up again. Two reports are of note. In 1948, three army officers who were

Figure 3.2
A species of langur monkey.

headed to Kolahoi Glacier in Kashmir had reached the snowline at 13,000 feet when they saw a large primate bounding toward them. At first, they thought that it might be the Yeti, but at closer range it turned out to be a large langur monkey (perhaps *Semnopithecus schistaceus* or *Semnopithecus entellus*), which can weigh nearly 50 pounds and walk bipedally for long distances (figure 3.2). Shortly thereafter, two Scandinavian prospectors reported large langur monkeys at 13,000 feet. The monkeys apparently attacked them and left large footprints in the snow[42] (though some sources doubt their story).[43]

The 1950s were the golden age of Yeti hunting, when, as John Napier suggests, "folklore started to deteriorate into fakelore" and there was a stampede to hunt for evidence of the Yeti. The Nepalese government offered special licenses at £400 per Yeti as an inducement to hunters.[44] The crucial event occurred during the Mount Everest Reconnaissance Expedition of 1951, led by Eric Shipton. The explorers were scouting a route to the top of Mount Everest (which still had not been climbed) and had split into three parties. While Edmund Hillary and another expedition member scouted one area, and two other climbers explored another, Shipton, team doctor Michael Ward, and Sherpa San Tenzing crossed a glacier at 18,000 feet.[45] There they found footprints in the snow—and, for once, someone took photographs. In doing so, they made history, for the Shipton print remains the single most famous and most persuasive piece of evidence for the Yeti. Like the film footage of Bigfoot shot by Roger Patterson and Bob Gimlin, it echoes hauntingly throughout the literature. As Napier put it, "Something must have made the Shipton footprint. Like Mount Everest, it is there, and needs explaining. I only wish I could solve the puzzle; it would help me sleep better at night."[46] But Shipton and Ward may not have anticipated how famous this print would become. By Shipton's own account, few of the tracks seemed impressive. "The tracks were mostly distorted by melting into oval impressions," he said, "slightly longer and a good deal broader than those made by our large mountain boots."[47] Elsewhere, Shipton elaborated:

> As we went down the glacier, however, the snow became less deep and the footprints more regularly shaped. At length we came to places, particularly near crevasses, where the snow covering the glacier ice was less than an inch thick. Here we found specimens of the footprints so sharply defined that they could hardly have been clearer had they been carefully made in wax. We could tell both by comparing one print against another and by the sharpness of the outline there had been no distortion by melting, and from this we inferred that the creatures (for there had evidently been two) had passed only a very few hours before. They were going down the glacier. The footprints were 12 inches long and about 6 inches wide. There was a big, rounded toe, projecting very much to the side; the middle toe was well separated from this, while three small toes were grouped close together. There were several places where the creatures had jumped over small crevasses and where we could see clearly that they had dug their toes into the snow on the other side to prevent their feet from slipping back.[48]

Shipton photographed what he and Ward took to be the clearest, least-distorted track (figure 3.3), but seems not to have taken pictures of any of the other "Yeti" prints, which would have helped determine the significance of the photographs he did shoot.

Or did he? Since the late 1980s, a school of thought has grown that Shipton may have intentionally faked his famous track, and that argument hinges on discrepancies in the accounts about how many photographs were taken.[49] Some cryptozoological sources claim that Shipton shot not only two close-up photographs of the famous footprint (one with an ice ax for scale, and the other with Ward's boot), but also two wider shots showing the same Yeti trackway trailing off to a moraine.[50] The problem is that the tracks in the wider shots do not match the footprint in the close-ups. When Napier asked Ward about this, Ward told him that the wider photograph was frequently misidentified, as an accidental result of having been kept in the same photo file as the shots of the "Yeti" footprint. According to Ward, that second trackway—probably from a goat—was photographed separately, earlier on the same day, and had nothing at all to do with the "Yeti" footprint photos. When Shipton confirmed Ward's two-trackway story, Napier felt that this explanation "clears up an eighteen-year-old mystery"—but it did not *stay* clear. Writing two decades later as the

Figure 3.3
An alleged Yeti footprint photographed by Eric Shipton in 1951, on a glacier of the Menlung Basin (near the border of Nepal and China). (Reproduced by permission of Royal Geographical Society [with IBG], London)

only surviving witness of the photographing of the famous Yeti print,[51] Ward spoke of

> a whole series of footprints in the snow. These seemed to be of two varieties, one rather indistinct leading to the surrounding snowfields, while the other had in places a markedly individual imprint etched in the two- to four-inch snow covering on the top of hard névé. We had no means of measuring these, so Shipton took four photographs, two of the indistinct prints with myself, my footprints and rucksack as comparison; the other two photos were of one of the most distinct and detailed prints, one with my ice axe and one with my booted foot for scale.[52]

Is "a whole series of footprints . . . of two varieties" one trail or two trails? Most early accounts seem to describe only a single set of tracks, while later versions (and later interpretations of the older records) either suggest two sets or one—or are ambiguous. The records and memories on this point are as conflicting as the comparison between the prints in the wide shots versus the print in the close-ups. As team member Hillary later reflected, "I was not aware of any second set of tracks."[53]

Whatever these photographs truly depict, the shot of the footprint compared with the ice ax was soon celebrated and debated in newspapers around the world, popular magazines, and even leading scientific journals[54]—and soon a rush of people were trying to mount expeditions to find more evidence of the Yeti. "The discovery of these footprints in 1951 stimulated great interest in the yeti," Ward recalled, "and, as a result, many expeditions of greatly varying degrees of competence have taken place in the Himalaya and Central Asia."[55]

But could the shot have been a hoax or prank perpetrated by Shipton and Ward? That was the suspicion of investigative journalist Peter Gillman, who found it "particularly striking" that the two-trackway story, in his opinion, "contradicted all previous accounts."[56] Potentially more damaging, Gillman noted that rare uncropped versions of the close-up photo with the ice ax may show part of a *second* footprint that seems to lack the distinctive features of the "main" footprint:

> Unlike the full footprint above it, it had only vague impressions of the smaller toes; and where you would expect a big toe to match the one in the main footprint, there was nothing at all. It was after pondering all the evasions and inconsistencies that we concluded that the most obvious explanation for the unique and anomalous single footprint was that it had been fabricated by Shipton. It would have been the work of moments to enhance one of the oval footprints by adding the "toe-prints" by hand, particularly a hand wearing a woollen glove.[57]

For his part, Ward denied that there was a hoax of any kind,[58] but Gillman suggested that Ward need not have known. "Under my scenario, this is what could have happened," Gillman speculates. "When Shipton and Ward came upon the line of prints, Shipton—unknown to Ward—added the crucial embellishments to the single footprint. He called Ward's attention to it, and joined in speculation that they had found evidence of the yeti."[59] The largely circumstantial case for a hoax then rests on Shipton's personality. Was he a prankster? Gillman cited examples uncovered by mountaineering writer Audrey Salkeld showing that Shipton sometimes indulged "in jokes of exactly this kind," making up far-fetched stories about other mountaineers—stories flatly denied by others who were there.[60] (Shipton apparently claimed that climber and geologist Noel Odell once tried to eat rocks in the mistaken belief that they were sandwiches and asserted that the body of climber Maurice Wilson had been found in possession of an incriminating "sex diary" and several items of women's clothing.)[61] Hillary found that his teammate Shipton was evasive when quizzed about the footprint and that he "tended to rather dodge giving too much of a reply."[62] Hillary told Gillman that Shipton "definitely liked to take the mickey out of people," and agreed that "he might have tidied it [the footprint] up, made it look fresh and new and photographed it." But would Shipton have let the hype get so out of hand for so long? Yes, thought Hillary, "he would think that was quite a good joke."[63]

FOOTPRINTS IN THE SNOW

Ever since the photograph of the footprint taken by Eric Shipton (see figure 3.3) became famous, the strongest line of evidence for the existence of the Yeti has been footprints in snow and ice. The Shipton prints *appear to resemble* gigantic human footprints, which has convinced many cryptozoologists and Yeti believers that they were made by a Yeti. But is the track evidence as compelling as cryptozoologists believe? Collectively, there are problems of consistency from one alleged trackway to another. Assessing this evidence in 1973, John Napier concluded that "so-called Yeti footprints" are collectively "useless in proving the existence of a Himalayan Bigfoot," but "do at least offer some hints of the origins of the tale. Unlike the Sasquatch, there is little uniformity of pattern, and what uniformity there is incriminates the bear."*

Trackways can also be deceiving on a case-by-case basis. "Most of the prints have been seen by mountaineers, the majority of whom are laymen," warned Ward, co-discoverer with Shipton of the most famous Yeti track find of all time.† Recall as well that footprints are not necessarily an accurate record of the shape of the sole of the foot, but more of a record of the way the foot strikes the substrate through the walking cycle.

If you look at your own footprints in the wet sand on a beach, you will see that—depending on how you were moving—the shape of the prints differs. If you run, the toes and the ball of the foot dig into the sand, the heel may never touch it, and the print may be shallow because you barely touch the ground as you run. If you walk normally, the heel impression is wider and deeper than the actual size of the heel because the foot touches heel-first and thus the weight of the body goes into the heel part of the footprint as you walk. In soft sand, you may leave a footprint that is wider than your foot because it sinks into the sand, which collapses easily into the impression. On muddy surfaces, similarly, the foot may slide forward and/or sideways, leaving a print that is larger than

* John Napier, *Bigfoot: The Yeti and Sasquatch in Myth and Reality* (New York: Dutton, 1973), 205.

† Michael Ward, "The Yeti Footprints: Myth and Reality," *Alpine Journal* (1999): 85.

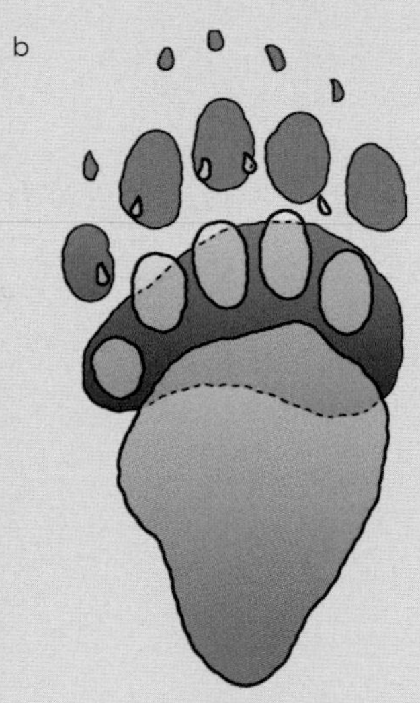

Figure B.1 The transformation of bear tracks (*a*), such as these photographed near Havre de Grace, Maryland, into human-looking footprints (*b*). ([*a*] Photograph by Lewis B. Glick; U.S. Fish and Wildlife Service; [*b*] illustration by Daniel Loxton)

your foot. In firm sand, though, you may leave only a partial print or a footprint that is narrower than your foot, as do people with high arches, since the curve of the instep may never touch the sand.

The same dynamics apply, but with slight differences, to footprints in the snow. If the snow is quite deep, you leave very large prints, since the ankle and leg as well as the foot plunge into the snow; the hole may get larger when you pull your leg out and forward on the next stride. Even in a thin coating of snow, the sides of the print often collapse inward after the foot is removed. Even the slightest bit of melting enlarges a footprint, and prints may melt unevenly if shaded parts of a trackway remain cool while the sun strikes another part directly. Ordinary tracks can become distorted, with some parts remaining sharp while others are erased.

Many of those who saw "Yeti" prints realized that hoofed animals and wolves can leave large paired trackways in the snow (especially deep snow). Most animals tend to walk single file and often step in the footprints of their leader, thus enlarging them. Many of the "Yeti" tracks discussed in this chapter have suspiciously narrow heels, which is typical of bears but not humans. All it takes for bear tracks to look more human is for the hind foot to step into the print made by the front foot (which bears do naturally), wiping out the hind-toe impressions and leaving only a much longer footprint with four or five oval front-toe impressions that resemble those of humans (figure B.1). Add a bit of melting to round out the heel region, and you have a very plausible-looking Yeti footprint.

Another thing to remember about human-like Yeti tracks: whatever made them, it was *not* a large ape-like creature similar to *Gigantopithecus*. All living and fossil apes have feet that are shaped very differently from those that made the tracks, with a much larger big toe farther back on the foot that allows for grasping. If the "Yeti" and "Bigfoot" tracks that have been photographed so far are from a real cryptid—not a hoax or a bear or another known animal—the creature would have to be very human-like, since completely modern bipedal posture is a feature found only in the more advanced hominids. The earliest known foot bones of the hominids, such as those of *Ardipithecus* from Ethiopia, dated to 4.4 million years ago, indicate that they did not have a modern human foot or walking gait.* But the famous footprints of *Australopithecus* from Laetoli, Tanzania, which are 3.5 million years old, show that a fully modern gait had evolved by that time.† Thus if the Yeti and Bigfoot footprints are real, the creature that made them should look much more human-like (comparable to *Australopithecus* or early *Homo*) than ape-like.

Plaster replicas of "Yeti" footprints are artistic reconstructions from photographs, not precise casts of the original prints. Just try to make a plaster cast of a footprint in the snow, and you will discover that you run into many problems. First, most of the people who have reported "Yeti" tracks rarely carry large quantities of plaster, a mixing bowl, and extra water to make a cast. Second, if the temperature is at or below freezing, only ice would be available to mix with the powdered plaster, so it would be impossible to make a liquid mixture. And third, plaster produces heat (an exothermic reaction) as it begins to set, which would melt the track and distort its original shape. Thus the "Yeti"

* Stone Age Institute, "New Pliocene Hominids from Gona, Afar, Ethiopia," January 10, 2005, http://web.archive.org/web/20080624005441/http://www.stoneageinstitute.org/news/gona_nature_paper.shtml (accessed October 22, 2011).

† David A. Raichlen, Adam D. Gordon, William E. H. Harcourt-Smith, Adam D. Foster, and Wm. Randall Haas Jr., "Laetoli Footprints Preserve Earliest Direct Evidence of Human-Like Bipedal Biomechanics," *PLoS ONE* 5, no. 3 (2010): e9769.

tracks exhibited on television shows that are touted by cryptozoologists are pure guesswork. They are replicas that match the outline of the prints as seen in a photograph, but the depth and three-dimensional detail are inferred from the pattern of light and shadow in the hollows of the prints as they appear in the photograph.

British explorers had been making repeated attempts to reach the peak of Mount Everest since the 1920s, when surveyors determined that it was the world's highest mountain. Many expeditions had climbed partway up and made observations on the best route, but none had succeeded in reaching the summit at 29,028 feet. The biggest obstacle is that the air is so thin (less than one-third the pressure at sea level) that most climbers require bottled oxygen, which was not available to the early climbers. The extreme altitude causes fatal heart attacks, blindness, and hemorrhaging of blood into the brain or lungs. More than 175 people have died trying to reach the summit, including all the members of the 1924 British Mount Everest Expedition (one of the first full-scale attempts), who vanished completely. The frozen body of one of the climbers, George Mallory, was found and recovered seventy-five years later!

During an expedition mounted in 1952 to explore routes to Everest, Edouard Wyss-Dunant's Swiss team discovered possible Yeti tracks at an elevation above 16,000 feet. Wyss-Dunant determined that the clawed tracks his team found had been made by a family of plantigrade quadrupeds—which is to say, bears.[64] The real zoological question, he felt, was *which* species of bear? He wrote, "The moment has come to attack the real problem and no longer seek at any cost to justify a myth. But there will doubtless be fewer followers for a zoological expedition than for one that searches for a mythical creature to the full accompaniment of the Press."[65] The Swiss expedition is also notable for an alleged Yeti attack on a porter (who claimed later in the year that he had been seized by a Yeti while on the trail, and then released). The Swiss team members pointed out that there was absolutely no evidence to corroborate this yarn: "It might have been the Snowman, it might have been a bear. Or, it might have been some freak of imagination."[66]

Finally, in 1953, Edmund Hillary, a New Zealander, and the experienced Sherpa Tenzing Norgay reached the summit of Mount Everest—and the world was electrified with the news. Yet, in the midst of the celebration for this remarkable human achievement, people wanted to know: Had Hillary or the team found footprints on the mountain, as Wyss-Dunant's team had the year before? Team leader Colonel John Hunt was a believer in the creature[67] (he had found tracks on an earlier expedition,[68] and called for an expedition in search of it)[69] but said, "I'm sorry to disappoint you, but we saw no trace of him at all. I think we were too large a party for any Abominable Snowmen. They like seclusion and solitude."[70] Tenzing Norgay reported in his first autobiography that he had never seen the Yeti, but believed that it existed because his father had seen it.[71] Later in his life, however, Tenzing became much more skeptical about the existence the Yeti.[72]

In search of a sensational story to capitalize on the Everest craze and attract readers, in 1954 the *Daily Mail* funded a big expedition to search the Himalayas for evidence of the Yeti. It included a number of mountaineers, journalists, and scientists—plus as many as 370 men to carry all the gear.[73] They tramped the Himalayas for fifteen weeks, but found nothing more than some possible Yeti hair, poor tracks, and questionable animal droppings. The lead climber of the expedition, John Angelo Jackson, photographed footprints in the snow of many known animals, as well as large tracks that could not be identified, although later he admitted that they could have been the result of the melting of a more conventional footprint. He also photographed symbolic paintings of the Yeti in Tengboche gompa, a fortified Tibetan monastery.

The team examined an alleged Yeti scalp found in the small Buddhist temple in the village of Pangboche in Nepal (figure 3.4). Team zoologist Charles Stonor at first assessed this object as "quite obviously the skin from the top of the head of the animal," adding that he was "utterly unable to identify it with any animal known to science."[74] The hairs varied from dark brown to fox red. However, hair samples and photographs sent to Frederick Wood Jones, an anatomist familiar with hair samples, revealed a very different story. As Stonor explained, Wood Jones "was able eventually to pronounce beyond all doubt that the object was not a true scalp at all, but fashioned from a piece cut off the shoulder of some unknown beast—certainly not a bear or an ape—and worked into the shape of a cranium while the skin was still fresh and soft."[75] (The "unknown beast" that contributed its skin to the manufacture of this and another alleged Yeti scalp relic has since been firmly identified, as we will see.)

Despite the lack of success, two members of the expedition published lengthy books on their participation and their impressions of the evidence for

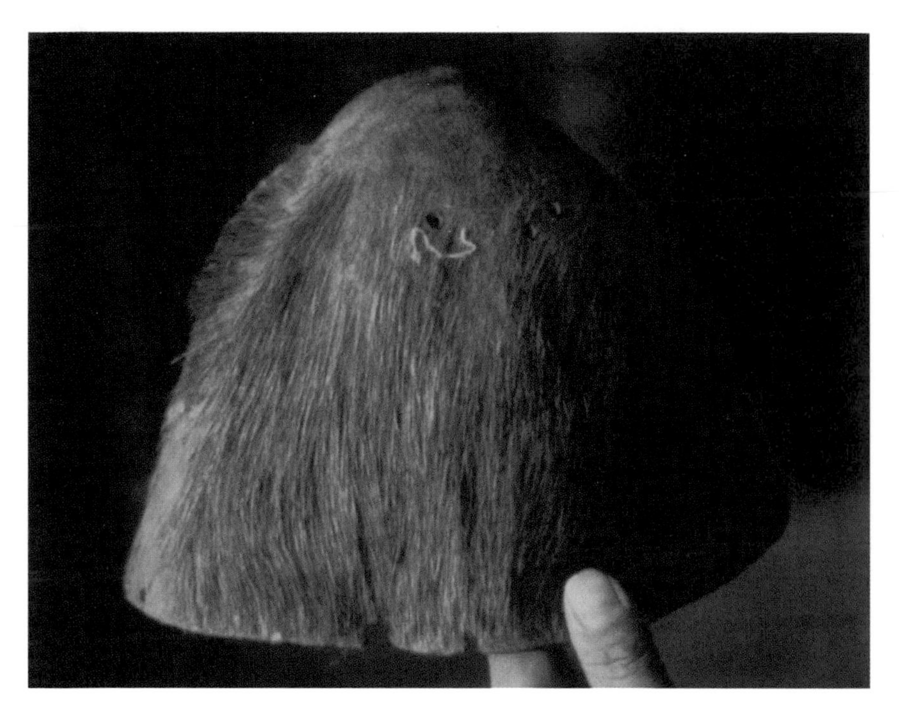

Figure 3.4 An alleged Yeti scalp examined at the Pangboche monastery by Charles Stonor, a member of the expedition to the Himalayas funded by the *Daily Mail* in 1954. (Reproduced by permission of Fortean Picture Library, Ruthin, Wales)

the existence of the Yeti. Both books were published in 1955, and each cites long passages from the other. One book, by newspaperman Ralph Izzard, is a gripping account of the expedition from the journalist's perspective.[76] Most of the book focuses on the details of each day of the expedition and the travails of trekking through such rugged country. Izzard spent two days following an 8-mile trackway of what he believed to be a pair of Yetis.[77] They made footprints about 8 to 9 inches long and 4 to 5 inches across, with a stride of about 2 feet, 3 inches between footfalls. Izzard detailed how the Yeti tracks made detours to avoid man-made paths and cottages, preventing contact with humans, and how the Yetis stumbled over a hidden hazard in the snow and left an imprint in the snowbank when they landed. Izzard and his partner Gerald Russell took only a limited number of photographs of this trackway after two days of following it, however. He also admitted that the footprints were about four days old and showed signs of melting, which enlarged them and made them less distinct.

Izzard also reprinted a letter originally published in the *Statesman* from Prince Peter of Greece, who lived in Kalimpong, Nepal, at the time.[78] Although the prince's letter asserts his belief in the existence of some sort of Abominable Snowman, he wrote:

> The real species, the flesh-and-blood Snowman is, I think, all the same, both a bear (perhaps a brown Himalayan bear) and a monkey. Tracks of both have been seen and the Swiss Everest Expedition's leader, Professor [Edouard] Wyss-Dunant, found what he considered to be the footmarks of a whole family of plantigrades [animals that walk on the soles of their feet, with their heels on the ground, as do humans, apes, and bears] on the slopes of the world's highest mountain. The monkey, however, seems to be the more prevalent of the two, and it is usually an anthropoid, ape-like creature that witnesses have reported having seen.[79]

The other book that emerged from the *Daily Mail* expedition was written by Stonor, one of the scientific officers of the group. As befits his training, his book is filled with descriptions of much of the wildlife and other scientific dimensions of the trek (including descriptions of Tibetan natural history and culture), rather than with the more sensationalist reports of Izzard. Stonor recounts the stories of the Sherpas and other tribesmen about the nature of the ape-like Yeti:

> It was the first that Pasang Nima had seen, and he described it as the size and build of a small man. Its head was covered with long hair, as was the middle part of the body and the thighs. The face and chest did not look to be so hairy, and the hair on its legs below its knees was short.
>
> The colour he described was "both dark and light" and the chest looked to be reddish. The Yeti was walking on two legs, nearly as upright as a man, but kept bending down to grub in the ground, he thought for roots. . . . After a time it saw the watchers, and ran off into the undergrowth, still on two legs, but with a sidling gait (which he imitated) giving a loud high-pitched cry heard by all who had been watching it.[80]

In his book and his reports for the newspapers, Stonor discussed different kinds of beasts that the Tibetans refer to as Yeti, and he determined from eyewitness accounts that the *Dzu-teh* of Tibet "is undoubtedly identical with the Himalayan red bear."[81] Like the journalist Izzard, the scientist Stonor found enticing "Yeti" footprints—but, as a scientist, was forced to admit that the tracks were too blurred with melting to "do more than record a general impression that they were made by a two-legged animal." Scientific honesty, he wrote, "required that we should claim no positive identification."[82] Newspaper stories claimed that there was a live Yeti in a zoo in Shigatse, a town in Tibet,[83] but Stonor concluded from a traveler's description that the animal was a gibbon.[84] Although some Yeti tales clearly described bears,[85] and despite the expedition's general failure to confirm the existence of the Yeti, Stonor concluded that "some unknown and highly intelligent form of ape does in fact maintain a precarious foothold in the alpine zone of the Himalayas."[86]

OILMAN SLICK, CIA CONNECTIONS, AND HILLARY RETURNS

In the late nineteenth and early twentieth centuries, British, French, Swiss, and German colonials and adventurers were among the leaders in the exploration of the Himalayas. Because they worked closely with Sherpas and other local peoples, European mountaineers also recorded most of the early reports of the Yeti. After the failure of the *Daily Mail* expedition of 1954 to produce conclusive evidence to support the claims for the existence of the Yeti, the next attempt was made not by British explorers, scientists, or journalists, but by a rich Texas oilman with the interesting name of Tom Slick, who financed three expeditions following his personal scouting in the area in 1956. According to his biographer Loren Coleman, Slick had been fascinated by Central Asia since his childhood and had traveled to the region many times and heard the Yeti legends. Although Texas-bred, he was a member of the eastern Establishment, having graduated from Phillips Exeter Academy and Yale University. He used the vast wealth he inherited from his father not only to fund expeditions to find Bigfoot, the Loch Ness monster, and the Yeti, but to establish a number of institutions for scientific and technological research. Slick was "a member of an elite circle of internationalists" whose "informal discussions on world peace" included Jimmy Stewart, Albert Schweitzer, Dwight Eisenhower, Winston Churchill, and John Foster Dulles. A moderate Republican, Slick rebelled against the isolationism and exceptionalism of the Republican Party in the 1950s and yearned for a more internationalist approach; these ideas were expressed in his book *Permanent Peace: A Check and Balance Plan* (1958). He argued that the United States and the Communist bloc should form a world police force that would prevent wars (much stronger than the modern United Nations peacekeeper forces) and urged the United States and the Soviet Union to dismantle their nuclear arsenals and work toward global peace. In 1960 and 1961, he even sponsored several large international meetings to further those goals.[87] Unfortunately, the weakness of the United Nations at that time and the hawkish attitudes of the Cold Warriors in both the United States and the Soviet Union prevented such goals from being realized.

According to Brian Regal, Slick made his first foray to find the Yeti in 1956. While in Central Asia, he met Irish explorer Peter Byrne, who became a critical player in both the Yeti and Bigfoot hunts.[88] Coleman suggested that both Byrne and Slick were working for the CIA during their exploits in Central Asia,[89] and, according to Regal, the connection between spying and hunting cryptids was a common pattern.[90] For example, anthropologists Carleton Coon and George Agogino, both of whom were active in trying to determine if the Yeti was real, also had CIA connections. Ivan Sanderson worked for British intelligence during World War II, and several other prominent scientists (such as S. Dillon Ripley and George Gaylord Simpson) also worked in intelligence in the Allied armies during the war. In the postwar world, there were few Americans in Central Asia who could keep tabs on the Chinese struggles with Tibet or the Soviet interests in the region. Thus people like Slick (with his money and powerful friends and his contacts with Dulles) were valuable resources for the United States to monitor the southern regions of Communist China and Soviet Russia.

In 1957, Slick and Byrne (with the backing of Luce, who wanted the rights to the first photographs of the Yeti) spent a month in the region.[91] Although they reported three sets of tracks that they attributed to the Yeti, little came of this trip.[92] "Like all the various monster expeditions before and since," wrote Regal, "the Slick project wandered around the region finding a few bits and pieces here and there: a few possible footprints, some hair samples, an apparent Yeti dung sample, lots of local stories, and not much else."[93]

Slick's sequel in 1958 was an altogether more Texas-size affair: bigger headlines, bigger budget, and bigger boasts. (Slick declared at the outset that the Yeti would be found by the end of the year.)[94] Led this time by Gerald Russell (a veteran of the *Daily Mail* expedition of 1954), the expedition of almost 100 people brought innovations as well as money and manpower—some sillier than others.[95] Their "pack of three Bluetick hounds" seemed like a good idea, as did their optimistic carting around of "two of a newly developed air-gun which propels hypodermic cartridges containing a temporary paralyzing drug."[96] But it is hard to take entirely seriously Russell's decision that "all the white hunters will be disguised as natives. We will wear rough woolen Sherpa vests, woolen hats, and felt Tibetan knee boots. Our faces will be stained brown." Even Byrne described the disguise plan as "seemingly ridiculous,"[97] which it may well have been in the light of the horde of folks stomping along in their party. They stayed in the field from February until June,[98] during which time they heard about Yetis catching frogs,[99] boasted of finding the cave of an Abominable Snowman,[100] and visited the Pangboche temple where the *Daily Mail* expedition had sampled hairs from the "Yeti scalp"[101] and tried to reanalyze the hairs or obtain the scalp. The temple also had a mummified hand that was claimed to belong to the Yeti. It was a sacred relic, so it was not supposed to be desecrated for scientific study.

Nonetheless, Slick's people got another shot at the Yeti hand at Pangboche in 1959, during Slick's last expedition (a much smaller affair, involving only Byrne and his brother).[102] According to Regal, Byrne ingratiated himself with the monks—and then, when he got the chance, stealthily removed some finger bones from the hand, replaced them with human bones, and rewrapped the relic so the theft was undetected.[103] In an interview with the BBC, Byrne confirmed that he had indeed swapped a human finger for a finger from the relic (though he said the substitution was done openly, in exchange for a donation to the temple).[104] Another story claims that Slick had Byrne give the bones of the "Yeti hand" to the famous actor Jimmy Stewart, a silent backer of their project, who smuggled it home in his luggage.[105] According to Byrne, the finger bones were actually smuggled in Gloria Stewart's lingerie case, so customs officials were reluctant to search it.[106] The bones were sent to anatomist, primatologist, and anthropologist William Charles Osman Hill, at the Zoological Society of London, who was puzzled by them and at first thought that they resembled Neanderthal finger bones; later studies by Carleton Coon showed that the finger bones were human.[107] Recently rediscovered among the Osman Hill material in the Hunterian Museum of the Royal College of Surgeons in London, the finger bones have now been definitively identified as human using DNA analysis. Said genetic expert Rob Ogden, who conducted the tests, "It wasn't too surprising but it was obviously slightly disappointing that you hadn't discovered something brand new. Human was what we were expecting and human is what we got."[108] This was the final epilogue of a book that long since closed on failure. Despite the audacious hype ("Expedition a Success, Proves Yeti Exists" was the title of Slick's summary for the 1958 expedition)[109] and all the time and money spent (at least $100,000 and perhaps double that),[110] the results of the Slick expeditions were ultimately dismal. Coon and the others who examined the hair and dung samples sent by Slick and Byrne concluded that there was no evidence of the Yeti in them and became disillusioned with the entire Yeti hunt, and a mummified "Yeti leg" tracked down during one of Slick's expeditions turned out to be from a snow leopard.[111] Looking over this body of evidence, Napier reflected that "the mummified paw of a snow leopard, a mummified human hand, and a footprint or two—add up to nothing at all. No single item contributed one jot or tittle of proof to the Himalayan Bigfoot legend."[112]

Meanwhile, the Chinese began to crack down after the Tibetan uprising of 1959, which led the Dalai Lama to flee into exile in India. The Chinese viewed the Yeti hunts as espionage, so they closed the country's borders and made it very hard for expeditions to travel beyond Nepal. One of the last to do so, in 1960/1961, was mounted by Everest conqueror Sir Edmund Hillary and America's favorite animal expert, Marlin Perkins. Perkins, a zoologist, was the director of the Lincoln Park Zoo in Chicago, in which role he had also hosted the hit animal program *Zoo Parade*.[113] (Perkins would go on to host *Mutual of Omaha's Wild Kingdom*, another immensely popular television show about wild animals and nature.) Backed by *World Book Encyclopedia*, the huge expedition team included a number of scientists (and 150 porters).[114] The Hillary-Perkins crew went straight to the Pangboche monastery, where they persuaded the monks to let them examine the "Yeti hand" (not knowing that Byrne had switched the relic's finger bones with those of a human). They even successfully arranged to borrow the "Yeti scalp" from the Khumjung monastery, and brought it back with them for expert examination in London, Paris, and Chicago. As Hillary explained, the scientists who studied the scalp found that "although it was an interesting and probably ancient relic, it was really a fake—molded from the skin of a serow. It had certainly never held the cunning brain of the elusive Yeti."[115]

These examinations confirmed what Hillary's team had discovered in the field: when they acquired three hides from serows (mid-size wild animals that resemble goats or antelopes), they suspected immediately that the hides were a match to the skin used to manufacture the Pangboche "Yeti scalp" and a very similar relic at the monastery at Khumjung (a village 8 miles from Pangboche). As team member Desmond Doig excitedly noted in his journal that night, "Almost certainly the Khumjung and Pangboche scalps are twins, made on the same mold of serow's hide. The shape, color, and essential dimensions are identical; two conical, balding caps of black-and-henna-colored bristles with distinct median crests and punctured with holes around their bases which look suspiciously like those made by nails used to hold the scalps onto their mold." To test this impression, "we set about making two of our own, sacrificing one of the three serow hides we had acquired, and employing an unsuspecting lama to carve the molds directly from a freshly felled pine log. . . . After four days of experiment we possessed two distinctly anthropoid 'scalps,' unlike those in the Khumjung and Pangboche monasteries only in their newness and wealth of hair." Finally, they sent hair samples from both of the allegedly authentic Yeti scalps and from both of the newly re-created fake scalps to Osman Hill for analysis, without telling him the origins of the samples. "If the two samples were declared identical," Doig reflected, "the scalps would be fakes and an important argument in favor of Yetis would have been defeated."[116] And that is exactly what happened. Hillary takes up the story: "Hill showed us microscopic sectional photographs he had taken of the bristles from the Khumjung and Pangboche scalps and from our own fakes, molded from serow. Until then he did not know which was which—we hadn't told him the origin of the bristles when we sent them from Khumjung. Dr. Hill was confident that all the specimens belonged to the same genus of animal."[117] The other physical evidence that supposedly argued for the existence of the Yeti was similarly disappointing. The alleged Yeti hand from the Pangboche monastery proved to be "essentially a human hand, strung together with wire, with the possible inclusion of several animal bones" (presumably those switched by Byrne).[118] The "Yeti skins" turned out to be the pelts of Himalayan brown bears, leaving Hillary's expedition with the "very strong inference that this bear might well have been the source of much of the Yeti legend." Looking over this evidence, and considering their failure to turn up "a single case of a lama who claimed personally to have seen a Yeti," Hillary concluded:

> There is no doubt that the Sherpas accept the fact that the Yeti really exists. But then they believe just as confidently that their gods live in comfort on the summit of Mount Everest. We found it quite impossible to divorce the Yeti from the supernatural. To a Sherpa the ability of a Yeti to make himself invisible at will is just as important a part of description as his probable shape and size. Part animal, part human, part demon—the Yeti is as calmly and uncritically accepted as we accept Father Christmas when we are children.
>
> Pleasant though we felt it would be to believe in the existence of the Yeti, when faced with the universal collapse of the main evidence in support of this creature the members of my expedition—doctor, scientists, zoologists and mountaineers alike—could not in all conscience view it as more than a fascinating fairy tale, born of the rare and frightening view of strange animals, molded by superstition, and enthusiastically nurtured by Western expeditions.[119]

The Hillary–Perkins expedition not only investigated remains of the Yeti, but also did research on high-altitude medicine, so the absence of proof of the existence of the Yeti did not make the expedition a complete failure. More important, Hillary was sensitive to and concerned about the plight of the Nepalese and especially the Sherpas, who had helped him climb Mount Everest and become world famous. In exchange for the loan of the "Yeti scalp," he and his expedition's sponsors paid for a village elder to "accompany the Yeti scalp wherever it may be taken in America and Europe" and for urgent renovations at the monastery.[120] Even more lasting were his efforts to raise funds for the construction of numerous schools and other public buildings in impoverished Nepalese villages.[121]

By this point, nearly all the Western Yeti hunters were thoroughly discouraged by the repeated failures of the successive expeditions. The more people looked for the Yeti and the more "evidence" they analyzed, the less likely the existence of the Yeti became. The shenanigans and CIA connections of the Slick–Byrne group not only had disillusioned many believers, but had alarmed the Chinese government. Because the Chinese viewed all the Yeti hunters as spies, they clamped down even further on them, even as they crushed Tibet, destroyed the Lamaist religion and political system, and drove the Dalai Lama into exile. After the Hillary–Perkins expedition, no effort was made to mount further trips into Tibet, since the Chinese were completely in control and did not allow foreigners into the country. The short golden age of Yeti hunting through the 1950s came to an end, and almost no one returned to the Himalayas for many years to find the Yeti. As John Napier put it, "The Hillary Expedition of 1960–61 had the effect of dampening the enthusiasm of all but the staunchest of Yeti supporters. Scientists smiled knowingly, mountaineers lost interest, and journalists found other fish to fry."[122]

WHILLANS, WOOLDRIDGE, THE CHINESE, THE RUSSIANS—AND A HOAX

For almost a decade, the closure of Tibet by the Chinese made it difficult to travel to the country to find evidence of the Yeti. The various expeditions undertaken before 1960 had found nothing: all the "Yeti scalps" in Tibetan monasteries were serow pelts; "Yeti" hands were human hands; "Yeti" dung and hair came from known animals; and "Yeti" trackways were still not convincing. Although the local peoples continued to report seeing the Yeti, the only "sighting" of a supposed Yeti by a Westerner had been by the British-Greek explorer and photographer N. A. Tombazi in 1925; he later decided that the figure he had seen was a hermit, although John Napier showed that the prints Tombazi reported are better explained as those of a bear.[123] But in the summer of 1970, the mountaineer Don Whillans, who was co-leading an expedition to climb Annapurna from the Nepalese side, reported footprints of Yeti-like appearance at 13,000 feet. However, his photograph shows a trackway that probably was made by a quadruped.[124] The same night, Whillans saw what he considered to be an ape-like animal bounding along on all fours a long distance from his tent. "I saw some black blobs up the slope in the shadows, then I saw a blob that hadn't been there a moment before," he said. "It was something moving 'round. . . . Then it bounded out of the shadows and headed straight up the slope in the absolutely bright moonlight. It looked like an ape. I don't think it was a bear."[125] He had been equally uncertain days earlier: "It could have been a bear but it looked much more like a monkey-type thing."[126] Whillans never saw the creature again, and it had been so far away from him ("about a quarter of a mile") that he could provide few details about his distant sighting in the darkness—except that it was apparently running on four legs. This detail suggests that it was a Himalayan brown bear or another known quadruped, rather than a supposedly bipedal Yeti.

The continued closure of Tibet only increased the desire of Western mountaineers and cryptozoologists to visit the country. It focused attention on the area, suggesting that the Chinese had proof of the existence of the Yeti and were hiding it. Indeed, the Chinese themselves examined the legend, sending more than 100 investigators into the region in 1977 to exhaustively track down every lead and look in every possible place where the Yeti might lurk. They found absolutely nothing. Eventually, in 1988, the Chinese government allowed a Western expedition into Tibet, but it, too, found no evidence of the Yeti. Chinese scientists have themselves investigated the legends of the Yeti and other Wildmen that fall within their national borders. For example, according to anthropologist Zhou Guoxing,

> A large-scale scientific investigation sponsored by the Chinese Academy of Sciences was carried out in these areas [northwestern Hubei and southern Shanxi provinces] in 1977. More than 100 people participated in it for nearly a year. . . . Although the investigation was unusual in its scale, number of participants, and duration, no direct proof was found of the existence of the Wildman, and only footprints, pieces of head hair, and feces presumed to be those of Wildman were recovered.

Although Zhou believed that it can be inferred on the basis of indirect evidence that "these unknown animals are not mere creatures of fiction," he conceded that the majority of Chinese scientists do not agree. Zhou also noted the long history of failed Chinese attempts to confirm the Wildman:

> Since the end of the 1950's, China has organized a series of on-the-spot investigations of Wildman in Tibet, and the provinces of Yunnan, Hubei, Shanxi, and Zhejiang. Among the participants in these investigations have been a number of professional scientists, such as anthropologists, geologists, zoologists, and botanists, as well as personnel in specific fields of zoological parks and natural history museums. Taking part in the investigation in the Shennongjia forest area are experienced huntsmen and skilled scouts. Up to the present time, apart from the abovementioned hand and foot samples obtained in the Jiolong Mountain areas of Zhejiang Province [found to be from a monkey, perhaps a macaque], no direct physical evidence has been found to support the existence of Wildman.[127]

In 1998, the head of the Regional Association for Wildlife Protection, Liu Wulin, told the Xinhua news agency that all the fur samples and footprints came from the Himalayan brown bear and that nothing gave evidence of the existence of the Yeti in Tibet[128] (although the Chinese are still hopeful of finding the creature, despite the complete failure of every effort).

Cold War tension even extended to monster hunting. Gerald Russell, leader of the American-financed expedition led by Tom Slick in 1958, said, "Our task this time has assumed all the more significance now that the Russians are already in the field on the Pamir plateau. It is now an international race for the yeti."[129] Led by Boris Porshnev, several Soviet expeditions were made to the Pamir Mountains, the northwestern extension of the Himalayas, which stretch from Tajikistan and Kyrgyzstan through China, Pakistan, and Afghanistan. Those who live in these mountains also have their share of legends about the "Wild Man of the Snows," known as the Almasti or Almas. Porshnev based his ideas on the earlier work of Badzar Baradiin, an ethnologist who claimed to have seen the Almasti when exploring and collecting cultural artifacts in Tibet in 1906. Porshnev and the Soviets were motivated by their Marxist ideology to view the search for the "Wild Man" as a way to confirm materialism and evolutionary explanations of human origins and behavior.[130] Cryptozoologists like Bernard Heuvelmans and Ivan Sanderson regarded *Gigantopithecus* as evidence for the survival of a giant ape in the form of the Yeti, but the Almasti does not fit that legend. Rather than being a ferocious beast that kills yaks and humans, the Almasti was kind to the Soviet peasants and does not have mythological stature or gigantic proportions, as the Yeti allegedly does. Porshnev thought that they were a relict population of Neanderthals who had survived the extinction of that species during the Ice Age of the Pleistocene epoch (2.5 million–11,700 years ago). He led an unsuccessful expedition to the Pamir Mountains, and soon his career went into decline and he fell out of favor with the Soviet regime. After Porshnev's downfall, some of his followers, including Dmitri Bayanov and Marie-Jeanne Koffman, tried to keep the Almasti hunt in the Pamirs alive in the 1960s and 1970s, but their expeditions met with failure.[131]

"DRAGON BONES" AND *GIGANTOPITHECUS*

In almost any Chinese city are open-air markets with many different things for sale. The Chinese pharmacists, or apothecaries, sell a wide variety of products that are thought to be effective medicines, from minerals and crystals, to roots and herbs, to parts of various animals—over 130,000 are known and used in traditional Chinese medicine. Some are common and harmless, but others—like tiger penises, rhinoceros horns, bear gallbladders, turtle plastrons, and seahorses—have caused vulnerable animal populations to be poached and hunted nearly to extinction. Chinese medicine also includes human products, including organs, bones, fingernails, hair, dandruff, earwax, feces, urine, and sweat. But many Chinese apothecaries also sell ground-up "dragon bones"—pieces of fossil bone and teeth that have been poached from caves and fossil sites, often before scientists were able to find them. Indeed, for centuries, the Chinese have been mining and destroying fossils to be ground up and turned into medicine.

Some paleontologists have learned that shopping in the apothecaries for "dragon bones" is the best way to find Chinese fossils and acquire them for science. However, it was often impossible to determine where there might be important deposits nearby because the "dragon bones" were extremely valuable and the Chinese carefully guarded the secrets of their sources. In the 1850s and 1860s, British travelers sent their specimens to London, where in 1870, paleontologist Richard Owen at the British Museum identified teeth of extinct rhinos, tapirs, elephants, hyenas, horses, and many other Ice Age mammals whose extant descendants are no longer found in China. Between 1899 and 1902, German travelers in China sent paleontologist Max Schlosser

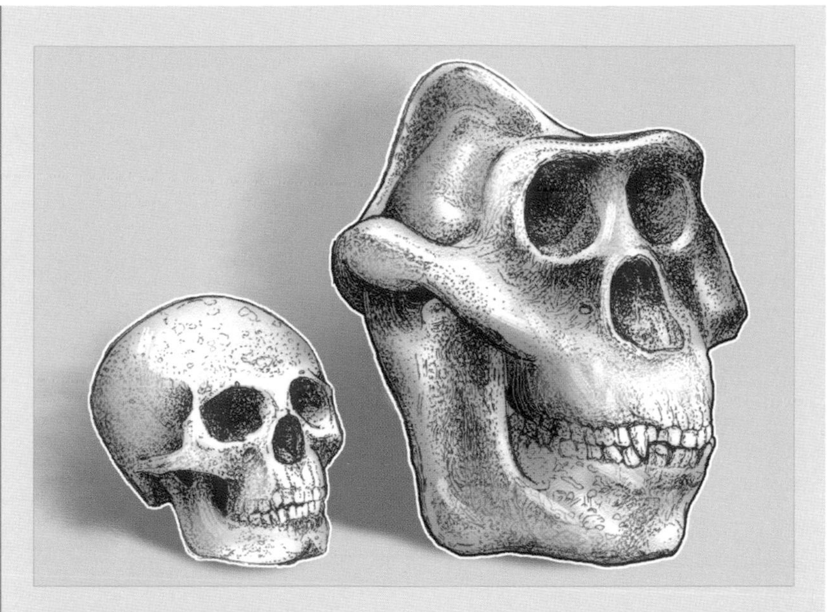

Figure B.2 A human skull compared with a reconstructed skull of *Gigantopithecus blacki*. (Illustration by Pat Linse, with Daniel Loxton)

many teeth from Ice Age mammals. Among them was the first tooth that led to the discovery of "Peking Man" (now considered to be *Homo erectus*) from the Zhoukoudian caves near Beijing. In 1921, paleontologist Walter Granger of the American Museum of Natural History managed to find the farmer who was supplying the apothecaries, and he excavated a cave that yielded many extinct mammals, including an Ice Age tapir that was as large as a modern rhino.

So fossil collecting in Chinese drugstores was a tried-and-true technique for scientists such as German geologist and paleontologist Gustav Heinrich Ralph von Koenigswald. He started his career by finding and describing the famous skullcap of "Java Man" (*Homo erectus*) at Mojokerto in 1937 and was the first to realize that "Java Man" was the same species as "Peking Man." He also did work on early humans and primates in Africa and many other regions.

In 1935, he found a huge primate molar in a Chinese apothecary shop. He recognized immediately that it was from a gigantic ape not too different from the modern orangutan, but much larger, and named it *Gigantopithecus blacki* (figure B.2). The genus means "gigantic ape" in Greek, and the species is named in honor of Davidson Black, a pioneer in Chinese paleontology who was especially important in the discovery of "Peking Man." Von Koenigswald spent four more years roaming Chinese marketplaces, but found only three more teeth of *Gigantopithecus*. He then talked to pharmacists who told him that the teeth came from Guangxi Province, in southern China. Using the kind of dirt clinging to the teeth and the gnaw marks of rodents as clues, he guessed that they were from cave deposits. When he finally found the caves, the teeth were mixed with middle Pleistocene elephant and panda fossils, so he estimated their age as between 125,000 and 700,000 years.

As he continued to track down the source of the teeth, the Japanese invaded China and von Koenigswald was taken prisoner. Even though he had been born in Berlin in 1902 (making him German by birth and supposedly a Japanese ally), von Koenigswald had acquired Dutch citizenship during his years in Java (then a Dutch colony), so the Japanese incarcerated him in a prisoner-of-war camp. He survived the ordeal and managed to hide all but one of his fossils from the Japanese; one skull was presented to Emperor Hirohito but was recovered after the war. His *Gigantopithecus* fossils had been buried in a milk bottle in a friend's backyard for safekeeping and were retrieved after the war. While he was imprisoned, his colleague Franz Weidenreich was at the American Museum of Natural History, where he published many of von Koenigswald's finds (including the idea that "Java Man" and "Peking Man" were just different samples of *Homo erectus*, and the best descriptions of the "Peking Man" specimens), assuming that von Koenigswald was dead. So even though von Koenigswald made most of the discoveries, the volumes were published with Weidenreich as the author. After the war, von Koenigswald joined Weidenreich in New York and worked with him again.*

Since the 1930s, many more *Gigantopithecus* specimens have been found in the original cave deposits, including some complete lower jaws (figure B.3). Unfortunately, no other non-dentary skeletal parts from this mysterious ape are known, despite decades of searching by the large number of Chinese paleontologists who now work on the deposits. More recently, anthropologist Russell Ciochon revisited the region around Guangxi Province and found more specimens of *Gigantopithecus*. He did so by shifting his focus to cave deposits in northern Vietnam, which are unspoiled by the fossil poachers who rob the Chinese caves to supply bones for apothecaries to grind up. Still, even after

* G. H. R. von Koenigswald, "*Gigantopithecus blacki* von Koenigswald, a Giant Fossil Hominoid from the Pleistocene of Southern China," *Anthropological Papers of the American Museum of Natural History* 43 (1952): 295–325.

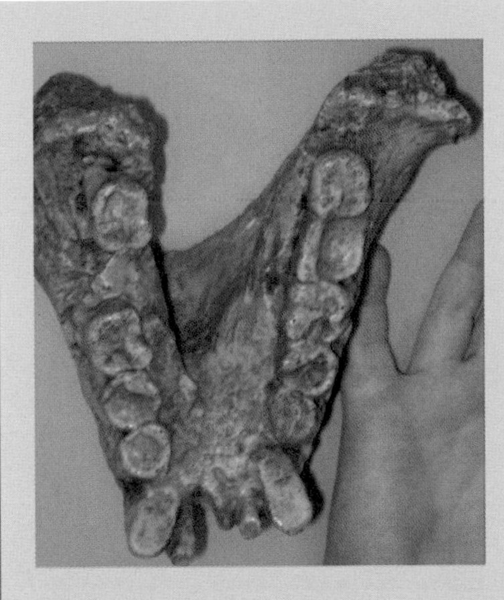

Figure B.3
A cast of the jaw of *Gigantopithecus blacki*, showing the robust thick-enameled and heavily worn molars. (Photograph by Donald R. Prothero)

more than seventy-five years since the first tooth was found, only three lower jaws and about 1,300 isolated teeth of this mysterious primate have been discovered. There is also a second species, *Gigantopithecus giganteus*, from India, which (despite its name) is about half the size of *Gigantopithecus blacki*. A third species, *Gigantopithecus bilaspurensis*, comes from much older beds (6–9 million years old) in India, suggesting that the *Gigantopithecus* line goes back to at least 9 million years ago and the evolutionary radiation of early apes such as the dryopithecines.

Because paleontologists have only the lower jaw to go on, it is hard to reliably estimate the size of the entire creature. Ciochon and his colleagues estimated that it was about 10 feet tall and weighed about 1,200 pounds. Elwyn Simons and Peter Ettel suggested that it was proportioned like a gorilla, standing about 9 feet tall and weighing about 900 pounds.* Either way, it was the largest primate that has ever lived, immensely bigger than a gorilla (the largest living primate) or even the biggest human giants.

What does remain of *Gigantopithecus* are the heavily built jaws with huge teeth, especially the molars, which have very thick enamel (see figure B.3). Both the molars and the cheek teeth in front of them (premolars) are very

* Russell L. Ciochon, John Olsen, and Jamie James, *Other Origins: The Search for the Giant Ape in Human Prehistory* (New York: Bantam Books, 1990); Elwyn L. Simons and Peter C. Ettel, "*Gigantopithecus*," *Scientific American*, January 1970, 77–85.

broad and low-crowned, often with their entire occlusal surface ground down flat, suggesting that these apes ate a very tough, gritty diet. Close microscopic analysis of wear facets on the tooth enamel, and the presence of phytolith fossils from plants, show that *Gigantopithecus* ate mostly bamboo, as does the living giant panda.*

Gigantopithecus lived in Asia since at least the middle Miocene, about 9 million years ago, and were found mostly in East Asia during the Ice Ages. Careful dating of cave deposits in Vietnam, which yield both *Gigantopithecus* and *Homo erectus*, show that early humans invaded China about 800,000 years ago and that *Gigantopithecus* died out about 500,000 years later, around 300,000 years ago.† Although this time line certainly disproves the idea that *Homo erectus* immediately killed off its distant cousin, there are other possible factors for its extinction, including competition with giant pandas for bamboo and bamboo die-offs, which occur every twenty to sixty years and may have stressed the ape population and made them more vulnerable to competition from pandas or people.

Or *did* they die out? As Brian Regal points out, in the 1950s and 1960s some anthropologists like Carleton Coon made the inference that the Yeti was a relict population of *Gigantopithecus*.‡ At that time, many anthropologists embraced the multiregional hypothesis, which suggested that different populations of *Homo sapiens* had evolved separately over a million years ago from different stocks of primates in different regions. Asians were descendants of "Peking Man"; Neanderthals, of an early European *Homo*; Africans, of African *Homo erectus*; and so on. Although a few holdouts still support a version of the multiregional model (like Milford Wolpoff at the University of Michigan), genetic evidence that has been amassed since the 1980s has overwhelmingly demonstrated that it is false. Instead, the human genome shows that modern

* Russell L. Ciochon, Dolores R. Piperno, and Robert G. Thompson, "Opal Phytoliths Found on the Teeth of the Extinct Ape *Gigantopithecus blacki*: Implications for Paleodietary Studies," *Proceedings of the National Academy of Science* 87 (1990): 8120–8124; Russell L. Ciochon, "The Ape That Was," *Natural History*, November 1991, 54–62.

† R. Ciochon, V. T. Long, R. Larick, L. González, R. Grün, J. de Vos, C. Yonge, L. Taylor, H. Yoshida, and M. Reagan, "Dated Co-Occurrence of *Homo erectus* and *Gigantopithecus* from Tham Khuyen Cave, Vietnam," *Proceedings of the National Academy of Sciences of the United States of America* 93, no. 7 (1996): 3016–3020.

‡ Brian Regal, *Searching for Sasquatch: Crackpots, Eggheads, and Cryptozoology* (New York: Palgrave Macmillan, 2011), 64–72.

Homo sapiens are all descended from African ancestors who spread across the Old World about 60,000 years ago, displacing any older populations of *Homo* (such as *Homo erectus*).* And the fossils plus the dating show that this "out of Africa" dispersal occurred more than once, since *Homo erectus* appears to have originated in Africa and then spread throughout the Old World (including China and Java) about 1.85 million years ago. Even more recent work in genetics shows that some populations (like Neanderthals) interbred with *Homo sapiens*, so when the invaders from Africa arrived, they incorporated the regional genome into theirs. Nonetheless, the hypothesis of multiregionalism and independent, isolated parallel evolution of *Homo sapiens* from local *Homo erectus* populations, as advocated by Coon in the 1950s (with its racist overtones), has long been discredited by anthropologists. Likewise, the idea that *Gigantopithecus* evolved in Asia into a human-like creature known as the Yeti is also discredited by the falsification of the multiregional hypothesis. So *Gigantopithecus* is no longer regarded as connected to the Yeti.

Not surprisingly, though, cryptozoologists beginning with Bernard Heuvelmans in 1952 have suggested that the Yeti (and, later, Bigfoot) is a surviving descendant of *Gigantopithecus*. The cryptozoological literature is full of unsupported speculations about the dispersal of these immense apes throughout Asia and North America from different primate stocks, and Bigfoot and Yeti are their relicts.† None of this amateur theorizing, based on outdated concepts of primate evolution, bears any relation to what anthropologists know about the real history of hominid fossils and human evolution. Many strong lines of evidence argue against the idea that either the Yeti or Bigfoot is a surviving *Gigantopithecus*:

- *Gigantopithecus* was a giant relative of the orangutan, not a close relative of humans. Although anthropologists do not have much evidence of its skeleton, it is reasonable to assume that its foot resembled that of an orangutan or another great ape, not that of a human, with its reduced big toe and inability to grasp. Thus its footprints should resemble ape footprints, not the human-like footprints allegedly produced by the Yeti or

* Spencer Wells, *The Journey of Man: A Genetic Odyssey* (Princeton, N.J.: Princeton University Press, 2002).

† "The Bigfoot–Giganto Theory," Bigfoot Field Researchers Organization, http://www.bfro.net/ref/theories/mjm/whatrtha.asp (accessed October 22, 2011).

Bigfoot. And the tracks should show the stooped knuckle-walking gait of the orangutan and all the other great apes, not the extremely human-like bipedal-walking posture allegedly shown by the Yeti and Bigfoot. (Indeed, one of the biggest problems with the film by Roger Patterson and Bob Gimlin that purports to have captured Bigfoot is that the walking posture of the creature is almost completely human, not ape-like in the least.).

- Although *Gigantopithecus* fossils are rare, an animal as large as that ape would still be expected to be fossilized at least a few times if it had survived anywhere in the world after 300,000 years ago. For example, one pro-Bigfoot organization claims: "No research group has ever made an attempt to look for Giganto bones in North America, so no one should be surprised that Giganto remains have never been identified in North America. Ironically, the most vocal skeptics and scientists who rhetorically ask why no bones have been located and identified on this continent are the last people who would ever make an effort to look for them."* This assertion is patently false and reveals ignorance about the fossil record and the practice of paleontology. Paleontologists do not look for particular fossils, but collect any and all deposits that yield decent fossils. For deposits of the past 300,000 years (middle and late Pleistocene), there are extraordinarily good fossil records in both China, where hundreds of paleontologists have been working for many decades, and especially North America, where fossils of larger mammals (especially from cave deposits) have been found in every state in the United States and in most provinces in Canada.† Hundreds of paleontologists have collected these fossils for more than a century and documented them in excruciating detail. Many extremely rare species are known, including an American cheetah and a camel that was built like a mountain goat. Yet not once has anything resembling *Gigantopithecus* been found—not even the smallest tooth fragment (which could be easily recognized by its thick enamel and distinct low-crowned cusps). Contrary to the conspiratorial thinking of cryptozoologists, paleontologists would be overjoyed to find such a fossil and announce it with great fanfare because such a discovery could make a reputation. They have no reason to hide such a fossil in order to thwart cryptozoologists.

* Ibid.

† Björn Kurtén and Elaine Anderson, *Pleistocene Mammals of North America* (New York: Columbia University Press, 1980).

One of the most famous "sightings" was by physicist and runner Anthony B. Wooldridge. On March 5, 1986, while doing a 200-mile run with just a rucksack in the upper Alaknanda Valley of northern India, he saw what he thought was a Yeti at an elevation of about 12,500 feet near Ghangaria:

> It was difficult to restrain my excitement as I came to the realization that the only animal I could think of which remotely resembled this one before me was the Yeti. My skepticism about the creature's existence was overturned by this all-too-real creature then in view. It was standing with its legs apart, apparently looking down the slope, with its right shoulder turned towards me. The head was large and squarish, and the whole body appeared to be covered with dark hair, although the upper arm was a slightly lighter color. The creature was amazingly good at remaining motionless, although the bush vibrated once or twice, and when I moved back to lower ground, it appeared to have changed its head position and to be looking directly at me.
>
> I took a number of photographs from an estimated range of 150 meters [490 feet]. . . .
>
> After about 45 minutes, the weather continued to deteriorate, and light snow began to fall. The animal still showed no sign of moving, although occasionally the shrub vibrated slightly. I moved back down the slope a short distance, and got the impression that the animal was now peering towards me around the other side of the shrub, but its feet had not changed in position.[132]

Wooldridge then left the area as the snowfall increased. When he returned to civilization, his account was published. His photographs were a sensation in the cryptozoological community. Enlarged and digitally enhanced as far as they could be, they were analyzed, discussed, and debated endlessly about what the large dark object was and how it should be interpreted. But a subsequent trek revealed why the mysterious Yeti had been able to hold still for so long. A year later, Wooldridge retraced his trail to the same mountain site and took stereo photographs of the location. Photogrammetric analysis then revealed that the large black "hominid" was just an outcrop of rock! Wooldridge conceded that "beyond a reasonable doubt that what I had believed to be a stationary, living creature was, in reality, a rock."[133]

Like Bigfoot, the Yeti has been hoaxed. In 1996, the television program *Paranormal Borderland* played a video sequence taken from March 12 to August 6, 1996.[134] Called "The Snow Walker Film," it was presented as genuine footage of the Yeti, but later investigation proved that the video had been a staged hoax, probably by the producers of the film. It was broadcast on the same network that produced the famous program *Alien Autopsy: Fact or Fiction . . . ?*, which became one of the most famous hoaxes in television history—although some UFO fans still believe that the footage is real. Cynically, the show was later recycled by the same network in a program called *The World's Greatest Hoaxes: Secrets Finally Revealed.*

THE MOUNTAIN MAN

If Sir Edmund Hillary and Tenzing Norgay were the first humans to scale Mount Everest, most mountaineers regard Reinhold Messner as the greatest climber ever. Born in the South Tyrol region of the Italian Alps in 1944, Messner started his career making record-breaking climbs in the Alps and the Andes in the 1960s and early 1970s. In 1978, Messner and his climbing partner Peter Habeler became the first and so far the only humans to have reached the summit of Mount Everest without bottled oxygen, and two years later he scaled it again, not only without carrying oxygen but climbing alone. He is the only person to have climbed all fourteen of the "eight-thousanders," the highest peaks in the world, which are more than 8,000 meters (26,000 feet) above sea level (all in the Himalayas). He and Arved Fuchs were also the first people to have crossed Antarctica on foot. He has made many sacrifices for these achievements, losing several toes to frostbite and suffering numerous other injuries.

He is famous for these many incredible climbs, but Messner had another interest as well: the Yeti. As he described one night in 1986, while traveling alone in a deep, forested valley in the Himalayan foothills:

> Making my way through some ash-colored juniper bushes, I suddenly heard an eerie sound—a whistling noise, similar to the warning call mountain goats make. Out of the corner of my eye I saw the outline of an upright figure dart between the trees to the edge of the clearing, where low-growing thickets covered the steep slope. The figure hurried on, silent and hunched forward, disappearing behind a tree only to reappear again against the moonlight. It stopped for a moment and turned to look at me. Again I heard the whistle, more of an angry hiss, and for a heartbeat I saw eyes and teeth. The creature towered menacingly, its face a gray shadow, its body a black outline. Covered with hair, it stood upright on two short legs and had powerful arms that hung down almost to its knees. I guessed it to be

> over seven feet tall. Its body looked much heavier than that of a man of that size, but it moved with such agility and power toward the edge of the escarpment that I was both startled and relieved. Mostly I was stunned. No human would have been able to run like that in the middle of the night. It stopped again beyond the trees by the low-growing thickets, as if to catch its breath, and stood motionless in the moonlit night without looking back. I was too mesmerized to take my binoculars out of my backpack.
>
> The longer I stared at it, the more the figure seemed to change shape. . . . A heavy stench hung in the air, and the creature's receding calls resounded within me. I heard it plunge into the thicket, saw it rush up the slope on all fours, higher and higher, deeper into the night and into the mountains, until it disappeared and all was still again.[135]

Messner fled in terror, scrambling along the rocky slopes until he reached a village and safety.

For the next ten years, he trekked around the Himalayas—sneaking into and out of occupied Tibet and hiding from Chinese patrols, breaking out of Chinese jails, hiking up and down one mountain range after another—in search of more evidence. He was one of the first Westerners to visit Bhutan after the king ended the country's self-imposed isolation. For all his searching, Messner found a few tracks, but they were inconclusive. He visited the famous Pangboche monastery, whose "Yeti scalp" turned out to be a serow pelt. At a Bhutanese monastery, Messner viewed a relic alleged to be the mummified pelt, head, hands, and feet of a "Yeti cub," but it was also a fabricated taxidermic fake (what sideshow operators would call a "gaffed" creature):

> The hands and legs really did seem human, like those of a child of eight or nine. It was obvious they were shaped with little sticks, leather, cloth, and thread—in other words, re-created. They were hanging on a thin hide that looked as if it might have been that of a monkey. The head was clearly handcrafted, its face a mask, and the hide stretched over it probably that of an animal. I couldn't tell if what was beneath it was wood or bone. . . . The Gangtey Gompa relic was nothing more than a doll used to cast out spirits—or to keep the legend of the yeti alive.[136]

As had Ernst Schäfer in the 1930s, Messner accompanied Tibetans to find the lairs of the *chemo* (Yeti), only to discover bear dens.[137] In 1997, he visited the Sosar Gompa monastery in Kham, where two animals hung over the entrance: a stuffed yak and a stuffed *chemo*, which turned out to be a Himalayan brown bear.[138] As he spoke with more and more Nepalese and Tibetans, he began to realize that the Yeti is a mythical composite of several animals, including not only an ape-like creature but also the Himalayan brown bear.

While trekking near the peaks of the Nanga Partains in 1997, Messner saw several *chemos* (or *dremos*, in the local dialect):

> One afternoon, after a long trek, we encountered another *dremo*. He fled when he saw us, but then seemed to stop and rest in a hollow. I approached the spot from behind some ridges so that he wouldn't pick up my scent. Rozi Ali [his trekking guide] followed me. When I began to climb down to where the animal was sleeping in the grass, Rozi Ali tried to stop me. I broke free of his grasp and came within twenty yards of the animal, where I took some good pictures. Rozi Ali, crouching some way back, begged me to make a run for it. He was sweating with fear.
>
> The animal woke up and looked at me in the way a startled child would a stranger. It was a young brown bear. It would only turn into a *dremo*, *chemo*, or yeti later, when the local people were providing their version of the encounter. . . . We saw only one more *dremo* during our trip to Kashmir. It was running away on two legs. From a distance it looked uncannily like a wild man [but it was also a brown bear].[139]

In an interview, Messner was asked if locals had actually shown him the animal that they called a Yeti:

> In the eastern part of Tibet they said to me, "This yeti is stealing women, he's killing yaks, and sometimes in the wintertime he comes and steals a goat. And he is a little bit like we are, and when he is whistling we have to run away"—exactly the stories the Sherpas tell about the yeti. Finally they brought me to a place. They said, "There is one! You see it?" And it was a Tibetan bear. And they said, "This is exactly what is stealing the women and killing the yaks."[140]

After a full decade of trekking into and out of Tibet and Nepal, Bhutan and Kashmir, and examining every bit of evidence, one of the best mountain men who ever lived concluded that "the pieces fit":

> *Yeti* is a collective term for all the monsters of the Himalayas, real or imagined. It is the abominable snowman, that Western fantasy, as well as the *chemo* and *dremo*. My perspective was no longer Western. I did not believe yetis were relics of prehistoric anthropoid species that had managed to survive undetected. The yeti was a living creature, not a figment of the imagination, that corresponded to the brown bear. . . . After all, in an ancient Tibetan dialect, *yeti* translates as "snow bear."

> I hasten to add that this is an extraordinary animal—fearsome and preternaturally intelligent, as far as possible from the cuddly image people of the West sometimes have of bears.[141]

THE HUNT CONTINUES

Even though the accounts from Ernst Schäfer, Reinhold Messner, and others seem quite conclusive that the Yeti is a mythical animal based on the Himalayan brown bear, and all the "evidence" for the Yeti is inconclusive or points to the bear, not an ape-like creature, many believers refuse to give up the search. In 2007, Joshua Gates, the producer of the television series *Destination Truth*, mounted an expedition to the Himalayas. His team found 13-inch tracks in the snow [142] and brought casts to a Bigfoot advocate, anthropologist Jeffrey Meldrum, who thought that the footprints had been made by a primate and resembled Bigfoot tracks. They also found some "Yeti hair," which, when analyzed, once again appears to belong to a Himalayan ruminant such as the goral or serow. On July 25, 2008, hairs brought back from the Garo Hills in northeastern India by Yeti hunter Dipu Marak were analyzed, and they also turned out to be from a serow.[143]

In 2008, the cryptozoology series *MonsterQuest* mounted its own expedition to the Himalayas.[144] The resulting episode features primatologist Ian Redmond and other Yeti advocates; includes cameo interviews with the elderly Peter Byrne, telling the story of the expeditions funded and led by Tom Slick, as well as with Messner and other bear experts who affirm the "bear origins" story; and a short segment on Schäfer and the Nazi connection. The presentation is like that of most other episodes of *MonsterQuest*: lots of spooky music and moody, dark shots; computer-graphic simulations of encounters; plenty of footage of the scenery and of the "Yeti hunters" huffing and puffing their way through the Himalayas (as their Sherpas barely sweat)—and nothing in the way of hard evidence. The tracks are inconclusive, and one "trackway" was made by a rolling snowball. The crew spends an inordinate amount of time trying to rig up a camera on a weather balloon, but finds nothing. The camera traps on the "Yeti trail" capture only ordinary wildlife. The hair samples, sent to a lab for analysis, once again are from a serow or goral. The hunters examine a carcass that vultures were scavenging, and Redmond even eats some of the rotten meat, all to speculate that this kind of food *might* sustain a Yeti. Programs like *Destination Truth* and *MonsterQuest* get high ratings on cable television and sustain the beliefs of viewers who want the mystery of unknown monsters, such as the Yeti, in their lives. But they offer no actual evidence—only much speculation—thus reinforcing the conviction that behind the false leads and inconclusive bits of evidence for the Yeti are only brown bears, rolling snowballs, and the tendency of non-Western peoples to conflate legend with reality in a way that Westerners cannot fathom and usually misunderstand.

Indeed, the most conclusive proof of the true meaning of the Yeti, as documented by Reinhold Messner and others, is that many of the peoples of the Himalayas *know* and *admit* that the Yeti is a myth, built on the foundation of the terrors of the Himalayan brown bear and then transmuted into their religious symbol. Messner describes a visit with the Dalai Lama, who asks him this very question:

> "Do you think that the migio, chemong [another name for the brown bear], and yeti might well be the same thing?" [asked the Dalai Lama].
>
> "I not only think so, I am completely convinced that they are," I said. Then I put my finger to my lips, as did the Dalai Lama, as if acknowledging that this must remain our secret.[145]

NESSIE

THE LOCH NESS MONSTER

OF ALL THE "REAL" MONSTERS that stir the Western imagination, there are few so romantic as the Loch Ness monster. I'm not even slightly immune to that romance. My love affair with Nessie blossomed early and strongly. What could be more wonderful than the idea that a living plesiosaur might slide undetected through the frigid waters of a Scottish lake? I sought out and devoured every book available in my elementary-school and community libraries. As a young boy wishing to learn more in those pre-Google days, I even asked the local reference librarian to track down the address of the Loch Ness Phenomena Investigation Bureau. (Alas, the organization was defunct by the time I tried to contact it.)

In the pulpy books on the paranormal, I studied the famous cases, marveled at the amazing photographs of arching necks and underwater flippers, and absorbed the standard arguments. "Loch Ness is connected to the sea through underwater tunnels," I told my classmates at recess. (I was unaware that the surface of Loch Ness is more than 50 feet above sea level.) "Do you know why Nessie wasn't reported until 1933?" I asked on the playground. "Because that's when they finally built the road beside the loch!" (I now know that the road predates Nessie by more than a century.)[1]

If my understanding of the Nessie literature was a bit uncritical, at least my research techniques were inventive. I remember crouching around a Ouija board in my fifth-grade classroom, pragmatically asking the spirits at which end of Loch Ness I should concentrate my search for the monster. (Or perhaps this was not so innovative. Seven years before my Ouija board consultation, mentalist Tony "Doc" Shiels allegedly led a team of psychics—successfully, he claimed—to summon Nessie and other "aquatic serpent dragons throughout the world.")[2]

Years later, my love of these wonderful stories led me to the skeptical literature. Eventually, I found myself a magazine writer, which offered me the professional opportunity to pursue my childhood dreams of investigating monster mysteries. And so, it is with great pleasure that I turn now to the enduring mystery of the Loch Ness monster.

LOCH NESS

Loch Ness is a long, deep lake that lies on a geological fault line—a country-spanning cleft called the Great Glen. Bisecting Scotland from coast to coast, the Great Glen features several large lakes, of which Loch Ness is the

Figure 4.1 Loch Ness is the largest of the lakes along the Great Glen, which bisects Scotland. Since 1822, these lakes have been linked by the Caledonian Canal, allowing water travel from the North Sea to the Atlantic Ocean across the middle of Scotland. (Image by Daniel Loxton)

largest (figure 4.1). At 22 miles long and around 754 feet maximum depth, Loch Ness is the United Kingdom's largest body of freshwater. It is also, according to legend, the home of an unknown species of large animal. If there is a Loch Ness monster, it is a recent arrival. The steep sides of the loch were scoured by glaciers during the Ice Age of the Pleistocene epoch (2.5 million–11,700 years ago). Indeed, the whole of Scotland was crushed beneath a half-mile-thick sheet of solid ice as recently as 18,000 years ago (or less)![3]

Today, Loch Ness is connected to the Moray Firth inlet of the North Sea by the 7-mile length of the River Ness (figure 4.2).[4] Since 1822, Loch Ness has been part of a shipping channel called the Caledonian Canal.[5] Composed of a series of canals, locks, and natural lakes, the Caledonian Canal allows ships to cross Scotland from coast to coast. The route runs by

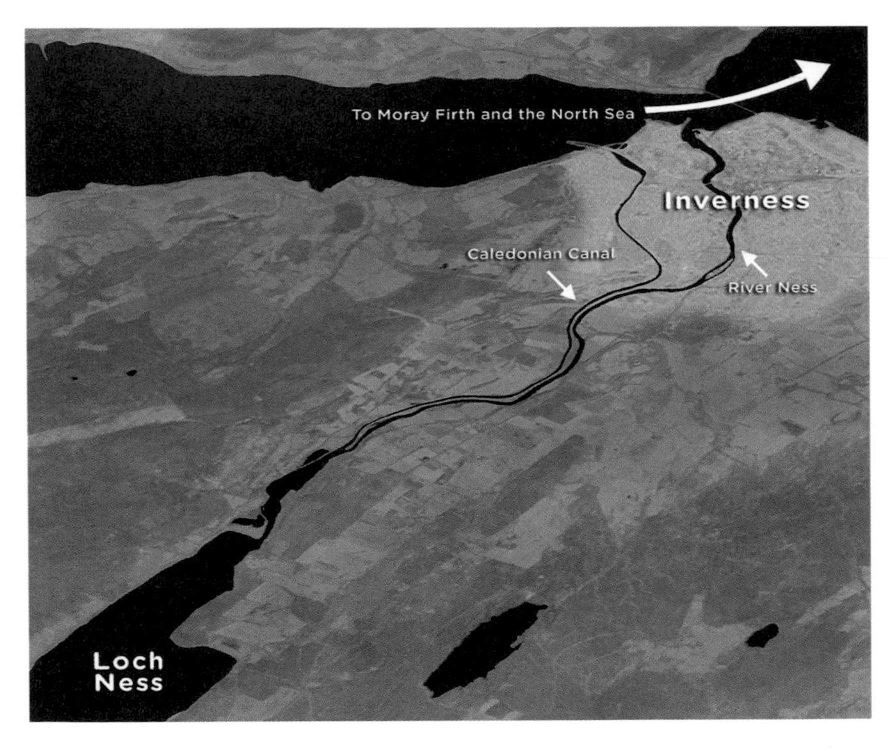

Figure 4.2 Loch Ness is connected to the Moray Firth and the North Sea by two short, parallel waterways: the River Ness and the Caledonian Canal—both of which run through the city of Inverness. (Image by Daniel Loxton)

canal through several locks from Moray Firth to Loch Ness (in parallel to the River Ness), continues down the length of Loch Ness, and then proceeds by canal to Loch Oich (and, eventually, to Scotland's west coast). Thus Loch Ness has been busy and well traveled for almost 200 years. Indeed, Loch Ness was well used and well populated even before the construction of the canal, crossed for centuries by boats and bordered by roads, towns, and villages (and, at the mouth of the River Ness, near the loch's northern end, the sizable city of Inverness). The opening of the canal ushered in a tourism boom, with daily steamship traffic running the length of Loch Ness. "Far from being a lonely, uninhabited spot before 1933," explains Ronald Binns, "Loch Ness was extremely popular with the leisured English middle-classes during the previous hundred years." Even Queen Victoria toured Loch Ness. Almost a century before the birth of the modern monster legend, Loch Ness was already overrun with recreational traffic, according to one disgruntled naturalist. He complained bitterly about the noisy, polluting steamboats "full of holiday people, with fiddles and parasols conspicuous on the deck."[6]

BEFORE NESSIE

Water-Horses

As we look toward the emergence of the Loch Ness monster in the 1930s, it is important to understand that a teeming menagerie of water-based supernatural creatures had already lived for centuries in Scottish folklore. In addition to the Great Sea Serpent—such as the Stronsay Beast, whose carcass washed up on the island of Stronsay in 1808 (chapter 5), and the sea serpent sighted in one of the inland freshwater lakes of the northern Isle of Lewis in 1856[7]—Scotland's feared folkloric monsters include the *boobrie* (a giant carnivorous waterfowl), the *buarach-bhaoi* (a nine-eyed eel that twists its body into a shackle around the feet of prey), the *biasd na srogaig* (a clumsy one-horned water beast with vast legs), and even the twelve-legged "big beast of Lochawe."[8]

Among this horde of folkloric creatures are the widespread traditions of kelpies (associated with running water), water-bulls, and water-horses (*Each uisge,* which haunted lochs and the sea).[9] Today, these related but distinct mythological creatures are harnessed in service of the legend of the Loch Ness monster, but there are strong reasons to think that this linkage is inappropriate. First, none of these creatures is anything like the modern Loch Ness cryptid. Second, none of them is indigenous to Loch Ness.

In Scottish folklore, water-bulls are small black bulls that are encountered when they venture onto land; they sometimes breed with terrestrial cattle before returning to the water. Water-horses (whether *Each uisge* or the distinct but similar kelpies) are lethal, shape-shifting demons. They are likewise encountered on land in the form of ordinary-looking horses, often with weeds in their manes and wet-looking, adhesive skin. If anyone is foolish enough to climb onto the back of a water-horse, he or she will become stuck in place—and the water-horse will carry the rider screaming into the water. Children are a favorite prey of water-horses. "Often he grazes in a field near the water," as one historian described the folklore in 1933, "and by his tameness tempts children to mount him. As they mount, he lengthens his back until all are accommodated in a line, when he rushes with them into the water." (A common folk tale describes a single surviving child who touches the monster and then

Figure 4.3
In a tale retold by W. Carew Hazlitt, a ravenous kelpie pursues a beautiful maiden. After a desperate flight, she escapes over the threshold of her house and collapses. Vampire-like, the kelpie is unable to enter the open door because it is protected by a branch from a rowan, a tree believed to offer protection against malevolent beings, but does capture the young woman's shoe, which extended slightly beyond the boundary of magical protection when she fell. (From W. Carew Hazlitt, *Dictionary of Faiths and Folklore: Beliefs, Superstitions, and Popular Customs* [London: Reeves and Turner, 1905], facing 334)

must cut off his own finger to escape. But all his friends are carried off to their deaths.)[10]

Water-horses are much closer to vampires or werewolves than to any modern cryptid (figure 4.3). "When killed, the water-horse proved to be nothing but turf and a soft mass like jelly-fish," explained an article about the mythical beasts of Scotland, just weeks before the dawn of the Nessie legend. "It could be shot only with a silver bullet, excellent proof of its supernatural character."[11] Even more vampire-like, water-horses can take human form: "In such form he very often goes courting a young woman, with the entirely unromantic object of eating her."[12]

Supernatural creatures with no true physical form, water-horses can be identified with modern cryptids only by badly distorting Scottish folklore. They do not act like or resemble Nessie in any meaningful respect. Moreover, they are part of global folklore and have no unique association with Loch Ness. Water-horses are said to lurk in most of the bodies of water in Scotland, including Loch Lomond, Loch Glass, Loch Awe, Loch Rannoch,[13]

Loch Cauldsheils, Loch Hourn, Loch Basibol,[14] Loch na Mna (on the island of Raasay),[15] Loch Garbet Beag, Loch Garten, and Loch Pityoulish.[16] Nor are they restricted to the United Kingdom. According to folklorist Michel Meurger, water-horses "are very widespread: the British Isles, Scandinavia, Siberian Russia, France, Italy, Czechoslovakia, and Southern Slavic countries."[17] Water-horses were even found in the New World. In 1535, while exploring a tributary to the Saint Lawrence, Jacques Cartier heard that "within this river are some fish shaped like horses, which at night take to the land and by day to the sea, as we are told by our savages."[18]

Hoaxes and Mistakes

The tradition of lake monster and sea serpent hoaxes also long predates the modern Nessie legend. One notable example occurred in upstate New York in 1904—an elaborate hoax involving a carved wooden monster submerged in Lake George by pulleys. (In 1934, the elderly hoaxer revealed the trick and expressed his opinion that Nessie was a similar prank.)[19] Indeed, it seems that monster hoaxing even occurred at Loch Ness in particular, several decades before Nessie. In 1868, a "bottle-nosed whale about six feet long" was found on the banks of Loch Ness. This "monster" caused quite a stir, drawing "large crowds of country people" before it was identified. The *Inverness Courier* concluded that the cetacean "had, of course, been caught at sea, and had been cast adrift in the waters of Loch Ness by some waggish crew" as a prank. "The ruse," said the newspaper, "was eminently successful."[20] (Adrian Shine, head of the respected Loch Ness and Morar Project, describes this story as "the earliest reference we've found to something more than the Water Horse of legend."[21] Similar hoaxes have occurred in more recent times: a dead elephant seal was dumped in Loch Ness in 1972 and dead conger eels in 2001.)[22]

Likewise, the history of water monster misidentification is long and well documented, throughout the world and at Loch Ness in particular. In 1852, as reported in the *Inverness Courier*, an armed mob prepared to battle a "sea serpent" at Loch Ness—literally with pitchforks!

> One day last week, while Lochness lay in a perfect state of calm, without a ripple on its surface, the inhabitants of Lochend were suddenly thrown into a state of excitement by the appearance of two large bodies steadily moving on the loch, and making for the north side from the opposite shore of Aldourie. Every man, woman and child turned out to be witness to the extraordinary spectacle. Many were the

> conjectures as to what species of creation these animals could belong; some thought it was the sea serpent coiling along the surface, and others a couple of whales or large seals. As the uncanny objects approached the shore various weapons were prepared for the onslaught. The men were armed with hatchets . . . the young lads with scythes, and the women principally with pitchforks. One fierce-looking amazon, wielding a tremendous flail about her head, commenced to flagellate a hillock by way of practice. At last a venerable patriarch . . . set to fetch an old [rifle] . . . took aim, and was just on the eve of firing when suddenly he dashed the gun to the ground.[23]

The man cried out in shock at what he took to be supernatural water-horses,[24] but the shock was short-lived. The creatures were "not actually the much dreaded 'kelpies,'" the newspaper continued, but perfectly ordinary, well-known animals. In fact, "they proved to be a valuable pair of ponies . . . indulging themselves with a dip in the cooling waters of Lochness." (Remarkably, the animals had swum "fully a mile" across the lake.)

This case brings two salient facts into sharp focus. First, it is possible for people to mistake ordinary animals for mysterious monsters—even crowds of people and even in broad daylight. Second, neither this group of locals nor the *Inverness Courier* gave any hint of an ongoing tradition of monsters in the loch. (All the explanations proposed in the story—sea serpent, whales, seals, and kelpies—are generic. None are specific to Loch Ness.)

Dressed with folklore and mischief, the Loch Ness stage had been prepared for decades—centuries—without any star to step on it. But in the 1930s, all that was to change.

A SERIES OF SIGHTINGS

"Three Young Anglers"

This brings us to a small news story from 1930 that may (depending on your point of view) be considered the first modern Loch Ness monster case. According to the *Northern Chronicle*, "three young anglers"[25] had a strange experience while fishing for trout on Loch Ness, as described by Ian Milne, one of the men:

> About 8:15 o'clock we heard a terrible noise on the water, and looking around we saw, about 600 yards distant, a great commotion with spray flying everywhere. Then the fish—or whatever it was—started coming toward us, and when it was about 300 yards away it turned to the right and went into the Holly Bush Bay above

> Dores and disappeared in the depths. During its rush it caused a wave about 2½ feet high, and we could see a wriggling motion, but that was all, the wash hiding it from view. The wash, however, was sufficient to cause our boat to rock violently. We have no idea what it was, but we are quite positive it could not have been a salmon.[26]

This story is in many ways a bit of a dud. The witnesses did not actually see a monster, and the story died in the news very quickly. (Interestingly, the newspaper also claimed to have spoken with an unnamed "keeper who dwells on the shores of the loch" who "some years ago saw . . . the fish—or whatever it was—coming along the centre of the loch, and afterwards stated that it was dark in color and like an upturned angling boat, and quite as big." Little was made of this alleged sighting, which was not recorded at the time it occurred.)[27]

But the "three young anglers" story is noteworthy in other ways. It is one of the vanishingly few reports of anything like a Loch Ness monster recorded before 1933. It also is one of the first to use (even if offhand) the term "monster" in relation to Loch Ness. In a skeptical letter to the *Inverness Courier*, someone going under the name Piscator paraphrased Milne's account as describing "a wave 2½ feet high caused by some unknown monster that, presumably, inhabits Loch Ness."[28] Retelling the tale in a small item two months later, the *Kokomo Tribune* claimed, "People in the vicinity of Loch Ness, in Scotland, are much mystified over reports of a monster having taken up its abode in the lake."[29]

"Strange Spectacle on Loch Ness"

The "three young anglers" case of 1930 failed to ignite a popular monster legend. But it was remembered by at least one person: a small-town, part-time reporter named Alex Campbell.

Then, in 1933, Campbell heard that his friends Aldie[30] and John Mackay (proprietors of the Drumnadrochit Hotel)[31] had spotted something in the water while driving along the shore of Loch Ness. Campbell wrote the story for the *Inverness Courier*, which ran it under the headline "Strange Spectacle on Loch Ness: What Was It?"

> Loch Ness has for generations been credited with being the home of a fearsome-looking monster, but, somehow or other, the "water kelpie," as this legendary creature is called, has always been regarded as a myth, if not a joke. Now, how-

> ever, comes the news that the beast has been seen once more, for, on Friday of last week, a well-known businessman, who lives in Inverness, and his wife (a University graduate), when motoring along the north shore of the loch . . . were startled to see a tremendous upheaval on the loch, which previously had been as calm as the proverbial millpond. The lady was the first to notice the disturbance, which occurred fully three-quarters of a mile from the shore, and it was her sudden cries to stop that drew her husband's attention to the water. There, the creature disported itself, rolling and plunging for fully a minute, its body resembling that of a whale, and the water cascading and churning like a simmering cauldron. Soon, however, it disappeared in a boiling mass of foam.[32]

Presto: Loch Ness was home to a "fearsome-looking monster" and suddenly had been "for generations"!

Campbell's story was a bit on the sensational side. The Mackays clarified their sighting later that year when they spoke with Rupert Gould. First, Aldie was the only one who saw any kind of object or animal; her husband saw only splashing. The article's "tremendous upheaval" was perhaps a little exaggerated: Aldie "thought at first that it was caused by two ducks fighting" (although she decided, "on reflection," that the splashing was "far too extensive to be caused this way"). When she finally saw the cause of the splashing, it was not one body "resembling that of a whale," but two dark humps in the distance. The two humps had a total length (she estimated) of about 20 feet.[33] (If accurate, this would make each hump about the size of a seal. Because Loch Ness is connected to the North Sea by both a river and a canal, seals play an important role in Nessie debates to this day—as we will see.)

Campbell was quick to tie the Mackays' sighting to the "three young anglers" case from three years earlier. In the original version of that story, the three fishermen did not see the cause of the splashing and had "no idea what it was"; in Campbell's revisionist retelling, they saw "an unknown creature, whose bulk, movements, and the amount of water displaced at once suggested that it was either a very large seal, a porpoise, or, indeed, the monster itself!" This was pure embellishment by Campbell. The newspaper account from 1930 contains no hint that the three fishermen suggested a seal, porpoise, or monster as an explanation for the spray or the wave.

In any event, the news of the Mackays' sighting (unimpressive even with the help of Campbell's purple prose) was met with considerable skepticism. In a response to the *Inverness Courier*, steamship captain John Macdonald expressed exasperation with the Mackays' amateur description of a tremendous upheaval on the loch. "I am afraid," he wrote, "that it was their imagination that was stirred, and that the spectacle is not an extraordinary one." During fifty years of navigating the loch ("no fewer than 20,000 trips up and down Loch Ness"), Macdonald had become familiar with an ordinary occurrence that very closely matched the Mackays' description: "sporting salmon in lively mood, who, by their leaping out of the water, and racing about, created a great commotion in the calm waters, and certainly looked strange and perhaps fearsome when viewed some distance from the scene." Proposed within days, this very plausible explanation for the Mackay case—by most accounts, the original Loch Ness monster sighting—was an event that the steamship captain had "seen many hundred times."[34]

The Mackays' sighting was a weak case by any reasonable standard, but this small, faltering spark would soon flare into perhaps the greatest monster mystery of modern times. Why did the notion of a monster in Loch Ness come to catch the public imagination in 1933, when the "three young anglers" case of 1930 did not? The Mackays' modest story and George Spicer's much more spectacular sighting, which would soon follow it, had something going for them that the "three young anglers" story had not: the release of one of the biggest blockbuster monster films in Hollywood history.

King Kong

Between the Great Depression and the rise of Nazi Germany, media audiences were ready for a diverting popular mystery. The centuries-old folklore traditions of water-horses and sea serpents had the potential to supply such a mystery, but something more was needed—a catalyst.

Hollywood supplied the perfect catalyst at the perfect time: the gigantic, long-necked water monster depicted in *King Kong* (and again in *Son of Kong*, later in 1933). I am not the first researcher to draw a connection between Nessie and *King Kong*. For example, Dick Raynor suggests a link, noting that, as a result of the film, the "entire western world was gripped by monster fever" in early 1933.[35] Ronald Binns agrees that "it is probably no coincidence that the Loch Ness monster was discovered at the very moment that *King Kong* . . . was released across Scotland in 1933."[36] But I think that a stronger relationship between the film and the myth can be asserted than has usually been argued in the past: in essence, that *King Kong* directly inspired the Loch Ness monster.

There is no question that the birth of Nessie correlates closely in time with the release of the film. *King Kong* opened in London on April 10, 1933, just four days before Aldie Mackay's sighting of the "disturbance" in Loch Ness.[37] The film was an instant box-office smash: "Thousands are being turned away from *Kong*," reported the *Daily Express* from Trafalgar Square. Those who did make it into the packed theaters came out "white and breathing heavily." It was a sensation—a monster thriller so real and so terrifying that moviegoers cried out in their seats.[38]

As the notion of a "fearsome-looking monster" at Loch Ness began to quietly percolate, the *Scotsman* marveled at *King Kong*'s success in "giving the impression that its monsters have newly emerged from the primaeval slime." Most important, the film created a viscerally believable illusion of "prehistoric monsters in contact with modern conditions."[39]

These two correlated factors—the sighting by the Mackays and the release of *King Kong*—soon inspired the most influential Nessie report of all time.

"The Nearest Approach to a Dragon or Prehistoric Animal"

A very few vague sightings followed the Mackays' story over the summer of 1933, but those first small embers of popular belief were fading. And then, in August, the legend suddenly burst into incandescence—and the influence of *King Kong* became unmistakable.

On August 4, the *Inverness Courier* published an astonishing letter from a Londoner named George Spicer. He had, he said, recently spotted a strange creature while driving along the shore of Loch Ness with his wife. His description of their spectacular sighting in broad daylight changed the legend forever: "I saw the nearest approach to a dragon or prehistoric animal that I have ever seen in my life. It crossed my road about fifty yards ahead and appeared to be carrying a small lamb or animal of some kind. It seemed to have a long neck which moved up and down, in the manner of a scenic railway, and the body was fairly big, with a high back."[40]

This account, essentially describing a dinosaur strolling across a road in modern Scotland, was (according to one respected Nessie researcher) "so extraordinary it taxed the imagination of even the most confirmed believers."[41] Yet the influence of the Spicer case simply cannot be overstated. As Adrian Shine explains, "There were no records of long neck sightings before the Spicers' encounter on land made the first connection to plesiosaurs."[42]

Figure 4.4 The water monster in a scene from *King Kong*. (Redrawn by Daniel Loxton)

Whereas the few previous witnesses had reported mere splashes or humps in the water, Spicer claimed a close-up view of a long-necked creature that could have been lifted right off of *King Kong*'s Skull Island. And that, I believe, is exactly what happened.

Among the most memorable scenes in *King Kong* is a night attack by a long-necked water monster. As crewmen from the *Venture* raft tensely across a fog-shrouded lake in pursuit of the abducted heroine, something sinister stirs in the water. A dark, swan-like neck arcs out of the water and then slides back out of sight. The men peer through the dense fog, when suddenly the looming neck attacks out of the darkness. The raft is overturned, spilling the men into the lake. In a series of dramatic shots, the huge, plesiosaur-like animal plucks men out of the water and kills them. This creature—with its rounded back, arched neck, and small head—is essentially identical to the plesiosaur-like popular Nessie that would grow out of Spicer's story (figure 4.4). As the remaining *Venture* crewmen scramble to the seeming safety of the shore, they learn a terrible truth: the creature is not an aquatic plesiosaur, but a *Diplodocus*-like sauropod! The monster pursues the men onto land—and, at this point, Spicer's sighting snaps sharply into focus.

In both his description and his sketch, Spicer almost exactly re-created this scene from *King Kong*. Spicer's creature crossed the road from left to right,

Figure 4.5 The "prehistoric animal" as described by George Spicer. (Illustration by Rupert Gould, under Spicer's direction; redrawn by Daniel Loxton from Rupert T. Gould, *The Loch Ness Monster* [1934; repr., Secaucus, N.J.: Citadel Press, 1976])

just as the *Diplodocus* on land crosses the movie screen (figure 4.5). As Spicer's beast "crossed the road, we could see a very long neck which moved rapidly up and down in curves . . . the body then came into view";[43] for its part, the somewhat implausibly writhing neck of the film's dinosaur enters first, followed by its huge body. The movie's creature gives the impression of having gray, elephant-like skin; Spicer's creature had gray skin, "like a dirty elephant or a rhinoceros."[44] The 30-foot Loch Ness beast is of roughly similar size to the movie monster: "When it was broadside on it took up all the road. . . . It was big enough to have upset our car. . . . I estimated the creature's length to be about 25–30 feet."[45]

A few other diagnostic details make the case especially compelling. In the shots from the movie, the sauropod's feet are not visible. (For ease of animation, they are shielded from view by bushes and enshrouding fog.) Likewise, in Spicer's version, "We did not see any feet." Especially striking is the profile that the two creatures have in common. The film's *Diplodocus* is shown with its tail curved out of view around the far side of its body. According to Spicer's description of his monster, "I think its tail was curved round the other side from our view."[46]

Finally, there is the troublesome description in Spicer's letter to the *Inverness Courier* that the monster "appeared to be carrying a small lamb or animal of some kind."[47] Later, paraphrased versions of his story suggested that the "lamb or animal" could refer to the end of the creature's tail sticking up above its shoulder (or perhaps something riding on the monster's back); but, at face value, this appears to be a direct description of the last shot in *King Kong*'s sauropod scene. Reaching into a tree, the dinosaur grabs a surviving crew member in its mouth and shakes him. In a shot that exactly matches Spicer's sketch, the doomed man looks exactly like a "small lamb or animal of some kind" in the monster's mouth!

This is not the first time that a similarity between the film and the sighting has been noted. Rupert Gould discussed *King Kong* with Spicer just months after the sighting: "While discussing his experience, I happened to refer to the diplodocus-like dinosaur in *King Kong*: a film which, I discovered, we had both seen. He told me that the creature he saw much resembled this, except that in his case no legs were visible, while the neck was much larger and more flexible."[48] But despite the red flag of Spicer's admission that he had seen *King Kong* and that his creature looked like the monster in the film (and even though the obscuring of the creature's legs and feet was a striking detail that the sighting and the film version shared), Gould seems to have moved on without further comment. More recently, Shine's always thorough research zeroed in on both the pivotal importance of Spicer's story (being, once again, the original report of a long-necked monster at Loch Ness) and the significance of Spicer's discussion of *King Kong*.[49] "I believe personally that *King Kong* was the main influence behind the 'Jurassic Park' hypothesis at Loch Ness," Shine confirmed, when asked about the likelihood of a connection. "Before Spicer's land sighting there were no long neck reports at all and it was the long neck that was so crucial."[50] Shine does not seem to have specified the details of the similarities in print, however, and most researchers overlook this smoking gun altogether.

What we are left with is familiar to critical researchers of other paranormal topics, such as UFOs: a feedback loop among popular entertainment, news media, and paranormal belief. The Loch Ness monster grew out of an existing genre of fictional encounters between modern humans and prehistoric creatures (plesiosaurs and sauropods, in particular). Audiences for the hit silent film *The Lost World* (1925) watched a *Diplodocus* rampage through the streets of London and those for *King Kong* saw another stop-motion sauropod dinosaur attack a raft full of men. These fictional stories prepared the public imagination to accept similar "true" stories that the press happily publicized. The press hype ensured further public interest, which inevitably generated more reports. Some were hoaxes, of course, but many were mistakes generated by expectant attention. In 1934, Gould gravely noted "the undoubted fact that, in proportion, as more people looked for the 'monster,' more people saw it."[51] Then, popular fiction stepped in to recursively capitalize on the media-based monster of Loch Ness. The first feature film version, *The Secret of the Loch*, hit theaters less than a year after Spicer's sighting! (Almost inevitably, the lead character declares the Loch Ness monster to be—what else?—a *Diplodocus*.)

"Horrible Great Beast"

Triggered by the Spicer case, a wave of new sighting reports poured in. The sheer number of accounts not only seemed to show that the monster was real, but also exposed a critical flaw in the newly minted legend. "There is one vital question regarding it which must always cause warrantable doubt," one writer nailed it in 1938. "Why have we heard of it only within the last five years or so, when there is no authenticated record of its existence in the centuries which have gone?"[52]

While news coverage asserted a "traditional Loch Ness 'monster'—a superstition which has existed for generations,"[53] knowledgeable locals refuted these media claims. Scolding the *Inverness Courier* in 1933, steamship captain John Macdonald wrote, "It is news to me to learn, as your correspondent states, that 'for generations the Loch has been credited with being the home of a fearsome monster.'"[54]

A genuine monster should imply a history of monster sightings. Where were they? Seeking such a history, people turned to old sources—and old memories—in search of support. Newspapers carried many letters like this one from the Duke of Portland:

> Sir,—The correspondence about "the Monster" recently seen in Loch Ness reminds me that when I became the tenant in 1895, nearly forty years ago, of the fishings in the River Garry and Loch Oich, the ghillies, the head forester, and several other individuals at Invergarry often discussed a "horrible great beast," as they termed it, which appeared from time to time in Loch Ness. None of them, however, claimed to have seen it themselves, but each one knew individuals who had actually done so.[55]

It is difficult even to know what to say about this story. After the media hype started, the duke came forward to say that he remembered that forty years earlier, he had heard some hearsay about a monster? Rupert Gould rather understated the provenance problem when he said of this claim, "Although interesting, such stories are of no great value as evidence."[56] Yet, to this day, many Loch Ness monster sources include this story, without comment, as a firsthand sighting in 1895 by multiple witnesses.[57] Other similarly backdated tales, about alleged but unrecorded events or hearsay from decades earlier, are interspersed with older stories of uncertain or unlikely relevance to create a fictional time line of Nessie prehistory. Some early references included on the canonical time line are invented from whole cloth. For example, one full-page article in the *Atlanta Constitution* is rumored to have described a modern Loch Ness monster (complete with woodcut illustration) in the 1890s. This would be important if true, but it appears that the article never existed. (All attempts to locate it have failed.)[58]

Despite this fictive history, the truth is that unambiguous Nessie sightings did not exist before the release of *King Kong* and the subsequent media hype. Some monster proponents confront this state of affairs honestly. With the possible exception of the "three young anglers" case, writes Henry Bauer, "I can produce nothing written before 1933 that unequivocally refers to large, nonmythical animals in Loch Ness."[59]

Still, as 1933 wore on and reports of sightings of the Loch Ness monster accumulated, people found apparent corroboration in the form of supernatural folklore. To begin with, there were the kelpie traditions, notwithstanding that they are hardly exclusive to Loch Ness. But enthusiasts soon unearthed a more specific tale that appeared to establish Nessie's antiquity.

THE STORY OF SAINT COLUMBA

Within an early medieval biography called *The Life of Saint Columba*, there is a brief but exciting description of a Catholic saint's confrontation with a "water beast" at the River Ness. Now considered the canonical "first recorded sighting" of the Loch Ness monster, this anecdote is said to establish a provenance for Nessie that dates back 1,300 years. Unfortunately, there are good reasons to suspect that this encounter never occurred.

Columba (ca. 521–597) was an Irish monk who in 563 traveled to Scotland as a missionary and established a monastery on the island of Iona. About a century after Columba's death, Adomnán, the ninth abbot of Iona, wrote a biography of his predecessor. This fascinating work is considered one of the most important windows into early Scottish culture, so it is not surprising that monster proponents wish to claim it.

The story goes like this: Columba and his monks encountered some locals conducting a burial at the river's edge. When he asked what had happened, Columba was told that a terrible beast had killed a swimmer. Hearing this,

Columba directed one of his companions to swim across the river and fetch a boat from the other side. The monk obediently leaped into the water:

> But the beast, not so much satiated by what had gone before as whetted for prey, was lurking at the bottom of the river. Feeling the water above it disturbed by the swimming, and suddenly coming up to the surface, it rushed with a great roaring and with a wide-open mouth at the man swimming in the middle of the streambed. On seeing this, the blessed man, together with all who were there—the barbarians as much as the brethren being struck with terror—drew the sign of the saving cross in the empty air with his upraised holy hand. Having invoked the name of God, he commanded the ferocious beast, saying, "Go no further! Nor shall you touch the man. Turn back at once." Then indeed the beast, hearing this command of the holy man, fled terrified in pretty swift retreat, as if it were being hauled back with ropes; though just before this it had approached the swimming [monk] so closely that, between man and beast, there had not been more than the length of a small boat-pole.[60]

This does sound like a powerful endorsement for the legend of the Loch Ness monster. But this superficial resemblance between the river monster and Nessie is a coincidence and is profoundly useless as evidence. The problem is that Adomnán's *Life of Saint Columba* is a hearsay-based biography of a man whom Adomnán never met—a larger-than-life historical figure who was, by Adomnán's day, already shrouded in legend. Drawing from monastic documents, local folklore, second- (and third- and fourth-) hand testimony, and travelers' tales from faraway lands, Adomnán collected hundreds of unrelated and mostly implausible anecdotes—some as short as two unsupported sentences. He had no way even to put them in chronological order. Each story simply begins "At another time . . ." or "On another day. . . ."

Like other early medieval hagiographic literature, *The Life of Saint Columba* is packed with magic, monsters, and spine-tingling supernatural forces. In addition to making prophecies and performing healing miracles, Columba is said to have calmed storms, turned water into wine, summoned water from a stone, driven out "a Demon that Lurked in a Milk-pail," raised the dead, enchanted a stick so that wild animals would impale themselves on it every night, comforted a weeping horse as it "shed copious tears"—and so on. The water beast story is but one of several animal-related anecdotes, each showing the saint's holy power over wild beasts (a boar, a water beast, poisonous snakes). In this context, concluded historian Charles Thomas, the monster tale emerges as just "one minor literary trope within a deliberate and overt piece of religious propaganda."[61] Worse, the story was a cliché. As Thomas points out, medieval hagiographies very often include saintly adventures "in which snakes, or serpents, or dragons—terrestrial or aquatic, with or without wings, silent or bellowing—figured as stock properties in every variety of resuscitation or repulsion miracle."[62]

The bottom line is that no one knows what Columba saw. Indeed, there is no reason to think that this encounter happened. No one knows where the story came from, and it cannot be used as evidence of anything.

Unfortunately, Thomas was right when he wrote that it was "too much to hope that future writers on the topic of the Loch Ness Monster will abandon this reference as irrelevant and misleading."[63] Today, more than twenty years later, virtually all Loch Ness monster sources continue to showcase Saint Columba's as the "first recorded Nessie sighting"—the centerpiece of a revisionist time line cobbled together from "sightings" that entered the record (as decades-old or even millennia-old hearsay) only after the start of the Nessie media circus.

THE EMERGENCE OF THE "PLESIOSAUR HYPOTHESIS"

Inspired by the sauropod dinosaur in *King Kong*, George Spicer created the legend of a long-necked monster at Loch Ness. In turn, his yarn gave rise to many other dinosaur-like and plesiosaur-like sightings. Promoted in news reports and supported by photographs of alleged monsters, these eyewitness accounts made the "plesiosaur hypothesis" a favorite explanation for Nessie throughout the twentieth century.

The groundwork for this notion had been laid as early as 1833, when naturalists began to suggest that a surviving population of prehistoric marine reptiles (known from newly discovered fossils) could be the best explanation for sea serpent sightings.[64] The idea received a big boost from an immensely popular science writer named Philip Henry Gosse in 1861.[65] The concept of surviving plesiosaurs was further popularized in widely read science-fiction stories, such as Jules Verne's *Journey to the Center of the Earth* (1864). Science-fiction writers even relocated the marine reptiles to freshwater lakes. One short story, "The Monster of Lake LaMetrie" (1899) by Wardon Allan Curtis, features a violent encounter between a scientist and an *Elasmosaurus* that has been cast up from the hollow center of Earth (figure 4.6).[66] Familiar

Figure 4.6 The skull of an elasmosaur at the Courtenay and District Museum and Palaeontology Centre, Courtenay, British Columbia. (Image by Daniel Loxton)

as we are with the conceit of a Loch Ness plesiosaur, these stories now read as clear foreshadowing of the Nessie narrative and the cryptozoological hopes that it embodies. Consider this passage from Arthur Conan Doyle's runaway hit *The Lost World*:

> Here and there high serpent heads projected out of the water, cutting swiftly through it with a little collar of foam in front, and a long swirling wake behind, rising and falling in graceful, swan-like undulations as they went. It was not until one of these creatures wriggled on to a sand-bank within a few hundred yards of us, and exposed a barrel-shaped body and huge flippers behind the long serpent neck, that Challenger, and Summerlee, who had joined us, broke out into their duet of wonder and admiration.
>
> "Plesiosaurus! A fresh-water plesiosaurus!" cried Summerlee. "That I should have lived to see such a sight! We are blessed, my dear Challenger, above all zoologists since the world began!"[67]

Perhaps not coincidentally, the speculation that a population of plesiosaurs could have survived in the oceans was advocated in 1930 by Rupert Gould,[68] who went on to write the first and most influential book about the Loch Ness monster! The notion that Loch Ness, in particular, might shelter plesiosaurs was raised as early as August 9, 1933, immediately following Spicer's sighting of a long-necked monster. Linking Nessie to sea serpents, the *Northern Chronicle* argued that there can be little doubt that the Loch Ness creature "is a surviving variant of the Plesiosaurus."[69] Strangely enough, it was Alex Campbell, the author of the *Inverness Courier*'s original Aldie Mackay sighting story, who pushed the plesiosaur idea the hardest. Having written the canonical first Loch Ness monster news story and conjured up the fakelore that "Loch Ness has for generations been credited with being the home of a fearsome-looking monster," Campbell went on to claim that he personally had seen a plesiosaur in Loch Ness—multiple times.

In October 1933, Campbell and his neighbor told reporter Philip Stalker that they had experienced spectacular sightings of the monster.[70] Stalker wrote that these "men whose honesty and reliability have never been called in question" described the creature as "being of the same form as a prehistoric animal: resembling most nearly the plesiosaurus. That is to say that they described its form, and when shown a sketch of a plesiosaurus stated that was the kind of animal they had seen."[71] Stalker went on to report Campbell's detailed claim that "one afternoon a short time ago he saw a creature raise its head and body from the loch, pause, moving its head—a small head on a long neck—rapidly from side to side, apparently listening. . . . While it was above water, he said, he could see the swirl made by each movement of its limbs, and the creature seemed to him to be fully 30 feet in length."[72] This was, Stalker admitted, "a description which, to anyone who is at all sceptical, must appear to be very fantastic." Skepticism was, we know now, entirely appropriate: Campbell retracted his story just days after Stalker publicized it! In a letter, Campbell explained that his spectacular 30-foot plesiosaur was just a group of ordinary waterbirds:

> I discovered that what I took to be the Monster was nothing more than a few cormorants, and what seemed to be the head was a cormorant standing in the water and flapping its wings as they often do. The other cormorants, which were strung out in a line behind the leading bird, looked in the poor light and at first glance just like the body or humps of the Monster, as it has been described by various witnesses. But the most important thing was, that owing to the uncertain light the bodies were magnified out of all proportion to their proper size.[73]

Campbell's role in the development of the Nessie legend was absolutely pivotal, but it soon descended into farce. Having admitted that the long-necked creature "fully 30 feet in length," with its head held "fully 5 feet above the water," was just a flock of cormorants (figure 4.7), Campbell then went on in 1934 to claim a nearly identical sighting of a creature "30 ft. long" with a

Figure 4.7 Waterbirds, such as the plesiosaur-like cormorant, are a frequent source of false-positive sightings of cryptids. (© Stockphoto.com/Jesús David Carballo Prieto)

"swanlike neck reached six feet or so above the water"![74] By itself, the sheer implausibility of this coincidence seems to disqualify Campbell's testimony (and, by extension, all later plesiosaur sightings following his example) from serious consideration. But Campbell was not done: he eventually claimed a whopping eighteen sightings of the Loch Ness monster—often close up and sometimes of multiple creatures at one time. (In one of these entertaining adventure tales, his rowboat was lifted into the air on the monster's back!)[75]

We will return to the "plesiosaur hypothesis" later in this chapter. For now, we should pause to discuss the variability of eyewitness descriptions of the monster. "It seems quite clear, from the inquiries which I made between Inverness and Fort Augustus," wrote Stalker, "that the many observers . . . are divided, broadly-speaking, into two sections." On the one hand, there were those "who have seen a long greyish-black shape, evidently the back of a creature"; on the other, "there are those who have . . . described its head and neck and form as resembling a plesiosaur."[76] In dividing witnesses into these two camps, Stalker both captured and dramatically understated this critical problem. As F. W. Memory, writing in the *Daily Mail*, explained, "Hardly two descriptions tallied, and the monster took on both curious and fantastic shapes—long neck, short neck . . . one hump, two humps, even eight humps, and no humps at all! In fact, it rivals the most versatile quick-change artist of the vaudeville stage in the appearances it was able to assume between one viewing and another."[77]

This variability of description leads researchers in radically divergent directions. If it is true that "the most striking feature about the Loch Ness 'monster'—one which differentiates it from all other known living creatures—is a very long and slender neck, capable of being elevated very considerably above the water-level,"[78] as Gould concluded, then this leads the investigation in one direction. If, instead, the creature has "a massive bull-like head set on a very thick neck, not unlike that of a seal or sea lion,"[79] then the search takes an entirely different turn.

MONEY AND THE MONSTER

The commercial potential of the Loch Ness monster was obvious from the first—so obvious, indeed, that the Scottish Travel Association found it necessary to issue a denial: "[C]ontrary to rumours which are circulating, the Loch Ness 'monster' was not 'invented' by this Association as a means of publicity for bringing people to Scotland."[80] Yet, such is Nessie's value as an attraction that rumors of deliberate tourism-related fraud continue to circulate. (It seems that even Joseph Goebbels, the Minister of Propaganda in Nazi Germany, argued that Nessie was a hoax created by British tourism agencies.)[81]

There seems little need to conjure a central conspiracy when good humor, expectation, and simple human error could so easily provide ample fuel to spark a monster myth. The Nessie legend seems to have emerged in a sporadic, haphazard, and organic manner, with free enterprise simply seizing the opportunity. And what an opportunity it was! As the *Daily Express* explained in December 1933:

> It's an industry. It is Scotland's answer to the Taj Mahal, New York Empire State Building, Carnera, and Ripley, believe it or not. Many newspapers in many countries are doing all they can for it. Scotland is in the front-page news day by day. . . . When the winter sun pales on the Mediterranean the Riviera has no attraction left. Loch Ness can defy all weather, seasons, ages with this fellow. He's a monster advertisement—whether he's there or not.[82]

Travel companies leaped into the Nessie industry, aggressively promoting commercial train and bus tour packages.[83] Loch-side bus traffic became so excessive in 1934 that special safety rules had to be introduced.[84] And while Nessie brought crowds of visitors to Scotland, the monster could also be exported as popular entertainment. In addition to appearances on radio programs and in comic strips, Nessie splashed immediately onto the silver screen. For example, a British Pathé variety short presented a pop song called "I'm the Monster of Loch Ness" in January 1934.[85] Amazingly, "a new 'talkie'" feature film, *The Secret of the Loch*, was announced, shot, and released before the end of that year.[86]

Nessie was put to work as a sales monster as well. Advertisers raced to print Nessie-themed campaigns for consumer products from mustard to floor polish to breakfast cereal.[87] Nessie merchandising also exploded immediately. In January 1934, a parade featuring a 17-foot-long monster escorted by bagpipers ushered three sizes of a cuddly velveteen toy called Sandy, the Loch Ness Monster, into Selfridges, a posh department store.[88] A series of large advertisements and cash contests urged children to "Join the Monster Club" by purchasing a wooden puzzle toy called Archibald: The Loch Ness Monster.[89] Rubber beach toys were rushed into production in anticipation of the 1934 tourist season.[90]

The humor of the Nessie industry was not lost on commentators. As one headline put it, "Monster Bobs Up Again . . . Hotels Doing Fine."[91] Another noted, "There have, of course, been unworthy suggestions that, as a major and unique tourist attraction, she is being very strictly preserved from any undue scientific de-bunking, by those who recognize her value as an invisible export."[92]

A PARADE OF PHOTOGRAPHS

The Hugh Gray Photograph

The era of Loch Ness monster photography began in November 1933, when a British Aluminium Company employee named Hugh Gray was walking home from church along Loch Ness. Spotting the monster close to the shore, Gray apparently took five photographs, four of which showed nothing.[93] But the fifth photo showed—something wiggly? Or maybe not?

When Rupert Gould wrote that Gray's photo, "although indefinite, is both interesting and undoubtedly genuine," he rather understated the "indefinite" part.[94] The alleged monster photo is genuinely terrible (figure 4.8). Its status as the first photograph of Nessie makes it worth mentioning, but it could show anything. To appreciate the true ambiguity of this photograph, consider

Figure 4.8 The first photograph alleged to depict the Loch Ness monster, taken by Hugh Gray in November 1933. (Reproduced by permission of Fortean Picture Library, Ruthin, Wales)

that at least one book has printed it upside down.[95] Some researchers wisely refuse to argue on the basis of the photo. "I believe the picture is probably a genuine photograph of one of the aquatic animals in Loch Ness," said Roy Mackal (on the strength of Gray's verbal testimony). "However, objectively, nothing decisive can be derived from this picture. There is no apparent basis for determining which is front or back, and any such decisions must depend largely on what preconceptions one may have."[96]

While no source can claim any reliable insight into the identity of Gray's monster, I confess that I can see only one thing: a yellow Labrador retriever, swimming toward the photographer. Once this interpretation was pointed out to me,[97] I found it impossible to un-see it. However, I cannot remotely prove that the photo depicts a dog, and this interpretation has been critiqued as pareidolia (the human tendency to perceive bogus patterns—especially faces—in random noise). At the same time, neither can anyone else prove anything on the basis of this picture. In and of itself, Gray's photograph has negligible evidential value.

Is there circumstantial evidence that could shed light on the matter? Perhaps. One reason for suspicion is that Gray reported at least six Nessie sightings,[98] of which his photo was the second.[99] When one witness announces multiple, even habitual, sightings of an elusive cryptid, I regard this as a huge red flag. Think of the millions of people who have visited Loch Ness, the thousands who live and work around it, and the organized observation campaigns that have failed to catch any glimpse of the monster. Yet, inexplicably, certain names turn up repeatedly in the Loch Ness sighting record. Of these witnesses, some have admitted to misidentification errors (Alex Campbell disavowed at least one of his eighteen sightings), while others have been caught cheating. For example, Nessie researcher Frank Searle produced an absurdly lucky streak of photographs before his fellow monster hunters proved that he was a hoaxer. (Some of his fake photos showed parts of a painted sauropod dinosaur blatantly cut out of a postcard.) After Searle was exposed, tensions rose to such a point that he allegedly firebombed a Loch Ness and Morar Project expedition. No one was hurt, and he soon left the area.[100] Today, Searle is universally remembered as a fraud.

For his part, Gray also sighted Nessie with suspicious frequency—and there is a seemingly implausible convenience to his photograph as well. Walking home after church, he said, "I had hardly sat down on the bank when an object of considerable dimensions rose out of the loch two hundred yards away. I immediately got my camera into position and snapped the object which was two or three feet above the surface of the water."[101] Then, having successfully captured, at close range, history's first photograph of the Loch Ness monster, Gray apparently went home and left the film in a drawer for almost three weeks.[102]

Was Gray's photograph a hoax? I don't know, but the smart money would not bet any other way.

March of the Hippopotamus

In December 1933, the *Daily Mail* dispatched a special investigator to Loch Ness to get to the bottom of the mystery: "big-game hunter" Marmaduke Wetherell. Almost as soon as he arrived, Wetherell announced that he had discovered (and cast in plaster) the monster's footprints on the shore of the loch. This struck some observers as a little convenient. "I am aware," Wetherell acknowledged in an interview, "that it is suggested that I have had phenomenal luck in finding such definite traces in two days, although others have failed after a long search."[103] Still, it was an astonishing find. In a BBC interview (produced, as it happened, by Peter Fleming, the brother of the creator of James Bond),[104] Wetherell said, "You may imagine my great surprise when on a small patch of loose earth I found fresh spoor, or footprints, about nine inches wide, of a four-toed animal. It prints were very much like those of the hippo."[105]

Recalling this interview, Fleming described Wetherell and his assistant as "transparent rogues" whose "account of their discoveries carried little conviction." Wetherell struck Fleming as "a dense, fruity, pachydermatous man in pepper-and-salt tweeds."[106] These ad hominem attacks are in keeping with the portrait painted by researchers David Martin and Alastair Boyd. Wetherell was "an eccentric, a likeable rogue, a master of hoaxes, a vain attention seeker." They found that he was not, however, "as the literature would have it, merely a big-game hunter. He was primarily a film director and actor."[107] A mid-level showman, in short—and, with his monster footprints, a showman in the spotlight.

But it was not to last. The cast of the footprints was sent to the Natural History Museum in London, where it was studied by the head of the Department of Zoology and other scientists. These experts identified it easily enough: "We are unable to find any significant difference between these impressions and those made by the foot of a hippopotamus." Nor did the prints of the monster match those of a living hippo that the museum had cast at the zoo for comparison. Wetherell's tracks had been made by using a "dried mounted specimen."[108] The footprints were a hoax, and Wetherall himself had offered the first hint.

In 1999, Martin and Boyd revealed, "Marmaduke Wetherell planted the hippo footprints himself" by using a "silver cigarette ashtray mounted in a hippopotamus foot." The ashtray still exists, in the possession of Wetherell's grandson Peter.[109]

Saying that he "could not account" for the fake footprints, Wetherell soon announced that the Loch Ness monster was a seal—and left the area. But his role in the Nessie story was far from over.

The Surgeon's Photograph

On April 21, 1934, the *Daily Mail* published a stunning new photograph of the Loch Ness monster, allegedly taken by a gynecologist named Robert Kenneth Wilson. Referred to today as the Surgeon's Photograph, this is unquestionably

Figure 4.9 The hauntingly indistinct, close-cropped standard version of the Surgeon's Photograph, taken in April 1934. (Reproduced by permission of Fortean Picture Library, Ruthin, Wales)

the most famous Nessie image ever produced (and an icon for cryptozoology in general) (figure 4.9).[110] With a few exceptions, Roy Mackal wrote in 1976, "every student of the Loch Ness phenomena . . . has accepted this picture as depicting the head-neck of a large animal in Loch Ness."[111]

Contemporary critics were quick to propose alternative explanations. "I got a copy of the original from the *Times*," wrote legendary paleontological explorer Roy Chapman Andrews, "and it showed just what I expected—the dorsal fin of a killer whale."[112] Others suggested, plausibly enough, that the photograph depicted a diving otter or waterfowl. However, it is now known that the photograph was a hoax. In 1975, the *Sunday Telegraph* printed a short article in which Ian Wetherell, the sixty-three-year-old son of Nessie hippo-foot hoaxer Marmaduke Wetherell, revealed that the Surgeon's Photograph was actually another Wetherell family hoax, created by using a small model monster built around a toy submarine: "So my father said, 'All right, we'll give them their monster.' I remember that we drove up to Scotland. . . . I had the camera, which was a Leica, and still rather a novelty then. . . . We found an inlet where the tiny ripples would look like full size waves out of the loch, and with the actual scenery in the background. . . . I took about five shots with the Leica . . . and that was that."[113] After the photo shoot with the model, Marmaduke Wetherell handed off the film to Maurice Chambers, a collaborator who passed it to Wilson, who submitted it to the newspaper—and history was made.

Strangely, Ian Wetherell's public confession in 1975 remained little known for many years.[114] In 1990, Adrian Shine dug out the forgotten article and set researchers David Martin and Alastair Boyd on the trail. By this time, both Marmaduke and Ian Wetherell were dead, but the search led them to Marmaduke Wetherell's stepson, an elderly gentleman named Christian Spurling. Spurling confirmed Ian Wetherell's claim: "It's not a genuine photograph. It's a load of codswallop and always has been."[115] Spurling and Ian Wetherell had built the monster model together.

For his part, Wilson was always cagey about "his" famous picture. He hinted to researchers that "there is a slight doubt or suspicion as to the authenticity of the photograph."[116] He insisted, "I have never claimed that this photograph depicts the so-called 'Monster.' . . . In fact I am unconvinced and intend to remain so."[117] Moreover, some witnesses have said that Wilson had confessed to the hoax. One such was a friend of Wilson's named Major Egginton, who had served under Wilson in the military. In 1970, Egginton told a young Nessie researcher named Nicholas Witchell,

> I always recall the occasion when in 1940 my late Colonel (of the Gunners) Lt. Col. R. K. Wilson, who was before the War a Harley Street specialist, told three of us, quietly in the mess how he and a friend had hoaxed the local inhabitants of Loch Ness. His friend with whom he used to fish the loch from time to time . . . had apparently superimposed a model of a monster on the plate. . . . [T]he resulting publicity according to "R. K." so scared them that it was kept very quiet.[118]

The "friend with whom he used to fish" was Chambers, the link between Wilson and Wetherell. Wilson's relatives confirmed that Chambers and Wilson used to hunt together in Scotland,[119] and Witchell related that the two men leased a "wild-fowl shoot . . . close to Inverness."[120] (Egginton's widow likewise confirmed that Wilson "used to fish and shoot up in Scotland.")[121] Egginton's statement that Wilson said his "friend" (Chambers) had supplied the Surgeon's Photograph as a ready-made image independently confirms the first-person testimony from Ian Wetherell and Christian Spurling. And, as Egginton emphasized, "The story I have told you was that given to me by Lt. Col. Wilson and not second hand."[122]

There is also a great deal of tantalizing hearsay evidence regarding skepticism on the part of Wilson's friends and family. Researcher Maurice Bur-

Figure 4.10 Less tightly cropped versions of the Surgeon's Photograph reveal the small size of the model of the monster used to perpetrate the hoax. (Reproduced by permission of Fortean Picture Library, Ruthin, Wales)

ton was told that "everyone" in a London club to which Wilson had belonged "knew the picture was a hoax." Following up this lead, Burton wrote to a friend of Wilson, who "evaded my question but answered rather in the nod-is-as-good-as-a-wink manner."[123] According to Wilson's sister-in-law, Wilson's younger son "was very skeptical over the whole Loch Ness question, and he told a cousin it was a hoax."[124] A friend of Wilson's surviving son related that "from all accounts R. K. was a great prankster with a wicked sense of humour. . . . R. K. didn't discuss the photo with his family. . . . [T]he generally held opinion amongst the family is they are sure that it's probably a fake."[125] Finally, Egginton's widow recalled that Wilson had told the hoax story for years. "Everybody knew," she said, "we knew up there. We all laughed about it, we have done so for years. We knew the story, my husband did, I did too. I find it incredible that the hoax lasted so long."[126]

The case for the photograph's being a hoax is very strong, while that for its being real amounts, essentially, to the impression that the silhouette resembles a monster. Yet, even this gut test fails badly. Looking at the scale of the ripples in the photo—especially in the recently publicized less tightly cropped version—the image seems obviously to depict something very small (figure 4.10). As early as 1960, commentators assessed the object as under 1 foot in height, based on the ripple size. Today, Martin and Boyd explain, "Avid believers in the authenticity of the photograph centre their faith around [Paul] LeBlond and [M. J.] Collins' more recent size determination. . . . Using the wind waves visible in the uncropped photograph to deduce a scale for the object in the center, they concluded this to be about 1.2 m [4 feet] in height."[127] Addressing this embarrassingly small upper-range estimate, Ronald Binns retorted, "Even if the object was 1.2 meters high, so what?"[128] That scale is obviously consistent with a small model, not with a dinosaur-size monster.

Despite all the evidence to the contrary, this vague, evocative image remains the most powerful icon for Nessie and continues to shape the public imagination. Such is its power. But serious students of the mystery have little choice but to face the historical truth: the Surgeon's Photograph is a known fake.

The Lachlan Stuart Photograph

World War II suppressed interest in the Loch Ness monster, and sightings fell off correspondingly. But interest began to revive in the 1950s. Key to that cryptozoological renaissance was a new photograph, taken by a forester named Lachlan Stuart (figure 4.11). Among dedicated Nessie researchers, it became arguably as influential as the Surgeon's Photograph. And, like that famous icon, Lachlan Stuart's photo was a hoax.

According to Stuart, he got up early on July 14, 1951, to milk his cow. Spotting something strange out his window, he shouted for his wife and houseguest, and grabbed his camera. Running to the shore of the loch, they watched three humps and "a long thin neck and a head about the size and shape of a sheep's head. The head and neck kept bobbing down into the water." As the creature cruised—astonishingly, just 40 yards offshore—Stuart snapped his famous photo. He developed the film on his lunch break, and Nessie researcher Constance Whyte saw the negative and photo later the same day.[129]

I must stop here to say that this photograph bothered me from the first moment I saw it. Even as a monster-loving child, I felt that the weirdly angular humps did not look like as though they belonged to an animal. They also conflicted badly with the plesiosaur I understood Nessie to be. (Of course, humps-in-a-line sightings have always composed a significant subset of reports.) But

Figure 4.11 The photograph taken by Lachlan Stuart in July 1951. (Reproduced by permission of Fortean Picture Library, Ruthin, Wales)

the photo was received as a major breakthrough in the 1950s, partly on the strength of Stuart's apparent sincerity. "I could not," Whyte raved, "put forward this photograph with more confidence if I had taken it myself."[130] This confidence was misplaced. As researcher Nicholas Witchell learned, "Lachlan Stuart's photograph, of three angular 'humps' a short distance offshore, was a hoax: Mr. Stuart's account of what happened, a fabrication. He evidently intended no great mischief, and was both surprised and amused that his picture—in reality of three partly submerged bales of hay covered in tarpaulin—should have been taken seriously." Shortly after admitting to having set up the hoax, Stuart took another Loch Ness local (author Richard Frere) to see the props. They were "hidden in a clump of bushes."[131]

The revelation that the photograph is a crude fake may not seem surprising. It looks like a crude fake. Yet Ronald Binns argues that hindsight obscures the photo's historical influence. Consider that in 1976, the scientific director of the Loch Ness Investigation Bureau characterized Lachlan Stuart's photo as "direct photographic evidence as to the probable shape of the central back region of the animal."[132] Looking back, Binns reflects, "What has probably been lost sight of over the years is the impact which the Wilson and Stuart photographs had on monster-hunters back in the 1960s and 1970s. In those days we all firmly believed that they were genuine photographs and that the monster was indeed a very big animal with a long giraffe-like neck, capable of transforming itself into a three-humped object."[133]

The Tim Dinsdale Film

If asked to point to the Loch Ness mystery's equivalent to the Roger Patterson–Bob Gimlin film, which allegedly caught a glimpse of Bigfoot, most people would select the famous Surgeon's Photograph. Loch Ness researchers probably would give a different, deeper answer: the Tim Dinsdale film. This little piece of black-and-white movie footage, shot in 1960, shows a tantalizingly indistinct, far-off blob moving on the surface of the lake. Nessie researchers find it compelling and evocative. As Roy Mackal, scientific director of the Loch Ness Investigation Bureau, explained, when all other evidence seemed unconvincing, "the one thing that would not go away was the Dinsdale film sequence. Although it was grainy in quality because of the great distance involved in the photography, I could not explain it, try as I would. It alone was sufficient for me."[134]

Dinsdale was an engineer who developed a sudden and powerful fascination for Nessie, after reading a magazine article late one night in 1959:

> I kept turning the story over in my mind, and, late that night in bed, fitfully asleep, I dreamt I walked the steep jutting shores of the loch, and peered down at inky waters—searching for the monster; waiting for it to burst from the depths just as I had read, and as the wan light of dawn filtered through the curtains, I awoke and knew that the imaginary search beginning so clearly in my dream had grown into fact.[135]

Reviewing the literature on the monster, Dinsdale hatched a very simple plan: he would drive to Loch Ness and film Nessie. And that, he soon reported, is exactly what he did. Indeed, Dinsdale spotted the monster just minutes after his first view of the loch: "Intrigued, I drove up closer . . . and then, incredibly, two or three hundred yards from shore, I saw two sinuous grey humps breaking the surface with seven or eight feet of clear water between each! I looked again, blinking my eyes—but there it remained as large as life, lolling on the surface!"[136] Alas, this was a false alarm. Upon closer examination through binoculars, the monster "turned out to be a floating tree-trunk after all."[137] But the excitement of his vacation was just beginning. Four days later, after driving about, visiting authors and witnesses, and periodically stopping to observe the loch—bingo! The Loch Ness monster!

> The light began to fail, and then, quite suddenly, looking down towards the mouth of the river Foyers, I thought I could see a violent disturbance—a churning of rough water, centering about what appeared to be two long black shadows, or shapes, rising and falling in the water! Without hesitation I focused the camera upon it and exposed twenty feet or so of film. . . . I went straight to bed quite sure of the fact that the Monster, or part of it at least, was nicely in the bag![138]

Nor was Dinsdale's luck, it seemed, exhausted. On the sixth day of his vacation, he filmed Nessie a second time! Coming over a hill,

> I saw an object on the surface about two-thirds of the way across the loch. . . . I dropped my binoculars, and turned to the camera, and with deliberate and icy control, started to film; pressing the button, firing long steady bursts of film like a machine gunner. . . . I could see the Monster through the optical camera sight . . . as it swam away across the loch it changed course, leaving a glassy zigzag wake; and then it slowly began to submerge.[139]

Flush with the knowledge that he and his camera had "reached out across a thousand yards, and more, to grasp the Monster by the tail,"[140] Dinsdale returned home. When his motion-picture film was developed, however, he discovered with disappointment that his first sequence of "Nessie" footage was "no more than the wash and swirl of waves around a hidden shoal of rocks; caused by a sudden squall of wind. . . . I had in fact been fooled completely!"[141]

Two of his three "monster" sightings had proved to be simple misidentification errors. But what of the last? Viewing the remaining film sequence, Dinsdale was convinced that he had truly captured proof of the existence of the Loch Ness monster—and felt that science would soon agree. Unfortunately, even he had to concede that this "shabby little black and white image that traced its way across the screen" looked fairly unimpressive.[142] Shot at extremely long range, the film depicts an indistinct blob moving on the surface of the water a mile in the distance. Not surprisingly, it was greeted with indifference by the scientists whom Dinsdale approached. However, television audiences proved much more receptive. Dinsdale's television debut was covered by major newspapers as far away as Chicago,[143] securing him a Nessie book deal (the first of several) and launching his new career as a Nessie researcher, author, and lecturer.[144] Dinsdale became a monster celebrity. (He even appeared on *The Tonight Show* with Johnny Carson during a lecture tour of the United States in 1972—a six-week "whirlwind of appointments, interviews and rehearsals for TV, radio and the newspapers.")[145]

But what exactly did Dinsdale film? Certainly, his footage depicts a moving object on the surface of the loch; beyond that, it's difficult to say. This "fuzzy, distant, inconclusive" film, as the *Boston Globe* described it, [146] lends itself to various highly uncertain interpretations. For example, consider the conclusion of the Royal Air Force photographic experts from the Joint Air Reconnaissance Intelligence Centre (JARIC) who studied the film in 1966. Dinsdale felt that the investigators had "vindicated" his film,[147] but this is a significant overstatement. In fact, they found that the appearance and speed of the blob in the film are consistent with those of a motorboat ("a power boat hull with planing hull"). However, the report continued, if the blob were a motorboat, this "would scarcely be missed by an observer." That is, because Dinsdale *said* that it was not a boat, "it probably is an animate object."[148] The assumption that Dinsdale would have recognized a distant motorboat is extremely shaky. After all, Dinsdale mistook other inanimate objects (first a log, and then some rocks) for the Loch Ness monster two other times during the same week!

Essentially, the experts at JARIC found that Dinsdale's blob is consistent with a monster—or with a boat. This conclusion is not terribly helpful. Since the film was analyzed, various authors, experts, and television productions have attempted to enhance the film, with results supporting (variously) either a monster or a boat. For example, pro-Nessie author Henry Bauer asked "a faculty member in the Computer Science Department at Virginia Tech . . . [to] scan a number of frames at higher resolution and to examine them under various types of enhancing techniques"; from this, he concluded that the film does not show a boat.[149] By contrast, Adrian Shine found frames of Dinsdale's film that suggest a manned boat. When he asked members of the original JARIC team to have another look at the film in 2005, they found that the object "has the overall appearance of a small craft with a feature at the extreme rear, consistent with the position of a helmsman."[150]

The Dinsdale film has quite a bit in common with the Patterson-Gimlin film. Both are influential among core proponents. Both filmmakers took a short trip with the intention of filming a famous cryptid, and then immediately did. And both films ultimately frustrate the search for answers. Could Dinsdale really have filmed a monster? The film's mysterious blob cannot tell us—not for sure.

Nonetheless, there are circumstantial hints that suggest a more prosaic explanation. For example, Dinsdale's blob travels in plain sight of a motor vehicle on a road just 100 yards away. The vehicle's driver and the "monster" would have had clear views of each other, but neither slows or reacts to this close encounter. To critic Ronald Binns, "The fact that the driver just kept going suggests that this was because the object passing by was something perfectly ordinary—a motor boat."[151]

As well, there are some red flags about Dinsdale himself. Although he had a reputation among critics and friends alike as "a man of great sincerity,"[152] he sometimes let his imagination run away with him, as it did during the two false-positive Nessie sightings he made in the days immediately before he shot his famous film. Given those misidentification errors, it is hard to know how to interpret his spectacular claim of a later, close-up sighting in 1971:

> I glanced to starboard and instantly recognized a shape I had seen so often in a photograph, the famous "Surgeon's Photograph" of 1934—but it was alive and muscular! Incredulous, I stood for a moment without moving. All I could do was stare. Then I saw the neck-like object whip back underwater, only to reappear briefly, then go down in a boil of white foam. There was a battery of five cameras within inches of my right hand, but I made no move towards them.[153]

Inexplicably, his later books reduce this close encounter to a sort of aside or footnote: "The strangest radio interview I had was in 1971 at Inverness: moments after seeing the Monster's head and neck from [a boat] I had switched on a tape recorder to make a commentary, and BBC radio wanted to use it for a science programme." After barely mentioning having seen the Loch Ness monster, Dinsdale spends the next two paragraphs on a pointless anecdote about the "peculiar experience" of technical glitches for an interview.[154] What gives?

Dinsdale was also given to flights of supernatural fancy. He was, for example, "conscious of some dark influence" at one secluded beach where he often staked out the monster:

> Four of the five expeditions based there resulted in illness or injury, or accident of a most unpleasant nature, and I became aware of some strange influence which seemed to be malevolent. . . . I do subscribe to the belief that . . . purely psychic influence can have a physical effect on people, and sometimes on material objects. I have witnessed such phenomena, which absolutely defy the physical laws as we understand them.[155]

In the same vein, he told the *Washington Post* that he sometimes experienced "dread" while on his boat on the loch: "That's the only word I have to describe it. Dread. I knew I wasn't alone. I knew it."[156]

Finally, there is the Dinsdale family's puzzling approach to stewardship of the film. As longtime Nessie researcher Tony Harmsworth complains, Tim Dinsdale long refused to allow researchers with Adrian Shine's reputable Loch Ness and Morar Project to study the film—a situation that did not improve after Dinsdale's death. "Wendy Dinsdale still refuses to allow the film to be properly examined by the Loch Ness Project's evidence analysts," Harmsworth writes, "which is both sad and implies that there may be something to hide."[157] While television productions are sometimes granted permission to show portions of the film, scholarship would be much better served if all researchers had ready access to high-quality copies. As it is, many Loch Ness researchers have never seen the complete film, which has been called "the single best piece of photographic evidence of Nessie"![158]

A MIXED BAG

The Loch Ness monster is not a species—it is a social phenomenon in which, over many decades, a wide variety of witnesses have described a wide variety of things for a wide variety of reasons. It is not one overarching mystery waiting for one generic solution, but a collection of many unique cases. Among those many cases, some are certainly hoaxes. Others are certainly the result of mistaken identification of known phenomena for an unknown monster. The question is: Could some remainder or subset of "Nessie" cases describe genuinely unknown animals? It's possible (if unlikely).

Most sources agree intellectually that the Nessie database is a mixed bag. However, the temptation to propose a kind of Grand Unified Nessie Theory has often proved irresistible to skeptics and proponents alike. This urge flies in the face of the highly variable descriptions that eyewitnesses have provided. As one article joked in 1934, "From the evidence, he must be a mixture of all that walks, flies, swims—and makes the hotels do miraculously."[159] Decades later, Tim Dinsdale characterized Nessie witnesses as "consistently inconsistent" on the critical matter of basic body plan: "Everyone seemed to have noted a different number of 'humps' in the water—one, two, three and even more on occasions, and sometimes no humps at all, just a huge back like an upturned boat. It was difficult to understand."[160]

Figure 4.12 An elasmosaur-type plesiosaur. (Illustration by Daniel Loxton)

This difficulty evaporates once we accept that "Nessie" is a catchall term for whatever witnesses happen to find puzzling when they look out over the loch. But interesting questions remain: What are the various things in the mixed bag? And is it possible that a true monster could hide in that mix? Could there be a signal, or signals, hidden in all that noise?

The "Plesiosaur Hypothesis" Redux

We have seen that the "plesiosaur hypothesis" was based on three exceedingly wobbly pillars: George Spicer's *King Kong*-inspired sighting, a report that Alex Campbell recanted as a mistake, and the fraudulent Surgeon's Photograph. Given that unstable foundation, are plesiosaurs worth considering at all? Prominent cryptozoologist Loren Coleman suggests that they are not:

> Most American cryptozoologists, from the beginning, and, indeed, most critical thinking cryptozoologists today reject The Plesiosaur Hypothesis. The mammalian focus and other schools of thought have won out long ago. We realize that these extinct marine reptiles are extinct, and to promote or use them as candidates for Loch Monsters is done mostly by "true believers" (on both sides of the aisle, whether they are pro- or anti-Nessites).[161]

In addition, plesiosaurs are a terrible explanation for reports of a multi-humped, sea serpent–type Nessie (figure 4.12). Still, many witnesses have described creatures that clearly do resemble plesiosaurs—pivotal early witnesses and recent witnesses as well. In 1997, for example, a journalist claimed ("apparently seriously," according to Coleman)[162] that he had seen five plesiosaurs:

> I was on the banks of Loch Ness the other morning, sunning my long legs, when one of those Nessies—those plesiosaur-type monsters that inhabit the loch—surfaced in front of me. She was 10 or 15 yards away. She had three or four humps and a greenish-brownish hide. Small, imperious head. Long, graceful neck. . . . Within a few seconds, a second monster surfaced, and bobbed beside the first one. Then a third and a fourth one—followed by a juvenile. . . . But my camera was in my car.[163]

In the context of a lighthearted travel story, this outrageous sighting seems like an obvious joke—especially given the writer's aside: "Perhaps it was just a brain storm; increasingly, I seem prone to them." Be that as it may, this man and other witnesses clearly describe "plesiosaur-type monsters." The ethos of cryptozoology (essentially, "take eyewitness testimony seriously") seems to demand that cryptozoologists accept this option. And while the "plesiosaur hypothesis" may have fallen on hard times (especially in the wake of the revelation that the Surgeon's Photograph was a hoax), there is no getting around its decades-long popularity—not only in the public imagination, but in the cryptozoological literature as well. As legendary cryptozoology pioneer Bernard Heuvelmans pointed out in 1977, "the plesiosaur has remained the candidate of choice for Anglo-Saxon researchers favorable to the creature's existence."[164] Writing in 2003, cryptozoologist Karl Shuker agreed that the hypothetical relict plesiosaur "remains the most popular identity among cryptozoologists for the Loch Ness monster."[165] Many Nessie books and Nessie researchers have gone to bat for this idea or at least included it as a distinct possibility. Tim Dinsdale advocated plesiosaurs as a more parsimonious explanation than any large undiscovered species. At least we know that plesiosaurs used to exist! Besides, he said, "the Monster looks exactly like a type of long-necked Plesiosaur. . . . If it is not a type of Plesiosaur, then I have no idea what it can be."[166]

Notwithstanding this long (and continuing) history of reports of plesiosaur-like creatures in Loch Ness, this idea was always spectacularly unlikely. Plesiosaurs have been extinct for 65 million years, while Loch Ness was a solid glacier just a few thousand years ago. Even if we were to imagine that plesiosaurs had survived undetected somewhere in the oceans, it is unlikely that they could have colonized Loch Ness. They were tropical animals, unsuited for the cold waters of the loch—and most plesiosaurs were marine animals, unsuited for freshwater in general. Plesiosaurs would also require more food than the loch can provide. Indeed, extensive biological sampling and sonar surveys conducted by Adrian Shine and other researchers of the Loch

Ness and Morar Project have revealed that the loch has a total fish population of only about 22 tons.[167] This is simply not enough food for a breeding population of plesiosaurs. As Shine explains,

> Before hopes are raised too high . . . it should be borne in mind that predators upon this biomass should not amount to more than approximately a tenth of the gross weight. Thus we have available a total of approximately two tonnes [2.2 tons] of "Monster" . . . equivalent to scarcely half the weight of a 36-ft (13 m) Whale Shark *Rhincodon typus*. In fact, two tonnes divided into an absolute minimum viable population of, say, ten creatures, would give an individual weight of only 200 kg [440 pounds].[168]

Moreover, hypothetical plesiosaurs must not only eat, but also die. A population would leave bones on the loch floor. Dredging and submarine searches have not found bones or other physical remains. Finally, plesiosaurs were air breathers. Any plesiosaurs in Loch Ness could be photographed several times an hour, each time they surfaced to breathe. (Their air-filled lungs would also light up sonar like nobody's business—which fails to happen in survey after survey.)

Thus the plesiosaur-type Nessie, based on popular culture and hoaxing, can be definitively ruled out on biological grounds. But, then, what are witnesses seeing out on the loch, if not Nessie?

Misidentification Errors

Skeptics argue that simple misidentification accounts for many reports of a monster in Loch Ness, and proponents often agree. As pro-Nessie author Henry Bauer forthrightly asserts, "At Loch Ness the opportunities to be deceived are legion."[169] Under the right viewing conditions (such as the calm water and long distances typical in Nessie sightings), known objects or events can create illusions that seem compelling even to careful observers. Birds, splashing fish, otters, logs, boats and their wakes, and waves are especially common culprits. Reviewing the sighting records, Roy Mackal concluded that "careful examination of the reports tells us that a large proportion of these observations, perhaps 90%, can be identified as errors, mistakes, misinterpretations, and, in a few cases, conscious frauds."[170]

Hearing this, monster fans sometimes feel incredulous. Can people really mistake, for example, common birds for plesiosaurs? Perhaps surprisingly, the answer is unequivocal: yes—and they often do.

There are many specific, documented cases in which witnesses mistook other objects or animals for the Loch Ness monster. Tim Dinsdale describes seeing Nessie "as large as life," only to subsequently identify this faux-monster as a "floating tree-trunk after all."[171] Similarly, cryptozoologist John Kirk relates a personal misidentification experience during his "pilgrimage to Loch Ness, a shrine for cryptozoological enthusiasts the world over." Gazing hopefully over the loch, he "thought the monster had surfaced when a long slender neck popped up out of the water. . . . I can recall how, for a moment, the hair on the back of my neck stood up." However, Kirk had binoculars ready, which allowed him to positively identify his monster: "a crested grebe, a common waterfowl frequently responsible for causing false alarms in the loch."[172] Similarly, influential early Nessie witness Alex Campbell admitted that at least one of his sightings of a plesiosaur-like creature was really a sighting of cormorants. Mackal has referred to standing waves (caused when boat wakes reflecting from the sides of the loch intersect one another long after the boat is gone) as "the most ubiquitous of all Loch Ness mirages." So common were false positives caused by standing waves that "quite early, the [Loch Ness Investigation] Bureau began keeping a record of any and all boats passing through the lake, so as more readily to discount sightings that were really reports of the standing waves."[173]

As soon as the monster sightings began, people rushed to propose various living animals as the solution to the mystery: a crocodile,[174] a sturgeon,[175] an oarfish,[176] a beluga, a giant squid, a giant eel,[177] and so on. Some suggestions were rather fanciful. One wag joked that the monster could be a "somewhat rare" breed of camel remarkable "for the abnormal length of time it can remain under water while swimming."[178] Others proposed, inexplicably, that the "Loch Ness monster is a sunfish, a native of Australia,"[179] or even that "a pair of sharks might account for the mystery, if there are sharks without the green dorsal fin, which would betray them?"[180] Indeed, there is an entire book dedicated to the thesis that Nessie is a super-gigantic version of *Tullimonstrum gregarium*—a soft-bodied, 1-foot-long marine invertebrate known only from 300-million-year-old fossil beds in Illinois.[181]

Other early suspicions clustered around more plausible suspects: Could witnesses have spotted otters, dolphins, porpoises, or (especially) seals in the loch?

Legendary skeptical investigator Joe Nickell favors otters as a general explanation for many monster sightings in lakes.[182] Otter habitat geographically correlates well with lake monster distribution (especially in North America), and otters certainly can resemble monsters. Alone or in pairs, their long sinuous necks and snake-like heads resemble those of plesiosaurs. In larger groups, the rolling dives of otters can create a powerful illusion of undulating coils.[183] With that in mind, it is notable that otters do live around Loch Ness.[184] Expectant attention operating as it does, we can infer that some Nessie reports must describe otters. Mackal was more direct, asserting that "a good number of monster sightings reported by tourists do result from the observations of otters."[185] Furthermore, a case can be made that otters may lie behind specific sightings. Notably, one witness from the initial rush of 1933 reported a Nessie that "certainly looked like three or four otters together, going in a bunch."[186] Similarly, according to one local at the time, George Spicer's crucially important sighting could be explained as a misidentified otter.[187] (This is the explanation favored by many critics, including Ronald Binns and Nickell. In my opinion, the fingerprints of *King Kong* are unmistakable in Spicer's account, but it is possible that he filtered an actual animal sighting through his impressions from the film. This may explain how his initially modest 6- to 8-foot creature grew into a 30-foot giant in subsequent tellings.)

Because Loch Ness is connected to the North Sea by a short hop up the River Ness, Scotland's local porpoises and seals seemed like obvious suspects from the first—so obvious that Campbell took pains to deny the possibility in his news story about the original sighting by Aldie and John Mackay: "It should be mentioned that, so far as is known, neither seals or porpoises have ever been known to enter Loch Ness. Indeed, in the case of the latter, it would be utterly impossible for them to do so, and, as to the seals, it is a fact that though they have on rare occasions been seen in the River Ness, their presence in Loch Ness has never once been definitely established."[188] However, both types of local animal were more plausible than Campbell's article claimed. To begin with, they were consistent with many reports. Indeed, according to one of the earliest (in June 1933), "the creature seemed rather like a porpoise or seal."[189] As a matter of practicality, medium-size sea animals can swim up the river and into the loch during high water; and, as Campbell conceded, seals were sometimes seen in the River Ness. From day one, these facts made seals the leading skeptical hypothesis. Although he personally rejected this explanation, Rupert Gould acknowledged that seals were "the most plausible theory of all—one which, at present, holds the field. That is, that the 'Loch Ness monster' belongs to the Pinnepedia—and is, in all likelihood, a large grey seal."[190]

For our purposes, looking back over the Loch Ness mystery, the key fact is this: it is now known, definitively, that seals do enter the loch. Furthermore, there is documentary evidence that porpoises or dolphins also venture into Loch Ness. Campbell's claim that "it would be utterly impossible" for porpoises to enter the loch is flatly refuted by an article published in the *Daily Mail* two decades before the Mackays' sighting: "A rare phenomenon is now to be observed in Loch Ness, where a school of porpoises have got enclosed. They entered from the Moray Firth when the River Ness was in high flood, and now that the river is almost unprecedentedly low, even a baby porpoise would find it hard to pass the shallow stretches, while the adults would be hopelessly stranded."[191]

Even more to the point, seals are a documented fact of life in Loch Ness (as they are in many Scottish lochs with access to the North and Irish Seas and the Atlantic Ocean). All debate on the presence of seals in Loch Ness was settled in 1985, when a harbor seal made the loch its home for seven months (figure 4.13). During this time, "about 30 people reported about separate sightings of the seal," and clear photographs were taken by Gordon William-

Figure 4.13
Harbor seals thrive throughout the Northern Hemisphere, found everywhere from the coast of the North Sea in Scotland to the coast of the Pacific Ocean in Canada. (Photograph by Daniel Loxton)

son.[192] In 1999, Nessie researcher Dick Raynor captured new video footage of seals in Loch Ness.[193] In decades of investigation, Raynor writes (he has been involved since 1967), "*none of us* have turned up one iota of scientifically valid evidence for any biological 'monster.' The most significant development is the realization that seals frequently enter the loch, and probably account for many of the sighting events." (It is not necessary to dwell on the eyewitness testimony of seals that Williamson collected from water bailiffs and fishermen, but those reports did offer support for the idea that seals are frequent visitors to the loch. Locals claimed many seal sightings and said that salmon fishermen had shot several seals over the years.)[194]

The presence of seals in the loch had been well attested all along. Consider, for example, the monster captured on film in 1934 by members of a team organized by a businessman, Sir Edward Mountain. When screened for "many leading scientists of the day in zoology and natural history . . . it was the general opinion of the scientists present that from the movements and manner of swimming the creature depicted was in all probability a member of the seal family, possibly a grey seal."[195] The seal hypothesis was also apparently supported by key monster hunters, including Malcolm Irvine, who shot film footage of the alleged monster in late 1933 and again in 1936. Regarding the latter film, the *Scotsman* reported that Irvine believed that the creature was a seal.[196]

Finally, several witnesses during the initial rush of Nessie sightings in 1933 and 1934 either described a monster in seal-like terms or flat-out asserted that they had seen a seal. For example, one "Highland Minister, who for various reasons cannot allow his name to be published," reported a creature with a head "not unlike that of a seal or sea lion."[197] Mrs. Cranston of Foyers was more explicit. She told the *Daily Mail* that she had seen a salmon leap out of the water 200 yards away (practically point-blank range by Nessie sighting standards), followed immediately by a large round head that bobbed vertically in the manner of a seal. Cranston asserted that she was, sensibly, "convinced that it was a large seal chasing salmon in the loch." At the Natural History Museum in London, Cranston's explanation seemed obvious. As one curator explained, "We have thought that to be the probable explanation from the beginning. Grey seals have been known in the past in Loch Ness. They follow the salmon up the river."[198] Interestingly, Marmaduke Wetherell (of hippo-foot infamy) reported that he personally had seen a seal in the loch: "I have not the slightest doubt that what I saw was a very big seal. The head leaves no room for doubt on that point. What appeared to be a hump was actually the creature's back as it lunged forward to dive. I am now quite satisfied that there is not a prehistoric animal in Loch Ness, but a very big seal."[199] Based on the testimony of these witnesses, the *Daily Mail* concluded its investigation of the Loch Ness monster with the screaming headline "There *Is* a Seal in Loch Ness." While Wetherell is obviously not a trustworthy source, this headline—at least in general terms, and at least some of the time—is completely correct.

Memory Distortion

Reports of mistaken or ambiguous sightings can also become embellished over time, as a result of distortion of the witnesses' memories, creative enhancement, retelling by the media, or any combination. It is easy to understand how this happens. Consider cryptozoologist John Kirk's false-positive sighting of Nessie, which turned out to be a Great Crested Grebe. If he had not had binoculars to penetrate the illusion on the spot, how might the "hair on the back of [his] neck stood up" emotional charge have influenced his later memories of the sighting?

Nor is this problem hypothetical. Important Nessie witnesses have embellished their tales over time. For example, the witnesses in the early "three young anglers" story specified in 1930 that they "could see a wriggling motion, but that was all, the wash hiding it from view." By 1933, the story had already evolved to describe "a dark back, appearing as two or three shallow humps."[200] By 1974, the original hidden-from-view story had been so elaborated that "the length of the part of it we saw would be about twenty feet, and it was standing three feet or so out of the water."[201]

Such distortion is not unique to cryptid cases and certainly is not due to any unique flaw in those who claim to have seen monsters.[202] Psychologists emphasize that the plasticity of memory is a defining factor of all firsthand testimony. Saints, scoundrels, geniuses, and idiots all face the same reality: the normal operation of human memory involves both creative reconstruction and loss of accuracy over time.

THE ASTONISHING HISTORY OF ORGANIZED SEARCHES

The sheer longevity of the Nessie legend makes it difficult to appreciate the number and scale of the decades of serious, sustained efforts to penetrate the mystery. From gyrocopters to giant Nessie traps, almost any imaginable scheme has been tried—long ago and many times. The history of the hunt is too extensive to cover here in depth, but let's have a look at a few of the projects you might have thought of yourself.

- *Why not seek evidence in old books and newspapers?* Archival-sleuthing efforts hit the ground running in 1933 and have never stopped. Yet even today, with centuries' worth of national and regional newspapers scanned into searchable databases, the archival evidence remains the same: Nessie was born in 1933, in the wake of *King Kong*. There is no sign of a local cryptid tradition before that time (and certainly no record of a long-necked monster in the loch before George Spicer claimed to have seen it). Worse, what records there are suggest that such a tradition did not exist. When reporting that a group of Loch Ness locals mistook swimming ponies for kelpies in 1852, the *Inverness Courier* gave no hint of an existing Nessie legend beyond the generic international water-horse mythology. Four years later, the same newspaper covered claims of a 40-foot eel in a loch on the Isle of Lewis without ever hinting that Loch Ness might host a monster of its own.[203]

- *Why not hire an army of paid observers and station them around the loch with cameras?* This approach was tried in the summer of 1934, when Sir Edward Mountain paid twenty men to keep vigil around Loch Ness for a month. Working from dawn to dusk with "almost military precision, with a careful distribution of watchers at places most likely to yield satisfactory photographs," the men dutifully generated twenty-one photographs of the alleged monster.[204] Roy Mackal asserts that they were false positive photos of waves and boat wakes (with one possible exception, showing an ambiguous hump), and this seems to be the consensus view: Mountain's observation campaign was a bust. As Ronald Binns explains,

  > Unfortunately there was a flaw in Mountain's great idea. His watchers were all unemployed men, drawn from Inverness labour exchange. In the hungry thirties to be paid 2£ a week just to sit at the side of Loch Ness on a summer's day must have been an attractive proposition. To be offered a bonus of ten guineas each time anyone photographed Nessie must have provided a great incentive to pander to the noble baronet's whimsy that there was a monster in the loch.[205]

Figure 4.14 A battery of cameras overlooks Loch Ness in 1966. (Reproduced by permission of Fortean Picture Library, Ruthin, Wales)

Still, the strategy of an "army of observers" seems like an obvious one. For that reason, it has been tried many times since, using teams of volunteers—not to mention the informal coverage by thousands of camera-armed tourists!

- *Why not keep watch with super-high-powered telescopic movie cameras?* This scheme was tried, too, with person-years of observation from custom-built stationary platforms (figure 4.14) and mobile truck-mounted film rigs. For example, starting in 1964 the Loch Ness Phenomena Investigation Bureau kept a significant percentage of the loch under observation by using giant camera rigs, which, as F. W. Holiday recalled, "were almost certainly the most formidable photographic tools that had ever been used for natural history purposes in Britain."[206] With several months of continuous watches, the searchers expected success that very year. As Holiday explained, "It was not unreasonable to suppose that dawn to dusk watching, seven days a week for five months, would produce results. After all, if you watch a given area of sky for long enough, you are bound to see a rainbow."[207] And, indeed, this strategy should have worked, if eyewitnesses were really seeing a monster. As Mackal reflected, "The number of recorded reports during the 30-year period following 1933 was roughly 3,000. This figure, taken at face value, would mean that about 100 observations were made annually. From this it is clear why it was reasonable to expect photographic surveillance of the loch surface to produce evidence rapidly."[208] But it was not to be—not that year, or any year thereafter. In 1972, the camera batteries were finally dismantled. The failure of this heroic effort of sustained photographic observation indicates that most eyewitness reports must be wrong—with grim implications for a legend based on eyewitness sightings.

- *Why not search the loch using submarines?* This approach seems obvious, if expensive. It was, of course, tried. In 1969, for example, the Loch Ness Phenomena Investigation Bureau fielded a custom-built submersible called *Viperfish*, armed with biopsy harpoons (figure 4.15). Sponsored by *The World Book Encyclopedia*, the mission was a failure.[209] Other subs were deployed at Loch Ness in 1973 and (as a commercial-tourism concern) throughout 1994 and 1995.[210] The critical obstacle for this search strategy is that the water of Loch Ness is very opaque (typically described as "inky" or "murky"), and thus visibility under the water is limited to only a few feet.

- *Why not search the loch using underwater cameras?* The opacity of the water of Loch Ness likewise severely limits the usefulness of underwater photography. Despite this hindrance, a team led by lawyer Robert Rines claimed to have taken underwater photographs of a plesiosaur-like creature in 1972 and again in 1975. The first set of shots, purported to show a dramatic diamond-shaped flipper, was apparently taken with the assistance of a dowser—a local psychic who "detected" the monster using paranormal means and directed the camera placement.[211] (She also claimed at least fifteen

Figure 4.15 Dan Taylor in *Viperfish*, his yellow submarine, at Loch Ness in 1969. (Reproduced by permission of Fortean Picture Library, Ruthin, Wales)

personal sightings of Nessie.)[212] The resulting photos were extremely indistinct, but were altered by an artist to depict an unmistakable flipper (figure 4.16). Exactly how that modification was done is controversial to this day. After computer enhancement, the photos remained ambiguous—only to later achieve shocking clarity through some additional undocumented process of creative compositing or retouching.[213] According to Nessie author Tony Harmsworth, "paint brush marks" are clearly visible on large blow-ups of the revised versions. "That is how it became so clearly a flipper," Harmsworth explains. "The retouching is not sophisticated air brushing or modern Photoshop effects, but relatively crude paint brush effects." Confronted with this evidence at a "tense meeting" with Adrian Shine, Rines himself admitted that there may have been retouching by a magazine editor.[214] In any event, these photographs were followed in 1975 by a shot alleged to show a plesiosaur-like body and neck and another said to show the monster's head. Unfortunately, as pro-Nessie author Nicholas Witchell explained, "the camera and strobe light were photographing through the heavy gloom of the water, strongly stained with the peat brought down into the loch by the mountain streams. It was like a green fog in which the individual particles of this liquid suspension were almost large enough to see."[215] Critics argue that the images in Rines's photos

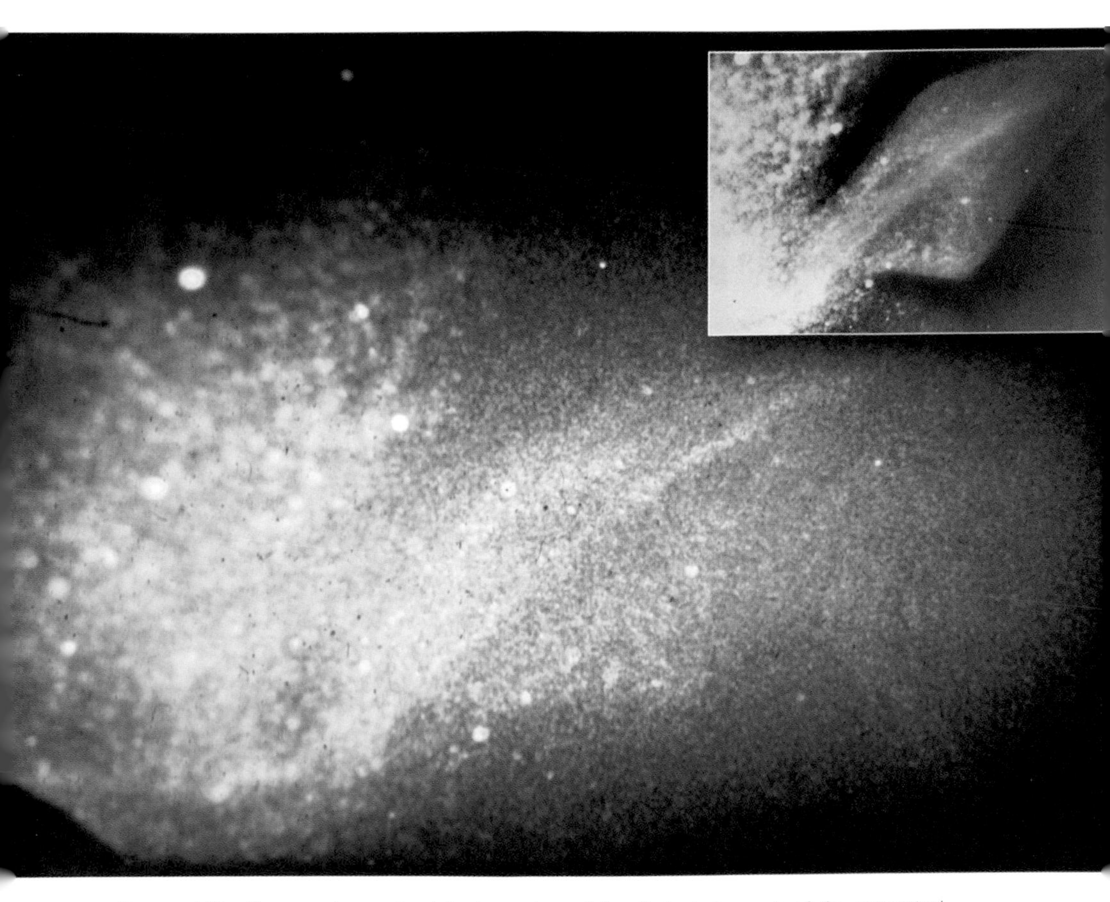

Figure 4.16 The unenhanced original version of the first photograph of the purported flipper of a plesiosaur-like monster, taken by Robert Rines, compared with the retouched version (*inset*). A scan of the unenhanced version made from a 35mm Kodachrome transparency was provided by Rines to Loch Ness investigator Dick Raynor around 1972 (stated to be copied from "camera original" 16mm Kodachrome 11, with an ASA speed rating of 25). (Reproduced by permission of Mirror Syndication International; photograph courtesy of Dick Raynor)

from 1975 are consistent with those of silt or wood debris. (The "head" photo turned out to be an underwater stump that has been identified and recovered.)[216] Thus, as Witchell wrote, "it would be foolish to cling to the photos in the face of such uncertainties. Bob Rines, I think, now agrees."[217] (I might add that the double-jackpot coincidence here would also stretch credibility. Rines's underwater cameras watched just a handful of the loch's 261 billion cubic feet—and for very short periods.)

- *Why not dredge the loch for bones?* In 1934, Rupert Gould pointed out that a population of giant animals should leave clear signs of its presence. Consider, for example, the "rather startling consequences" if the loch were home to a breeding colony of plesiosaurs: "What, on this assumption, has become of the bones which should, by now, have carpeted the entire floor of the Loch?" Could that biological treasure be just sitting there for the taking? Unfortunately, no. Gould noted that "no trawl-net—and many have been put down for biological purposes—has ever brought up any fragment of the kind."[218] Yet this was, on the face of it, a promising strategy. As Binns pointed out, "The bed of Loch Ness is, in most places, as flat as a billiard table. If the loch contains an unknown species of giant animal with a bone-structure then evidence of skeletal remains ought not to be hard to find."[219]

  In fact, Loch Ness has been sampled extensively using dredges, scoops, and core samplers, including serious scientific efforts undertaken from the 1970s through the 1990s.[220] (In 2003, for that matter, a diver named Lloyd Scott *walked* the entire length of the floor of Loch Ness in a charity fund-raising stunt.) No monster bones were discovered, although a great deal was learned about the life of the loch, from fish to plankton to an invasive species of American flatworm thought to have been introduced on monster-hunting equipment![221] The researchers with the Loch Ness and Morar Project also discovered that the deep, even, and undisturbed sediment on the floor of Loch Ness is a delicately stratified geologic record of the region. Thus any future dredging efforts will have to be weighed against the damage that dredges cause to this historical resource.

- *Why not search the loch using sonar?* This question, of course, is the doozy. It won't surprise you to learn that sonar has been tried, but few realize how intensively and how often Loch Ness has been scanned. In addition to attempts using a single boat, a submarine, or pier-based equipment, there have been many systematic grid- or dragnet-style sonar surveys of the entire loch (figure 4.17). In 1962, for example, a Cambridge University team used a four-boat fleet to sweep the entire loch. According to the team leader, "Nothing was detected by any of the boats. This seems to rule out the possibility that a large animal exists in the body of Loch Ness."[222] Exhaustive sonar trawling in 1968, 1969, and 1970 (by a team from the University of Birmingham) produced disappointing results. (The survey of 1968 recorded two unusual contacts, but the work of the following year produced

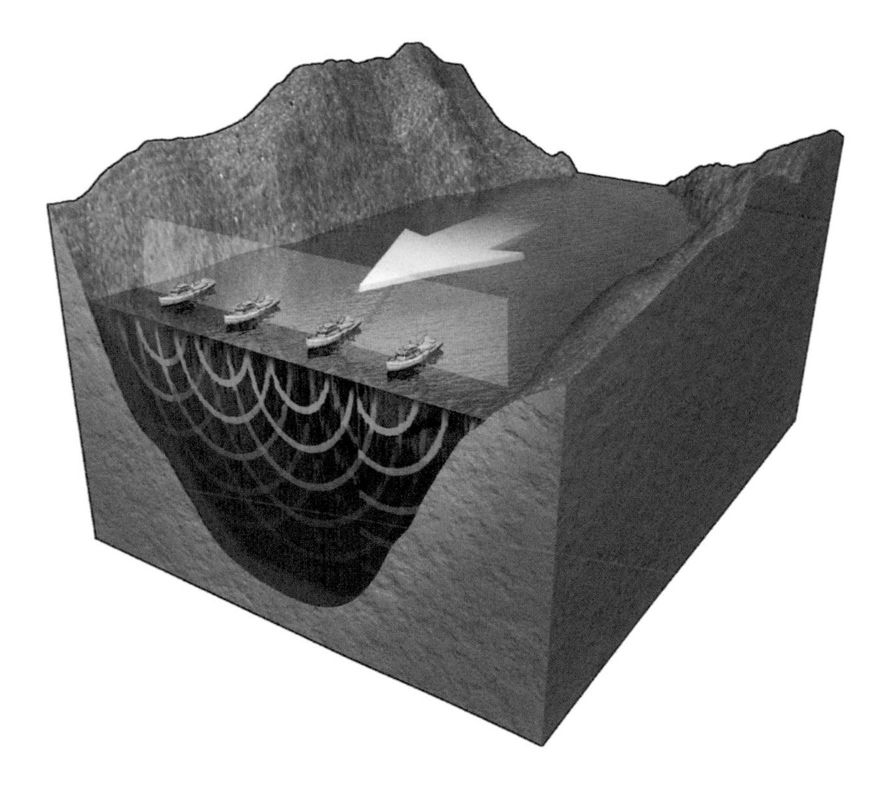

Figure 4.17 Several systematic, dragnet-style sonar surveys of the entire loch were conducted over twenty-five years. (Illustration by Daniel Loxton; not to scale)

nothing. The intensive survey of 1970 again, as Binns put it, "drew a complete blank.")[223] In 1981, the Loch Ness and Morar Project launched a custom-built sonar boat that scanned the loch from end to end, twenty-four hours a day. (In 1982, this effort racked up 1,500 hours of patrol time. While interesting sonar contacts were made, none indicated anything like the giant Nessie of eyewitness reports.)[224] Similarly, in 2003, the BBC sponsored a sonar search that used the Global Positioning System (GPS) to ensure complete coverage of the loch. "We went from shoreline to shoreline, top to bottom on this one, we have covered everything in this loch and we saw no signs of any large living animal in the loch," according to one of the BBC's sonar specialists.[225]

The granddaddy of all Loch Ness sonar searches was Operation Deepscan, conducted in 1987 under the command of Shine (figure 4.18). This coordinated, large-scale project mobilized twenty-four boats to sweep the

Figure 4.18 The fleet of boats participating in Operation Deepscan in 1987. (Reproduced by permission of Fortean Picture Library, Ruthin, Wales)

loch from end to end—multiple times. The fleet created a "sonar curtain" across the loch, with follow-up boats ready to investigate anything detected by the curtain. The results? "Sonar Search for 'Nessie' Reveals 3 Wobbly Scratches," ran the *Washington Post* headline.[226]

Results from the long history of sonar surveys of Loch Ness are mixed.[227] On the one hand, the sonar searches, conducted over forty years, are most notable for their complete collective failure to find giant monsters. Operation Deepscan and other surveys effectively rule out any possibility of Loch Ness plesiosaurs, giant newts, or 80-foot sea serpents. If they were living in the loch, sonar would have found them decades ago. On the other hand, some sonar contacts remain tantalizingly unexplained (including one of Operation Deepscan's "wobbly scratches"). These hits are much smaller than the gigantic creatures reported by Nessie witnesses, but some could suggest animals larger than salmon. Could these anom-ous contacts represent a genuine zoological surprise? It's possible. Shine as argued hopefully for the possibility of sturgeon in Loch Ness.

There are, however, many sources of false-positive sonar contacts. Sonar can refract and reflect off thermal boundaries in the water column. It can reflect off the walls of the loch. Sonar can record schools of fish as larger animals. Even boat wakes can create bogus sonar hits. Most important, sonar can record objects that are not animals. Logs are an obvious culprit, as is debris tethered to the loch bottom (such as from fishing gear). Amusingly, it is even possible that some sonar contacts record sunken equipment left behind over the decades by Loch Ness monster hunters!

UNDERWATER TUNNELS TO THE SEA?

The intensive searching has revealed none of the evidence we should expect if Loch Ness were home to one or more large monsters. This implies that eyewitness reports of such creatures must break down to a collection of hoaxes and (more commonly) errors of various types.

But what if Nessie only visits the loch? A part-time monster could explain the absence of sonar evidence. The Loch Ness and Morar Project concluded its review of the sonar evidence: "In the case of a single occasional migrant, detection would be virtually impossible."[228]

The rivers and canals that flow into Loch Ness can be confidently ruled out as commuter routes for large monsters, being shallow, well travelled, obstructed by weirs, broken up by shipping locks, or some combination. Indeed, the best route from Loch Ness to the North Sea—the River Ness—is so shallow that fishermen often wade across it.[229] (The river also runs smack dab through the city of Inverness.) But could there be an additional, secret route?

Many people believe (as I did, as a child) that Loch Ness is connected to the North Sea through underwater tunnels. This is an amazingly persistent myth, perhaps because it is so convenient. How else could a gigantic monster hide in a confined space, other than by slipping out through a back door? However, this explanation has been known to be silly since the very origin of the Loch Ness monster legend. Speaking to reporters in late 1933, Rupert Gould ruled out the rumors of "a subterranean passage connecting Loch Ness to the sea" on the sensible basis that "the surface of Loch Ness is above sea level, and obviously could not be if there were a passage by which it would be drained down to sea level."[230] Indeed, the surface of the loch stands more than 50 feet above sea level, and pressure of 50 vertical feet of water would propel a monster down a tunnel like a bullet.

It is worth noting that rumors of underwater tunnels to other lakes or to the sea are not unique to Loch Ness. Indeed, folklorist Michel Meurger has emphasized that hidden passageways are extremely common components of the legends surrounding monster-haunted lakes around the globe.[231] While cryptozoologists take pains to distance themselves from the supernatural origins of lake monster traditions (such as the shape-shifting demonic kelpies), Meurger argues that the myths of secret outlets to the sea are merely a revision of the idea that monster lakes conceal passages to hell or the underworld.

THE BOTTOM LINE

More than thirty years ago, Roy Mackal wrote that even unfalsifiable monsters must eventually fade away: "At least in theory no amount of failure could disprove our basic assumption. However, in practice, human nature being what it is, continued and total failure would be equivalent to disproof."[232] Has that day come at last?

For eighty years, serious researchers have thrown money, reputations, and long years of labor into the depths of Loch Ness—with no sign of a monster to show for it. This outcome was inevitable. For all the science and technology brought to bear, Nessie was born from magic: the kelpie folklore of Scotland and the movie magic of Hollywood. Today, the creature has been forced into a paradox as magical as its roots: a giant monster that many people see, but that cannot ever be detected by science.

As Adrian Shine and David Martin explained in 1987, "Work in the 1980s is not a quest for the dragon of popular expectation. The media-christened 'Monster,' by definition imaginary and by connotation prehistoric, has an existence in the realms of entertainment copy, quite independent of research findings."[233] After decades of fruitless searching, researchers had shifted from hunting the Loch Ness monster to studying the environment of the loch for its own sake.

And what of Nessie, that companion of my own childhood dreams? She swims on, swift and elusive, in the imagination of millions.

5

The Evolution of the Sea Serpent

From Hippocamp to Cadborosaurus

OF ALL THE WORLD'S STOMPING, slithering, and bloodsucking monsters, nothing else quite matches the Great Sea Serpent. Enduring for centuries, the sea serpent legend has a unique place in the history of cryptozoology—and a unique place in my own family history as well.

One of the lessons of cryptozoological investigation is that most cryptids are brand-spanking new. The Loch Ness monster (a Hollywood spin-off) was born after my own grandmother. The modern Bigfoot legend premiered the same year as *Leave It to Beaver*. If the chupacabra (goat sucker) were a person, it would be in high school, the story of this vampiric cryptic from Puerto Rico being only as old as the sci-fi horror movie *Species* (1995), on which it was based.[1]

But sea monsters are, if you will forgive me, a different kettle of fish. Descriptions of sea monsters are as old as written language. They appear in the Bible and in the writings of the ancient Greeks and Romans. Even by the very strictest definition, the fully formed modern version of the Great Sea Serpent goes back at least to 1755, when Bishop Erich Pontoppidan's book *The Natural History of Norway* made it a permanent part of global popular culture.

Over the past 250 years, sea serpents have enjoyed a unique level of support from the scientific community. They have been endorsed, discussed, or denounced by some of the great scientific minds in history, including such legendary pioneers of biology as Louis Agassiz, Richard Owen, and Thomas Henry Huxley. Developing alongside the evolution of paleontology and the discovery of dinosaurs and their marine-reptile cousins, the high-level debates about the reality of sea serpents that took place in learned journals and gas-lit parlors helped define the methods and borders of modern science.[2]

The idea of vast creatures sliding undetected through the abyss stirred my imagination deeply at a young age. How could it not? What kid could read a story like Ray Bradbury's "The Fog Horn" (about a huge, ancient sea serpent drawn to a lighthouse) and not thrill to the idea that it could be true? Imagine such a primeval monster, as Bradbury did, "hid away in the Deeps. Deep, deep down in the deepest Deeps. Isn't *that* a word now . . . the Deeps. There's all the coldness and darkness and deepness in the world in a word like that."[3]

Why, *anything* could be down there, couldn't it?

But I had a very personal reason to think that sea serpents were just waiting out there to be discovered: my parents saw one, when I was a young child. I was raised as something of a nomad. My family lived in a silver travel trailer and spent winters on the beaches near Victoria, British Columbia. We just pulled in the trailer, and presto: home. On one of those beaches, my parents saw the twentieth century's best-known regional version of the Great Sea Serpent: Victoria's own Cadborosaurus. I'll let my father tell the story as he remembers it—a story that has shaped my career to this day:

> We were right by the water. We were young and happy and life was good. We had a morning ritual of sitting up in bed and looking out to sea while we enjoyed our morning cup of tea.
>
> This particular beautiful summer morning we thought that we saw Cadborosaurus. We both saw it at the same time, and both said something stupid like "Wow, look, a sea serpent," followed by embarrassment at what we had just said. We jumped up in our bathrobes and ran barefoot down to the water's edge.
>
> There it was, swimming along just 40 feet or so from shore. It was swimming parallel to the beach and making pretty good time. We had to walk briskly to keep up with it, and we wished we had the camera.
>
> We were both saying things like, "It's definitely a sea serpent, see the head? I can even see its eyes!"
>
> It was about 30 ft. long, dark in color and clear as day. It had a head, a neck, and many humps. It was swimming along in perfectly coordinated undulating motion. It appeared to be very real, alive, efficient . . . and very clearly a sea serpent.[4]

What was it, really? Well, that's the question, isn't it? My parents had diverging interpretations of this experience, but as a kid, it seemed obvious to me: it was a sea monster! And I was going to catch it.

For years, I dreamed of just that. I scoured catalogues for monster-hunting equipment. I designed logos for my future Cadborosaurus expeditions. I got my local librarian to help me write to the Loch Ness Phenomena Investigation Bureau. And today, bizarrely enough, I find myself in the professional position of actually investigating monsters (albeit skeptically). Through it all, it seemed—and still seems—perfectly plausible that a sea serpent-like animal could exist. Why not? The ocean is certainly big enough, and it is an exhilarating fact that many of its inhabitants remain completely unknown.

Could some sea serpent stories be true? As with other cryptid mysteries, the best place to start is at the beginning. How did the idea get started in the first place? How did it take the form that it did? Why, for example (and this deeply puzzled me as a kid), are classic sea serpents so often described with horse-like manes?

Why, of all things, would sea serpents have heads that resemble those of horses (or cows, sheep, or camels)?

SEA MONSTERS IN ANTIQUITY

Sea Serpents in Ancient Literature and Classical Art

To find the roots of the sea serpent legend, we'll look back almost 3,000 years to the sunny shores of the Mediterranean. If sea serpents are real, it is likely that the sophisticated coastal cultures of the ancient world would have described and depicted them. After all, the ancients were no slouches, and trade goods and information flowed throughout Eurasia for thousands of years. Besides, sea serpents would tend to stick out. My parents' 30-foot Cadborosaurus was a minnow by sea serpent standards. Other witnesses have described sea serpents as long as a blue whale or larger: various reports have put them at 150, 200, 300, or even a staggering 600 feet long. A 600-foot sea serpent would be not so much a cryptid as Godzilla.[5] Surely, ancient sources would mention the largest animal ever to exist?

Sea monsters are as old as literature—and for good reason! Imagine the ocean through the eyes of the ancients: endless, dark, and deadly. When brave warriors, traders, and fishermen ventured out over that alien abyss, they could only imagine what vast and terrible dangers stirred below. One thing they did know: sometimes the sea swallowed men. To ancient mariners, explains classical scholar Emily Vermeule, the sea was "more like a tightrope over an open tiger-pit than a safe road home."[6]

Accordingly, the seas of the ancient imagination swarmed with a teeming variety of powerful, malevolent monsters. Many were inspired by true-life animals filtered through folklore, mythology, and the tales of travelers returned from distant lands. Even as these real and folkloric creatures were named, they also blended easily into one another. In particular, the dividing lines between sea monsters, whales, dragons, and snakes were fluid. Think of the biblical Leviathan: a primordial sea monster, yet simultaneously a dragon, but considered by many to be a whale. The book of Job emphasizes the fiery aspects of this vast (if vague) ocean-dwelling monster:

> Out of his mouth go flaming torches; sparks of fire leap forth.
> Out of his nostrils comes forth smoke, as from a boiling pot and burning rushes.
> His breath kindles coals, and a flame comes forth from his mouth. (41:11-13)

Isaiah amplifies this hybrid description, calling Leviathan "the dragon that is in the sea" (27:1).[7] At the same time, many scholars argue that Job's ver-

Figure 5.1 A *ketos* depicted on a Caeretan hydria black-figure vase, ca. 530–520 B.C.E. (Stavros S. Niarchos Collection). John Papadopoulos and Deborah Ruscillo note the "cetacean-like flippers . . . plus what looks suspiciously like a whale fin about two-thirds down the body." (Redrawn by Daniel Loxton, with reference to John Boardman, "'Very Like a Whale': Classical Sea Monsters," in *Monsters and Demons in the Ancient and Medieval Worlds*, ed. Anne E. Farkas, Prudence O. Harper, and Evelyn B. Harrison [Mainz: Zabern, 1987], and a photograph by S. Hertig [Zurich University Collection])

sion of Leviathan is based on a real animal, with the whale among the leading contenders.[8] (Such a distorted, mythologized view of a real animal would be consistent with the general lack of familiarity with marine life among the relatively land-locked writers of the Hebrew Bible. As archaeologists John K. Papadopoulos and Deborah Ruscillo note, the Hebrew Bible does not record even one species name to identify any particular sort of fish. Even Jonah's "whale" is identified simply as "big fish.")[9]

By contrast, writers and artists immersed in the coastal culture of the ancient Greeks faithfully described many dangerous, delightful, and delicious sea animals in detail. But whales, being relatively rare in the Mediterranean, were as hard to study as they were terrifying. As a result, no clearly identifiable, naturalistic Greek images of whales are known to exist.[10] Instead, "sea monster" and "whale" were interchangeable concepts in early classical literature, with the same word, *ketos*, used for both. *Ketos* is the basis for the word "cetacean," but it was not originally restricted to whales: it was an umbrella term for anything big and scary that lived in the sea, including whales, sharks, and tuna (figure 5.1).

Alongside the merman (Triton) and the hippocamp (a Capricorn-like creature whose body combines the front half of a horse with the back half of a fish, to which we will return), the *ketos* was one of the three standard sea monster types employed by Greek artists. As creative vase painters and sculptors explored the *ketos* form over many centuries, several varieties emerged. Many images of the *ketos* straightforwardly depict very big fish. Other artists imagined *ketos* monsters as grotesque hybrids: fishes with the heads of powerful terrestrial animals. Once this mix-and-match scheme became common, artists could combine creatures as creatively as they wished.[11] By the Hellenistic period (the three or four centuries following the continent-spanning conquests of Alexander the Great), sea monster art had become more light-hearted and fantastic. Artists were happy to try inventive depictions, as archaeologist Katharine Shepard notes: "We find for the first time in the Hellenistic period the sea-centaur, -bull, -boar, -stag, -lion, -panther and -pantheress."[12]

Papadopoulos and Ruscillo describe a third style of *ketos* illustration, which is much more intriguing from a cryptozoological perspective: "An alternative manner of representing the *ketos* is as a large serpent-like creature: a snake by any other name. Like the big fish, a suitably massive snake was one, relatively straightforward, way of giving iconographic substance to a massive sea creature that was, above all else, mysterious and frightening."[13] This last sort of fantastic, vertically undulating beast included elements from the mutually blurry notions of dragons, fish, whales, and pythons (among other animals),[14] but to the modern eye they scream "sea serpent." Just look at that vertical undulation! Does this mean that the ancients did have knowledge of a genuine sea serpent 2,000 years before the modern legend emerged in Scandinavia?

Not so fast. It is very easy for modern audiences to be misled when interpreting out-of-context artworks from other times and other cultures. For example, UFOlogists who see "evidence" of flying saucers in medieval or Renaissance paintings mistakenly project their own expectations onto stylized depictions of clouds, the Holy Spirit, or even broad-brimmed hats.[15] As it happens, the long, snaky, vertically undulating tails that make *ketos* images resemble our notion of sea serpents were not unique to those depictions. The same tails were used just as often for hippocamps and for mermen (figure 5.2). It is worth noting that while neither fish nor serpents move by means of vertical undulation, a wavy or an arched body is extremely handy for indicating sinuousness in profile images on vases or shallow relief sculptures. "Of course," notes legendary cryptozoological writer Bernard Heuvelmans (one of the coiners of the word "cryptozoology"),[16] "most primitive or childish pictures show snakes wriggling in the same way, but this is simply because it

Figure 5.2 Herakles wrestles the river deity Akheloios on an Attic red-figure vase attributed to Oltos, ca. 520 B.C.E. (Redrawn after London E437, British Museum, London)

needs some skill in perspective to draw the side view of an animal that wiggles horizontally."[17]

Is there reason to think that artists modeled the *ketos* on a genuine serpentine animal? No. The *ketos* may be widespread in classical art, but it cannot be used as evidence for the existence of a cryptid. As art historian and archaeologist John Boardman explains in an influential overview of the *ketos*, it is a wasted effort to speculate about "our beast's relationship to real or imagined sea monsters since we can show that his creation is an artistic amalgam not much troubled by what swam or was thought to swim in the Mediterranean."[18] Moreover, it is clear that images of the *ketos* changed with the fashions among artists—a pattern more consistent with the depiction of imagined creatures than real animals.

But consider the opposite: What if art inspired the notion of the sea serpent? The universality of vertical undulation as an artistic solution for portraying wiggly movement in static two-dimensional media (and the rarity of vertical undulation in nature) hints that the Great Sea Serpent of modern legend could owe something to art. We will return to the influence of art, but first let's explore the efforts during the classical period that we would call "science" today. Classical artists may not have recorded evidence for the existence of the modern sea serpent, but what of those ancient thinkers who attempted to accurately describe the animals of the air, land, and sea?

Sea Serpents in Classical Natural History

The cryptozoological hypothesis that immense sea serpents literally exist has an Achilles' heel: extremely little demonstrated support in the historical record before the Enlightenment (when sea serpents suddenly burst into prominence in Scandinavia). The pioneering book *The Great Sea-Serpent*, for example, was perfectly up front about the failure to identify sightings from antiquity. Author Anthonie Cornelis Oudemans wrote off the few possibilities he was aware of as misidentification errors and simply declared that he would "review only reports of no earlier date than the year 1500" C.E.[19]

Was Oudemans too hasty? It is unambiguously the case that the sophisticated scholars of classical Greece and Rome did, in fact, preserve accounts of sea monster sightings. For example, classical folklorist Adrienne Mayor has identified dozens of ancient references to allegedly real aquatic monsters.[20] But did these authorities describe creatures that are recognizably "sea serpents" in the modern sense? To qualify as sea serpents, creatures must (minimally) live in the sea and look like serpents. Some classical texts describe marine monsters that are not serpents. They include a giant man-eating sea turtle[21] and some clear references to whales (such as the orca enclosed in a harbor and battled by boats as entertainment under the Roman emperor Claudius).[22] Others describe serpentine creatures that did not live in the sea (such as a giant snake that allegedly fought Roman soldiers at a river in North Africa).[23]

Oudemans briskly dismissed the argument that the animals mentioned by Aristotle, Pliny the Elder, and other classical natural historians were sea serpents, considering them pythons, eels, and known species of sea-adapted snakes. This cavalier approach may seem hard to justify, but I think that he was correct: despite some serpent-like versions of the *ketos* in art, and despite many ancient references to sea monsters in general, classical history and literature can offer little to help the case for the sea serpent in particular.

Consider Aristotle. His meticulous *History of Animals* (350 B.C.E.) is often cited in support of the sea serpent, but this is a stretch too far. The most frequently quoted passage refers simply to "some strange creatures to be found in the sea, which from their rarity we are unable to classify." Fishermen apparently described some of these unidentified rarities as "sea animals like sticks, black, rounded, and of the same thickness throughout."[24] This representation may sound suggestive to those sufficiently willing to cherry-pick, but it is vague to the point of meaninglessness. Another passage is much more clear—in that it clearly does not describe a sea serpent:

> In Libya, according to all accounts, the length of the serpents is something appalling; sailors spin a yarn to the effect that some crews once put ashore and saw the bones of a number of oxen, and that they were sure that the oxen had been devoured by serpents, for, just as they were putting out to sea, serpents came chasing their galleys at full speed and overturned one galley and set upon the crew.[25]

Both Aristotle and Oudemans understood that this was an exaggerated, hearsay description of Africa's native pythons. Such stories of land-based, big snakes were common and had nothing to do with sea monsters. Yet, they sound so tempting! As Bernard Heuvelmans's respected book *In the Wake of the Sea-Serpents* explains, this hyperbole has led to the widespread misuse of "various stories of gigantic snakes, but living on land, which many authors have tried hard to include in the sea-serpent's dossier."[26]

As with stories of whales and sea monsters, ancient Greek lore about pythons was intertwined (and often interchangeable) with that of dragons. Consider an example from Greek epic poetry. When the legendary hero Jason approaches the grove of the Golden Fleece, the last challenge he must face is a vast serpent (figure 5.3):

> Directly in front of them the dragon stretched out its vast neck when its sharp eyes which never sleep spotted their approach, and its awful hissing resounded around the long reaches of the river-bank and the broad grove. . . . Women who had just given birth woke in terror, and in panic threw their arms around the infant children who slept in their arms and shivered at the hissing. As when vast, murky whirls of smoke roll above a forest which is burning, and a never-ending stream spirals upwards from the ground, one quickly taking the place of another, so then did that monster uncurl its vast coils which were covered with hard, dry scales.[27]

It is not surprising that Greek stories feature land-based mega-serpents. The Greeks had contact with some of the planet's largest snakes, but not such close contact as to easily distinguish folklore from biology. In modern Africa, large pythons range widely below the Sahara (and may have been encoun-

Figure 5.3
The serpent guardian of the Golden Fleece regurgitates Jason, as Athena watches, on a red-figure cup painted by Douris, ca. 480–470 B.C.E. (Redrawn after Vatican 16545, Vatican Museum, Rome)

tered in North Africa's Mediterranean regions in classical times, as Aristotle reports). The African rock python (*Python sebea*), in particular, which grows to lengths of 20 feet or more, is a "monster" even before the magnifying power of hearsay. Rock pythons occasionally hunt humans for food, with the largest specimens capable of swallowing an adult whole.[28]

The armies of Aristotle's student Alexander the Great conquered far to the east, invading India in 326 B.C.E. There, Alexander's officers reported seeing snakes as large as 24 feet long and heard claims of snakes many times larger.[29] By the time the Roman military commander and naturalist Pliny the Elder compiled his *Natural History* (ca. 77–79), stories of Indian mega-snakes had grown to describe "dragons"[30] fond of hunting elephants—indeed, of "so large a size that they easily encircle the elephants in their coils and fetter them with a twisted knot."[31] (By the Renaissance, the dragon-versus-elephant idea had been applied to African pythons as well. Edward Topsell wrote in 1608 of the "enmitie that is betwixt Dragons & Elephants, for so great is their hatred one to the other, that in Ethyopia the greatest dragons have no other name but Elephant killers.")[32]

Looking over this literature, Heuvelmans correctly concluded that classical tales "do not refer to the sea-serpent, even in the widest sense, but to other creatures which have nothing to with the case."[33] This absence is conspicuous when contrasted with the way that classical knowledge of other, known sea animals expanded and improved over time. Aristotle and other natural historians eventually hammered down an accurate understanding of whales, distinguishing these animals from other sea monsters with the term *phallaina* (the root of the Latin word *baleaena* and the English term "baleen"). By the fourth century B.C.E., Aristotle could describe how whales breathe air through a blowhole instead of gills, bear live young, and even provide milk for those young. No comparable understanding emerged for any sort of sea serpent. This suggests that there were no genuine sea serpents for classical informants to observe.

THE HIPPOCAMP: GRANDFATHER TO THE GREAT SEA SERPENT

Cryptozoology has tended to overlook what I believe to be the pivotal classical example of a sea serpent–like creature: the *hippokampos* or hippocamp (figure 5.4). An imaginary mer-creature, the hippocamp combines the foreparts of a horse with a looping, coiled tail inspired by fish or dolphins (like the Capricorn that many people know from astrology, whose foreparts are those of a goat). Hippocamps appeared first on vases, coins, jewelry, sculpture, and other artworks from ancient Greece and Italy, where they were often used as a purely decorative element. In the 2,500 years that followed, they became a fixture in European art and literature—especially in heraldry. For example, a hippocamp appears on the coat of arms of Belfast, Ireland.

Developed by the Greeks, embraced by the Romans, and passed from country to country during the Middle Ages, the image of the hippocamp slowly mutated into something more than a decoration for vases. Over time, this fantasy creature became an allegedly real cryptid. In my opinion, the modern

Figure 5.4
Poseidon, the Greek god of the sea, rides a hippocamp on Attic black-figure pottery, ca. sixth century B.C.E.

Figure 5.5
The classical hippocamp and the modern cryptid Cadborosaurus share a general body plan, as well as such specific anatomical details as a mane and whale-like tail. (Various artists have regularly depicted both creatures with between one and several humps or tail arches.) (Illustration by Daniel Loxton)

myth of the Great Sea Serpent (including the recent version, Cadborosaurus) is a cultural invention descended from the artistic tradition of the hippocamp.

It is not hard to spot a family resemblance.[34] Compare the typical hippocamp with a modern composite drawing of Cadborosaurus (figure 5.5). These creatures are a direct, one-to-one anatomical match—from the tips of their whale-like flukes, to their pectoral limbs, to their horse's manes.[35] Here is a clear answer to some of the peskiest questions about the sea serpent: Why the preposterous vertical arches? Why the mane of a horse? Why the head of a horse (or, in some variants, a cow, sheep, or camel)? It is a puzzle that bothered me as a kid and has bothered me throughout my career as a skeptical investigator. Why would sea serpents look so much like horses? It turns out that the answer is simplicity itself. Like Jessica Rabbit in the film *Who Framed Roger Rabbit*, sea serpents have their shape because they were, literally, "drawn this way."

The horse-headed hippocamp motif first appeared in Greece during the Orientalizing period at around the same time as the Triton (or merman) and the *ketos*-type sea monster. It is important to emphasize at the outset what hippocamps were not. They were not gods. They were not characters in myths. Most important, they were not believed to be real animals. The classical world did have a lively tradition akin to cryptozoology. Serious scholars like Aristotle and Pliny the Elder attempted to fold into natural history the

Figure 5.6
Poseidon, standing on a shell chariot, being drawn through the sea by hippocamps.

fantastic creatures described in rumor, legend, and travelers' tales from distant lands. Dragons, unicorns, and griffins fit into this tradition, as do ferocious *ketos* sea monsters. Even mermen were reportedly spotted by eyewitnesses, but hippocamps were not.

Hippocamps existed almost exclusively in art, where they often were depicted as steeds for mythological beings associated with the sea (figure 5.6). They appeared as an artistic motif at around the same time as the emergence of artwork that featured a similar-looking real-world fish: the spiral-tailed, horse-headed sea horse.[36] Indeed, these real and mythological creatures reflect each other so well that sea horses are classified as members of the genus *Hippocampus*, giving these delicate little fish the unlikely designation "horse sea monster." Recursively, the fantastical hippocamps may have been inspired by the fish—although the origin of the monster is unclear. Archaeologist Katharine Shepard reflected that interpreting the significance of the hippocamp is a "difficult problem, since he plays no part in any mythological tale":

> Some have thought that the horse was symbolic of the waves of the sea. Poseidon riding a hippocamp or Poseidon in a chariot drawn by hippocamps would then represent Poseidon borne by the waves. Another idea is that the monster is intended as a likeness of a real sea-horse, but our knowledge of sea-horses does not justify

> this theory. Probably the hippocamp is a purely fantastic monster, which served sometimes as a symbol of the sea. Often the animal seems merely to be used for decorative purposes.[37]

How did a decorative creature break free from the realm of art to take a central place in cryptozoology? Let's trace that strange transformation.

The Hippocamp Spreads Across Europe

The hippocamp motif was adopted by the Romans and eventually diffused across Europe. The Romans themselves may have carried it to England and Scotland (figure 5.7). They invaded Britain in 43 C.E. and ruled much of the island for almost four centuries. The cultural legacy of the Roman presence in Britain persisted long after the end of Roman rule. This influence is a plausible source of hippocamps found (for example) in Aberlemno, Scotland, where they add a decorative flourish to ninth-century Pictish relief sculpture.[38]

Figure 5.7 A section of the mosaic floor from the Roman Baths in Bath, England, ca. fourth century C.E. (Photograph by Andrew Dunn, via Wikimedia Commons. Released under Creative Commons Attribution-Share Alike 2.0 Generic license)

The enduring popularity of classical literature helped carry the concept of the hippocamp into the Middle Ages and beyond. For example, the Roman poet Virgil wrote that the mythological Proteus

> . . . rides o'er
> The sea, drawn by strange creatures, horse before
> And fish behind.[39]

Virgil was read everywhere in Christian Europe. (Likewise, the mosaics, paintings, and sculptures of antiquity were widely admired and copied long after the fall of the Roman Empire.)

An immensely popular Greek text called the *Physiologus* was instrumental in transmitting the concept of the hippocamp. Composed in Alexandria, Egypt, between the second and fourth centuries,[40] the *Physiologus* is a collection of legends about interesting and exotic animals from the fox to the unicorn—but it is not a work of natural history. The unknown compiler intended the *Physiologus* to be read for moral instruction. Drawing material from the legends and natural history of classical antiquity, the *Physiologus* reconceived each creature in the service of overt Christian allegory.[41] Appearing alongside such old favorites as the centaur, siren, and phoenix, the hippocamp (Hydrippus) became a symbol for Moses:

> There is also a beast called the Hydrippus. The front part of his body resembles a horse, but from the haunch backwards he has the shape of a fish. He swims in the sea, and is the leader of all fishes. But in the Eastern parts of the earth there is a gold-coloured fish whose body is all bright and burnished, and it never leaves its home. When the fish of the sea have met together and gathered themselves into flocks, they go in search of the Hydrippus; and, when they have found him, he turns himself towards the East, and they all follow him . . . and they draw near to the golden fish, the Hydrippus leading them. And, when the Hydrippus and all the fish are arrived, they greet the golden fish as their King. . . . The Hydrippus signifies Moses, the first of the prophets.[42]

The Greek *Physiologus* was soon translated into Ethiopian, Armenian, Syrian, Arabic, and Latin. From there, the evolution of the book got more complicated. (Translator Michael Curley dryly understates, "The influence of *Physiologus* on the literature and art of the later Middle Ages is too long a story to be fully recounted here.")[43] For our purposes, it's enough to note that a rich and

Figure 5.8 A hippocamp depicted in the Ashmole Bestiary, ca. 1225–1250. (MS. Bodl. 764, fol. 106r; reproduced by permission of the Bodleian Libraries, The University of Oxford)

varied ecosystem of versions and adaptations emerged, with hand-drawn manuscripts in many languages spreading throughout Europe.

"A major watershed in the later history of *Physiologus*," Curley continues, "is its incorporation into the encyclopedias and natural-historical compendia of the late Middle Ages." In particular, the influential *Etymologies* (ca. 623) of the archbishop and scholar Isidore of Seville drew heavily on the legends collected in the *Physiologus*, while dropping most of the Christian allegorical content. In Isidore's treatment, the hippocamp is rendered down from a Moses figure to an ordinary sea creature. Smack dab between whales and dolphins is the entry for "Sea-horses (*equus marinus*), because they are horses (*equus*) in their front part and then turn into fish."[44] This demythologized version brought the fantastical hippocamp a step closer to its modern reinvention as a cryptid.

The tradition of compiling animal legends continued in a genre of twelfth- and thirteenth-century books referred to generically as the "bestiary" (book of beasts). Larger collections than the *Physiologus*, they contain three times the content of the earlier books. Bestiaries include exotic monsters, but the central passages feature ordinary creatures made to serve Christian allegory. "For what is the good of a lesson that can only be taught by hearsay," asks translator Richard Barber, "relating to a beast that no one has ever seen in the flesh? The longest sermons are devoted to topics drawn from everyday life: the ant and the bee display the virtues of humility, obedience and industry, the viper warns against the sin of adultery."[45]

Bestiaries were hand-copied (and creatively improvised) from previous bestiaries, and they varied enormously. Some describe the hippocamp; in others, the hippocamp appears as an illustration. The lavishly gold-leafed Ashmole Bestiary, for example, features an illustration of the hippocamp in its general discussion of fishes (figure 5.8).

The Hippocamp in Nordic Culture

The modern sea serpent legend was born out of Nordic culture, with its origin in medieval Iceland and its florescence in Enlightenment Norway.

THE ICELANDIC *HROSSHVALR* By the twelfth century, the Norse society of Iceland had adopted a belief in a creature called the *hrosshvalr* (horse whale), which was depicted as an unmistakable hippocamp. We will see that this innovation—the Nordic reimagining of the Greek hippocamp as a maned, horse-headed "real" marine monster—is a key to solving the modern mystery of the Great Sea Serpent.

As a maritime culture, the Norse naturally told tales of a great many sea monsters—including a huge, kraken-like creature called the *hafgufa*—but even among these teeming monstrosities, the *hrosshvalr* was considered especially savage. It was described in a twelfth-century text called *The King's Mirror* (*Konungs Skuggsjá*), which was almost certainly intended to instruct the son of a Norwegian king.[46] Along with lessons in statesmanship and trade, *The King's Mirror* taught the unknown prince about useful species of whales and warned him of the dangerous monsters of the sea:

> There are certain varieties that are fierce and savage towards men and are constantly seeking to destroy them at every chance. One of these is called *hrosshvalr*, and another *raudkembingr*. They are very voracious and malicious and never grow tired of slaying men. They roam about in all the seas looking for ships, and when they find one they leap up, for in that way they are able to sink and destroy it the more quickly. These fishes are unfit for human food; being the natural enemies of mankind, they are, in fact, loathsome.

Scholars have tended to identify these creatures as walruses or sea lions, although *The King's Mirror* specifies that the *hrosshvalr* and *raudkembingr* grow to "thirty or forty ells in length," or about 67 to 90 feet. It's probable that the words *hrosshvalr* (horse + whale) and "walrus" (whale + horse) are etymologically related. (J. R. R. Tolkien, a philologist and the author of *The Lord of the Rings*, struggled with the etymology of the term "walrus" for the *Oxford English Dictionary* during his tenure with the dictionary in 1919 and 1920.)[47] It is possible that *hrosshvalr* started out as a word for "walrus" and was later applied to a fanciful monster. Nonetheless, whatever the original meaning of "horse-whale," the *hrosshvalr* took on a familiar form, a form that it keeps to this day: when Flemish mapmaker Abraham Ortelius depicted the *hrosshvalr*, he drew the creature as a classical hippocamp (figure 5.9).

Figure 5.9 The *hrosshvalr* depicted by Abraham Ortelius on his map of Iceland, published in 1585.

Figure 5.10
A life-like interpretation of a *hross-hvalr* on an Icelandic stamp issued in March 2009. (Illustration by Jón Baldur Hlíðberg; image courtesy of Iceland Post)

Ortelius made historic contributions to cartography, creating the first modern atlas in 1570 (and later, incidentally, proposing continental drift). His depiction of the hippocamp as a living monster also marked a critical milestone in the formulation of the sea serpent. Ortelius lavishly populated his maps with monsters, many borrowed from the work of Olaus Magnus or harkening back to the curly-tailed *ketos*. The map of Iceland that Ortelius published in 1585 teems with ferocious beasts, including our friend the *hrosshvalr*. He described it as "sea horse, with manes hanging down from its neck like a horse. It often causes great scare to fishermen." In his illustration, the hippocamp has long webbed toes and a finny frill on the back of its horse-like forelimbs—a form that it often takes in heraldry. (In other heraldic uses, the hippocamp may have front flippers instead of webbed feet.) The *hrosshvalr* still has essentially this shape, although now filtered through the sensibilities of modern fantasy art. It is depicted, for example, as a maned, flippered, hippocamp-like animal on a stamp issued by the Icelandic government in 2009 (figure 5.10).

THE MIDGARD SERPENT The hippocamp-inspired *hrosshvalr* of medieval Iceland is not the only relevant source for the Great Sea Serpent. We must also consider a vast entity from Norse mythology called Jörmungandr: the World Serpent or Midgard Serpent. Jörmungandr was a child of the Norse god Loki. Cast down into the watery abyss by Odin, the ruler of the gods, Jörmungandr grew into a serpent so large that his body stretched around Earth. According to an older story recorded in the medieval Norse *Prose Edda*, the god Thor successfully fished for Jörmungandr using a huge hook and an ox's head for bait (figure 5.11).

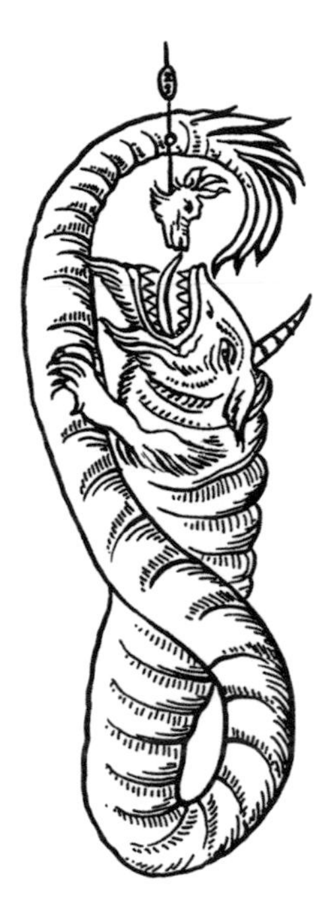

Figure 5.11
Jörmungandr, the World Serpent or Midgard Serpent, takes the ox's head bait dangled by the Norse thunder god, Thor. (Redrawn from a seventeenth-century Icelandic manuscript)

When the modern sea serpent legend eventually made its way into academic debates (starting in the 1750s with a treatment written by Bishop Erich Pontoppidan of Norway), scholars were quick to suggest that it could be the Jörmungandr myth repackaged for a scientific age. Discussing the tale of Thor fishing for the Midgard Serpent, Bishop Thomas Percy noted in 1770, "We see plainly in the . . . fable the origin of those vulgar opinions entertained in the north, and which Pontoppidan has recorded concerning the craken and that monstrous serpent described in his *History of Norway*."[48] A century later, this view was still invoked as an explanation for the sea serpent. In 1869, *The New American Cyclopedia* held:

> It is important to observe that the idea of a sea serpent certainly originated in northern Europe, and was clearly mythological in its first conception. The Midgard serpent, offspring of Loki, which girds the world in its folds and inhabits the deep ocean till the "twilight of the gods," when it and Thor will kill each other, plays a conspicuous part in the *Edda*; and the gradual degradation of the idea from mythology to natural history in its native seats may be traced in Olaus Magnus and the later sagas, till the Latin of Pontoppidan gave it currency in Europe with the natural additions of popular fancy.[49]

Is this view correct? Is the modern sea serpent descended from Jörmungandr? It is probably not a coincidence that a culture with a stupendously large mythological sea serpent later invented a stupendously large cryptozoological sea serpent, but the exact relationship is not entirely clear. (As early as 1822, attempts were made to turn the argument on its head: perhaps a species of serpentine cryptid gave rise to the Norse myth, rather than the other way around.[50] As A. C. Oudemans elaborated, "All fables have their foundation in facts, or in objects of nature, and it is plausible that the Norwegians had met with the sea-serpent before the fable of Thor's great Serpent was inserted into their *Eddas*.")[51]

It is certainly plausible, even likely, that the Jörmungandr myth could be among the roots of the Norwegian sea serpent legend. In turn, the Midgard Serpent can be plausibly interpreted as a regional iteration of primordial dragon myths, such as the Babylonian Tiamat and the biblical Leviathan. It's worth noting that our sources for the Midgard Serpent and other Norse myths date from well into the Christianization of the Nordic countries. Written by a Christian, *The Prose Edda* begins with the words, "In the beginning God created heaven and earth and all those things which are in them; and last of all, two of human kind, Adam and Eve, from whom the races are descended."[52]

Albertus Magnus

Whatever the true source and impact of the Jörmungandr myth, belief in the hippocamp continued to spread and develop. Some of Europe's greatest intellectual lights took up the topic of the mer-horse, including the natural philosopher Albertus Magnus (Albert the Great).

Born at the dawn of the thirteenth century, Albertus worked in a twilit age of both deep superstition and remarkable advances (such as windmills, ship rudders, and, at the end of his life, papermaking in Italy). Albertus started off slowly. "For the first thirty years of his life he appeared remarkably dull and stupid," according to Charles Mackay.[53] He became a Dominican friar and eventually a bishop—and, along the way, one of the medieval world's most celebrated scholars, remembered today as a towering prescientific thinker.[54] (The famed Catholic theologian Thomas Aquinas was a student of Albertus Magnus.)

Albertus Magnus is best known for his encyclopedic *De animalibus*, which built on the work of Aristotle (with material from other authorities, including insights from Albertus himself). Among the menagerie of real aquatic animals and mythical marine monsters is our old friend the hippocamp, described as a predator, "alive" and well at the cutting edge of scholarship that would one day give rise to zoology and cryptozoology: "equus maris (Sea horse) is a marine animal whose foreportion takes the form of a horse and rear parts terminate like a fish. It is a pugnacious creature, inimical to many marine species, and its diet consists of fish. It has an abject fear of man; outside the water it is utterly helpless and dies soon after being removed from its native element."[55]

Olaus Magnus

The now-familiar modern sea serpent had no existence as a cryptid during the Middle Ages. Medieval witnesses saw many other menacing monsters of the deep, including hippocamps, but not the Great Sea Serpent. On the contrary, the Great Sea Serpent turns out to be a fish story of relatively recent composition. An important milestone in its development came with the scholarship of Olaus Magnus, the archbishop of Uppsala in Sweden. For many commentators, Olaus is considered the original sea serpent author, although we'll see that the honor does not belong only to him.

Around the time that Nicolaus Copernicus proposed his heliocentric model of the solar system, Olaus Magnus created an advanced map of the Scandinavian countries and wrote a lengthy book about the people, customs, and animals of those lands. Known in English as *A Description of the Northern Peoples* (1555), the book became the standard reference on Scandinavia. It also provided the final stage set for the modern sea serpent (although the curtain would not fully open for a further 200 years). By then, the Scandinavian reimagining of the Greek hippocamp as a "real" sea monster was itself a centuries-old tradition. Like Albertus Magnus before him, Olaus Magnus described this *Equus marinus* as a genuine, living animal: "The sea-horse may be observed fairly often between Britain and Norway. It has a horse's head, utters a neighing sound, has cloven hoofs like a cow, and seeks its pasture as much in the sea as on land. It is seldom caught, even though it reaches the size of an ox. Lastly, its tail is bifurcated like a fish's."[56]

Olaus also presented a menagerie of other exotic sea monsters, including a "monstrous Hog" with "four feet like a Dragons, two eyes on both sides in his Loyns, and a third in his belly, inclining toward his navel." Another seemingly unlikely creature was "a Worm of blew and gray colour, that is above 40 cubits [60 feet] long, yet is hardly so thick as the arm of a child." Oddly enough, this worm (which he claimed to have regularly seen himself) appears to be a real animal: the thin nemertean bootlace worm (*Lineus longissimus*), which can grow longer than 100 feet.[57]

But it is to another Olaus Magnus monster that we must turn—a land-based Norwegian serpent that, Olaus said, sometimes entered the water to hunt. This has become one of the most-quoted passages in the vast sea monster literature:

> Those who do their work aboard ship off the shores of Norway, either in trading or fishing, give unanimous testimony to something utterly astounding: a serpent of gigantic bulk, at least two hundred feet long, and twenty feet thick, frequents the cliffs and hollows of the seacoast near Bergen. It leaves its caves in order to devour calves, sheep, and pigs, though only during the bright summer nights, or swims through the sea to batten on octopus, lobsters and other crustaceans. It has hairs eighteen inches long hanging from its neck, sharp, black scales, and flaming-red eyes. It assaults ships, rearing itself on high like a pillar, seizes men, and devours them. It never appears without denoting some unnatural phenomenon and threatening change within the state; the deaths of princes will ensue, or they will be hounded into exile, or violent war will instantly break out.[58]

This serpentine animal appears in virtually every book and documentary on the topic of sea serpents, cited as the canonical first important record of the Great Sea Serpent. It is easy to understand why so many authors fall into this trap. A maned snake that snatches screaming sailors from the rigging? Sure sounds like the monster of modern legend! But it isn't—not quite. Not yet.

Looking back at Olaus Magnus's account, pioneering sea serpent author A. C. Oudemans was happy in 1892 to take those elements that conformed to his own vision of the sea serpent—and to discard the rest arbitrarily, decreeing, for example, the testimony of "its devouring hogs, lambs, and calves, and its appearance on summer nights on land to take its prey to be a fable."[59] In constructing his cherry-picked version, Oudemans fundamentally misrepresented the dragon-like snake that Olaus had described. According to Olaus, this monster does not venture onto land to take its prey, but lives and hunts on land (in "cliffs and hollows of the seacoast") and occasionally enters the water to harvest seafood (such as the sailors it cracked "like sugared almonds," as a magazine article put it in 1843).[60] This lifestyle is firmly underlined by the

Figure 5.12 The serpent depicted in Olaus Magnus's *Description of the Northern Peoples* emerges from a cave on land to prey on the crews of passing ships.

accompanying illustration in *A Description of the Northern Peoples*, which clearly shows the serpent emerging from its cave on land (figure 5.12)!

A serpent, sure. The modern sea serpent? No. Olaus Magnus's snake should be viewed as a component legend, a root. It has the mane of the hippocamp, and it is a menace to ships, but Olaus Magnus was clear in describing it as a land-based monster. It is, in short, a *lindorm*—one of the standard gigantic snakes or dragons in Scandinavian folklore.

It's important to realize as well that supernatural phenomena were fundamental components of *lindorm* stories of this period. When Olaus Magnus wrote that the appearance of the serpent foretells a major upheaval, such as a royal death or a major war, he was not kidding—and he was not just tacking on some bit of superstitious gloss to the end of a naturalistic wildlife sighting. The supernatural significance is what those stories are about; the naturalism that cryptozoologists impose on them is the artificial gloss. Nonetheless, such revisionist demystification is a long-standing (and ongoing) tradition.[61] When science-minded Bishop Erich Pontoppidan repeated Olaus Magnus's snake story 200 years later, he felt it was obvious that Olaus "mixes truth and fable together . . . according to the superstitious notions of that age."[62] Bernard Heuvelmans likewise pooh-poohed the supernatural aspects of Olaus Magnus's account ("We may smile at His Eminence's naïveté"), but the truth is that such arbitrary editorializing is reckless. Folklorist Michel Meurger is scathing of the way that Heuvelmans and other cryptozoologists omit or downplay the supernatural, calling it a "gross distortion" of these stories.[63]

The Evolving Sea Serpent

Two more centuries would go by before the parallel currents of folklore described by Olaus Magnus—the *lindorm* and Scandinavian iterations of the hippocamp—would converge to form the modern sea serpent legend. During those 200 years, scholars continued to discuss the hippocamp by its many names: "hippocamp" or "hyppocamp" (English); *hippokamos* or Hydrippus (Greek); *sjø-hest* (Norwegian); *hroshvalur*, *hrossvalur*, or *roshwalr* (Icelandic); *Equus marinus*, *Equus aquaticus*, or *Equus bipes* (Latin); and *cheval marin* (French). Following those discussions is difficult, complicated by the fact that most European cultures had their own version of the hippocamp-style monster, plus related (yet distinct) supernatural, kelpie-type water-horses associated with bodies of freshwater. In most of the languages, the word for the hippocamp-monster is the same as the name for delicate little fishes: sea horse. And, to make things even more bewildering, many hippocamp-type monsters are interchangeable with walruses!

In all this confusion, the English-language cryptozoological and skeptical literatures have typically treated the hippocamp-type monsters as rather marginal creatures with no particular relationship to the sea serpent. Still, I am not the first to note that the old Scandinavian mer-horse traditions became part of the formulation of the new Scandinavian sea serpent. Referring to the work of Norwegian writer Halvor J. Sandsdalen, Michel Meurger explains that the modern sea serpent represents a hybridization, a "fusion of several fabulous creatures originally clearly differentiated in Norwegian folklore":[64] the terrifying half-fish, half-horse monster called the *havhest* (another Scandinavian synonym for "hippocamp") blended with the *lindorm*, "a land snake blown to gargantuan proportions."

We'll come to that moment of fusion—the moment the modern sea serpent was born—shortly. For now, let's look more closely at the evolution of the hippocamp as an allegedly real monster at the dawn of the scientific era.

Some authorities of the late Middle Ages and early modern period took the mer-horse to be a fact of natural history, on the sensible basis that people sometimes saw them. Others correctly linked the folk belief in horse-headed sea monsters to the classical concept of the hippocamp. French naturalist Pierre Belon and Swiss naturalist Conrad Gesner were among the skeptics. A specialist in marine animals, Belon straightforwardly identified the hippocamp as the "fabulous horse of Neptune" (figure 5.13).[65] As we have seen,

Figure 5.13 The hippocamp as depicted in both Pierre Belon's *De aquatilibus*, a treatise on fishes, and book 4 of Conrad Gesner's *Historiae animalium*.

Olaus Magnus argued in 1555 that the hippocamp was a genuine animal, but Gesner (an even more towering figure in zoological history) accepted Belon's debunking analysis. Gesner's encyclopedia *Historiae animalium* (1551–1558) was the most ambitious natural history text written since the fall of Rome. Gesner repeated Belon's argument that modern sea horse sightings are based on an invented classical fancy:

> The Ancients took great liberties with their charming fables, crafted to conceal the truth yet be believed and cloud the credulous minds of men in a haze of nonsense. . . . Those who put their faith in the silly pictures of the ancients are deceived. . . . It all comes from the desire of princes for the fame of their name and the wonder of scholars: since they wished to signify their dominion over land and sea, they joined together the two animals that symbolize these elements, horse and dolphin, which became a monstrous fusion of marine detritus they still call by the name of Hippocampus.[66]

Gesner was much less skeptical in his treatment of other monsters discussed by Olaus Magnus. He repeated the descriptions and illustrations of two species of sea serpent: a smaller type (about 30 or 40 feet long) and the dragon-like mega-serpent portrayed by Olaus:

> On the same map there is another sea-serpent, a hundred or two hundred feet long . . . or three hundred . . . which sometimes appears near Norway in fine weather, and is dangerous to Sea-men, as it snatches away men from ships. Mariners tell

> that it incloses ships, as large as out trading vessels . . . by laying itself round them in a circle, and that the ship is then turned upside down. It sometimes makes such large coils above the water, that a ship can go through one of them.[67]

Gesner's otherwise faithfully redrawn version of the illustration in *A Description of the Northern Peoples* makes the critical change of moving the monster to open water, rather than emerging from a cave on shore (figure 5.14). The illustration of the smaller serpent shows another modern feature: the serpent floats with its coils arching cartoonishly out of the water (figure 5.15). As Bernard Heuvelmans later noted, this now-canonical detail is fundamentally ridiculous: "Perhaps the great number of humps in the great sea serpent is the result of the naiveté or incompetence of the illustrators. Gesner's picture of his 'Baltic' sea serpent certainly leads one to think so, for the way the animal floats almost entirely out of the water would be mechanically impossible for anything but a balloon."[68]

Alongside the fledgling Scandinavian idea of the sea serpent, belief in the monstrous sea horse continued. Surgeon Ambroise Paré took up the topic in his book *Des monstres et prodiges* (1573). Caught between the superstition of the Middle Ages and the first stirrings of proto-scientific empiricism, Paré was intensely interested in medical marvels (gruesome birth defects, in particular) but also turned his attention to black magic, celestial portents, and

Figure 5.14 The land-based mega-serpent of Olaus Magnus becomes the ocean-going sea serpent of Conrad Gesner.

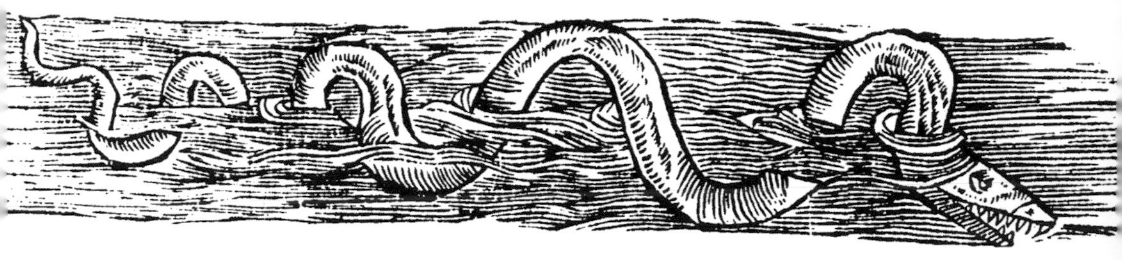

Figure 5.15 Conrad Gesner's smaller type of sea serpent, as reproduced in Edward Topsell's *Historie of Serpents*.

monsters. After discussing several sorts of mermen, Paré described the hippocamp as a living creature: "This marine monster having the head, mane and forequarters of a Horse, was seen in the Ocean sea; the picture of which was brought to Rome, to the Pope then reigning."[69] (It's interesting how these ideas cluster: after 2000 years, classical Greek Tritons and hippocamps were still swimming alongside each other in the European imagination.) It was around this time, in 1585, that mapmaker Abraham Ortelius depicted the Icelandic *hrosshvalr* as a classical hippocamp, "with manes hanging down from its neck like a horse."

As the infant scientific tradition began to wobble to its feet, so too did a related tradition that I work in: what is now called scientific skepticism, or the critical examination of popular beliefs, especially of paranormal claims.[70] Scientifically inclined writers began to ask which beliefs could be supported by empirical evidence and which were baseless superstitions. These attempts themselves often amounted to little more than reactionary scorn, and many of them proved to be as inaccurate as the claims they were critiquing. In England, physician Sir Thomas Browne's book *Pseudodoxia Epidemica: Or, Enquiries into Very Many Received Tenets and Commonly Presumed Truths* (1646; also called *Vulgar Errors*) blends religious reasoning, the new empirical spirit, and ancient wisdom to produce a wonderful hit-and-miss prototype of the skeptical genre. Browne was quite right to doubt the "very questionable" claims that chameleons consume nothing but air and that the "flesh of Peacocks corrupteth not," and he likewise nailed the falsity of the belief in hippocamps. In his day, the sea horse was cited as proof of the folkloric notion that all the animals of the land were reflected by counterparts in the sea. Browne would have none of it, retorting scornfully, "As for sea-horses which confirm this assertion; in their common descriptions they are but Grotesco deliniations which fill up empty spaces in maps, and meer pictorial inventions, not any physical shapes." He insisted that the folkloric sea horse was identical to the classical hippocamp.[71] (Browne emphasized that this mythical sea horse was distinct from the walrus and the curly-tailed fish, both of which were also called sea horse—as was, I might add, the hippopotamus in many medieval sources. I can't help but empathize with Browne's frustration: these several conflated creatures are a hassle to disentangle.)

Despite the skeptics, the belief in the monstrous sea horse persisted throughout European culture. For example, a quick-tempered Jesuit missionary named Louis Nicolas described the hippocamp as an animal living in Canada. Nicolas traveled extensively within Canada between 1664 and 1675 and produced his *Codex canadensis* (a hand-drawn, seventy-nine-page document depicting the animals and aboriginal peoples of Canada) around 1700.[72] Nicolas showed an unambiguous classical hippocamp, matter of factly drawn alongside a beaver (figure 5.16), as a "sea-horse which is seen in the fields on the banks of the Chisedek River, which empties into the Saint Lawrence."[73]

By that time, the hippocamp had endured for two millennia, evolving from a purely artistic creature of fantasy to a monster accepted as real by scholars and regular folks across Europe and beyond. And it was not done yet.

THE BIRTH OF THE SCANDINAVIAN SERPENT

The Scandinavian sea serpent was finally born at the end of the seventeenth century. It existed then only as a regional monster, but the moment of fusion had come. As the Scandinavian serpent matured, German scholar Adam Olearius recorded a third-hand sighting around 1676 ("on the Norwegian coast, he saw in the calm water a large serpent, which seen from afar, had the thickness of a wine barrel, and 25 windings"), and historian Jonas Ramus recorded another sighting in 1698.[74]

The fully modern reformulation of Olaus Magnus's shore-dragon as a gigantic, fully aquatic marine serpent with arching coils and a mane (the sea serpent form that survives as the cryptid Cadborosaurus) finally emerged as an established part of Norwegian folklore in 1694, when it was recorded in Hans Lilienskiold's hand-scripted and gorgeously painted four-volume *Speculum boreale* (*Northern Mirror*). Lilienskiold was a high government official whose writing explored the culture, geography, and wildlife of the northernmost parts of Norway.[75] The *sjø-ormen* (sea-worm), Lilienskiold wrote, "cannot be considered anything less than a bad vermin which often curves 236 feet in quiet seas, so it looks like a number of ox-heads have been thrown into the water."[76]

Figure 5.16 The hippocamp as *cheval marin* in Louis Nicolas's *Codex canadensis*. (Collection of the Gilcrease Museum, Tulsa, Oklahoma)

Crucially, Lilienskiold's sea serpent featured a "light-grey mane which goes a fathom down below the neck," thereby merging the parallel traditions of the hippocamp, the serpentine *lindorm*, and the unnamed monsters of the sea. In a remote corner of Norway, a hybrid folkloric monster had been born.

This new kind of sea serpent had what it took to become a star. All it needed was a promoter, a well-positioned publicist. That promoter was Bishop Erich Pontoppidan.

Erich Pontoppidan

Two hundred years after Olaus Magnus, the sea serpent was an established part of folk belief in Scandinavia—and only there. This point deserves boldfacing and underlining: the sea serpent was an exclusively Scandinavian creature, a cultural relative of the Icelandic *hrosshvalr* and *havhest* traditions. As Bernard Heuvelmans explained, "The sea-serpent was rarely heard of except in Scandinavia. There it was considered as a real animal and elsewhere as a Norse myth."[77]

Then everything changed. The man responsible was Erich Pontoppidan, the bishop of Bergen in Norway from 1747 to 1754. Pontoppidan was not only a highly placed clergyman, but also a reputable member of the Royal Danish Academy of Sciences. For this reason, it made a splash when his two-volume *Natural History of Norway* boldly argued for the flesh-and-blood reality of "the Mer-maid, the great Sea snake, of several hundred feet long, and the Krake[n] whose uncommon size seems to exceed belief."[78] His advocacy for these creatures caught the imagination of the world. In particular, Pontoppidan's sea serpent took off as a global popular mystery—igniting a scientific debate that would rage through the eighteenth and nineteenth centuries and continue among cryptozoologists in the twentieth and twenty-first.

When Pontoppidan turned his considerable talents to the discussion of the sea serpent, the Norwegian coast remained "the only place in Europe visited by this strange creature."[79] But there, sea serpents were already an entrenched part of popular culture. For example, Pontoppidan quoted a treatment by the foremost Norwegian poet of the previous generation, Petter Dass:

> The great Sea Snake's the subject of my verse,
> For tho' my eyes have never yet beheld him,

> Nor ever shall desire the hideous sight;
> Yet many accounts of men of truth unstain'd,
> Whose ev'ry word I firmly do believe,
> Shew it to be a very frightful monster.

Many Norwegians accepted the reality of the sea serpent, while skeptics, who "are enemies to credulity, entertain so much the greater doubt about it." Either way, by then everyone in Norway had heard of the legendary monster. As Pontoppidan reported, "I have hardly spoke with any intelligent person" who was not able to give "strong assurances of the existence of this Fish"—and to describe it.[80] That sea serpents had become a widely familiar, culturally available explanation for ambiguous or unfamiliar sights at sea should have given Pontoppidan pause, but he seems not to have realized the implications. Instead, he built his case on the aggregate of sighting reports, just as cryptozoologists do, citing "creditable and experienced fishermen, and sailors, in Norway, of which there are hundreds, who can testify that they have annually seen" sea serpents.[81] He was impressed by the way in which eyewitnesses "agree very well in the general description," adding that "others, who acknowledge that they only know it by report, or by what their neighbors have told them, still realize the same particulars."[82] This was a critical error. Pontoppidan treated cultural sea serpent lore as a *confirmation* of eyewitness testimony, rather than as a *generator* of sighting reports.

Be that as it may, some of the sightings were spectacular, leading the bishop to believe that sea serpents grew to a whopping 600 feet long. Witnesses also persuaded him that the necks of Norwegian serpents featured "a kind of mane, which looks like a parcel of sea-weeds hanging down to the water" (figure 5.17). Fifty years after Hans Lilienskiold first described the truly modern sea serpent, the *hrosshvalr-* and *havhest-*derived mane persisted as a key feature of the folk belief (and, by virtue of looking like seaweed, provided an obvious source for mistaken sightings).

This cultural DNA from the hippocamp is glaringly obvious in Pontoppidan's primary case study: the sworn testimony of naval officer Lawrence de Ferry. As a science-minded scholar, Pontoppidan was a voracious seeker of information who combed the literature, wrote letters, and questioned widely along the docks. "Last Winter, " he wrote,

> I fell by chance in conversation on this subject with captain Lawrence de Ferry . . . who said he doubted a great while, whether there was any such creature, till he had

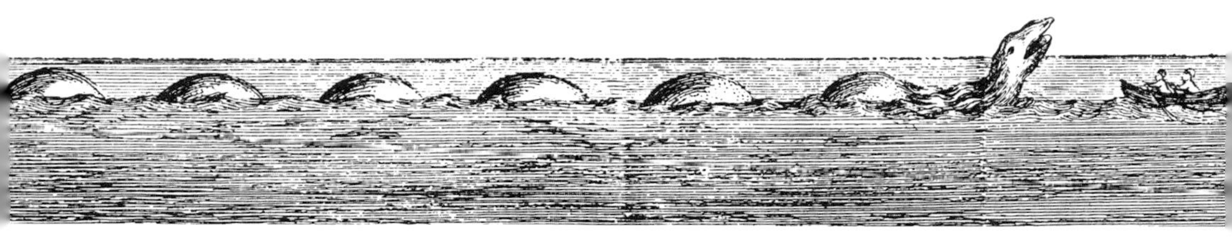

Figure 5.17 The maned *sjø-ormen* (sea-worm), drawn by Hans Strom, in Erich Pontoppidan's *Natural History of Norway*.

> the opportunity of being fully convinced, by ocular demonstration, in the year 1746. Though I had nothing material to object, still he was pleased, as a further confirmation of what he advanced, to bring before the magistrates, at a late sessions in the city of Bergen, two sea-faring men, who were with him in the boat when he shot one of these monsters.[83]

Ferry wrote a statement describing his encounter, and his men swore before the magistrates that the statement was correct. Ferry had, the statement said, been at sea on a calm day when his eight rowers told him that "there was a Sea-snake before us." He ordered the men to intercept the creature and "took my gun, that was ready charged, and fired at it." The animal dived, leaving blood in its wake. "The head of this snake," Ferry swore, "which it held more than two feet above the surface of the water, resembled a horse. . . . It had . . . a long white mane that hung down from the neck to the surface of the water. Beside the head and neck, we saw seven or eight folds or coils of this Snake, which was very thick . . . there was about [6 feet] distance between each fold."

What did Ferry and his men see? It is interesting to speculate, as did skeptical naturalist Henry Lee: "The supposed coils of the serpent's body present exactly the appearance of eight porpoises following each other in a line. This is a well-known habit of some of the smaller cetecea."[84] (I have seen this effect, and it is shockingly compelling. More likely suspects, however, would be the seals common to the Norwegian coast.) In any event, "the horse-like features of the sea-serpent's head were common knowledge," as folklorist Michel Meurger notes. "Therefore, Ferry's assertion is proof only of a traditional interpretation of a sighting he had six years prior to his official statement. This is too long a time to have a fresh recollection but more than enough time to blend memories with collective stereotypes."[85]

The real importance of Ferry's story is its subsequent role as a template. This widely publicized, often-translated, and precedent-setting sighting was the first credible eyewitness account to reach beyond Scandinavia, and it helped to lock in the canonical image of the sea serpent for the English-speaking world: multiple humps or coils, combined with the head and mane of a horse.

In the centuries since the publication of *The Natural History of Norway*, Pontoppidan has often been criticized for his credulity. "Indeed," wrote Heuvelmans, "the Bishop of Bergen was treated as a liar as arrant as Münchhausen."[86] Other critics have been kinder. "The Norwegian Bishop," granted Lee, "was a conscientious and painstaking investigator, and the tone of his writings is neither that of an intentional deceiver nor of an incautious dupe. He diligently endeavoured to separate the truth from the cloud of error and fiction by which it was obscured; and in this he was to a great extent successful."[87] I have argued that quite a bit of Pontoppidan's work can be regarded as early "scientific skepticism."[88] He advocated for science literacy, pointedly critiquing his clerical brethren for "supercilious neglect" of knowledge of the physical world. Pontoppidan even went out of his way to investigate and correct popular falsehoods, such as the idea that bottomless whirlpools penetrate through the entire Earth and the already ancient legend that the Barnacle Goose hatches out of trees or rotten wood.

Pontoppidan was no slouch, so it is not surprising that he zeroed in on a key problem for his sea serpent: its cultural specificity:

> Before I leave this subject, it may be proper to answer a question that may be put by some people, namely, what reason can be assigned why this Snake of such extraordinary size, &c. should be found in the North sea only? For, according to all accounts from seafaring people, it has never been seen anywhere else. Those who have sailed in other seas in different parts of the globe, have, in their journals, taken particular notice of other Sea-monsters, but not one of them mentions this.[89]

Many societies share the ocean, but only one encounters sea serpents. That is such a screaming, flashing, gigantic red flag that readers may be forgiven if they find Pontoppidan's answer to it unconvincing. He simply asserted that "when the thing is confirmed by unquestionable evidencc, and is found to be true," then armchair objections are beside the point. Moreover, Pontoppidan argued that "this objection requires no other answer, than that the Lord of nature disposes of the abodes of his various creatures, in different parts of the globe, according to his wise purposes and designs, the reason of his proceedings cannot, ought not to be comprehended by us." This dodge seems to me like the clergyman in Pontoppidan inappropriately pulling rank on the scientist.

THE GREAT SEA SERPENT

Erich Pontoppidan's book was the critical turning point for the sea serpent: a Scandinavian phenomenon that burst forth onto the world stage to become an enduring part of popular culture. (I can't help but think of the Swedish pop band ABBA.) To understand what a breakthrough this was, consider Bernard Heuvelmans's overview of the sighting database (figure 5.18). He estimated that in the centuries before the publication of Pontoppidan's *Natural History of Norway*, there were only nine dated and documented sightings of sea serpents in all of history. After Pontoppidan? "The number increases to twenty-three between 1751 and 1800 but they do not become really frequent until the first half of the nineteenth century: 166 from 1801 to 1850 and 149 from 1851 to 1900. The rate did not drop in the twentieth century, for there were 194 between 1901 and 1950." Heuvelmans noted that the trend was continuing in the 1960s, when he compiled the statistics. Assuming that one-third of sightings record misidentification errors or hoaxes, he estimated that "this makes 100 sightings every fifty years, an average of two a year, right up to the present."[90]

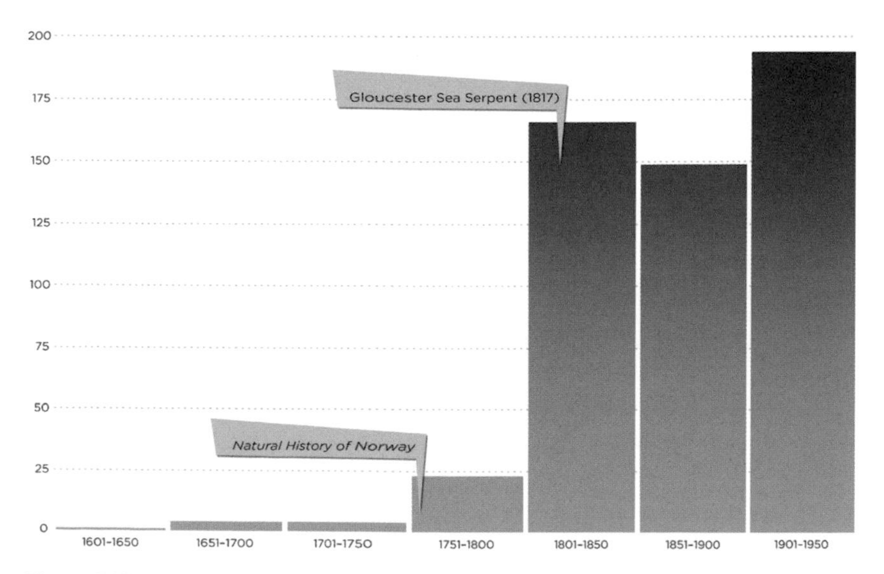

Figure 5.18 Sightings of sea serpents, 1600–1950, by fifty-year intervals, according to Bernard Heuvelmans.

This strikes me as exceptionally funny in two respects. First, Heuvelmans's assumption that two-thirds of claimed sightings are genuine was surely on the optimistic side; after all, it's not known that there has been even one genuine sea serpent sighting—ever. Second, the pattern he described is clearly that of a pop-culture phenomenon. By his own admission, sea serpents basically did not exist before Pontoppidan's book made them famous; since then, they have been seen regularly and increasingly. *The Natural History of Norway* launched the sea serpent to stardom; from there, it entered the self-sustaining fame cycle that perpetuates UFOs, Bigfoot, and other popular mysteries.

The Stronsay Beast

For as long as humans have walked beside the sea, they have marveled at the strange things that sometimes wash ashore. "During the rule of Tiberius, in an island off the coast of the province of Lyons the receding ocean tide left more than 300 monsters at the same time, of marvelous variety and size," recorded the Roman natural historian Pliny the Elder.[91] Similarly, a pamphlet published in 1674 describes a "Strange Monster or Wonderful Fish" that washed up on a beach in Ireland (figure 5.19). The description and illustration clearly identify the animal as a large squid, 19 feet in length (including the arms) "and in Bulk or Bigness of Body somewhat larger than a Horse." (The anonymous author of the pamphlet cheekily added that "some Zealots hearing of a strange Creature . . . took it for the Apocaliptical Beast, and fancied the Pope was landed in person.")[92]

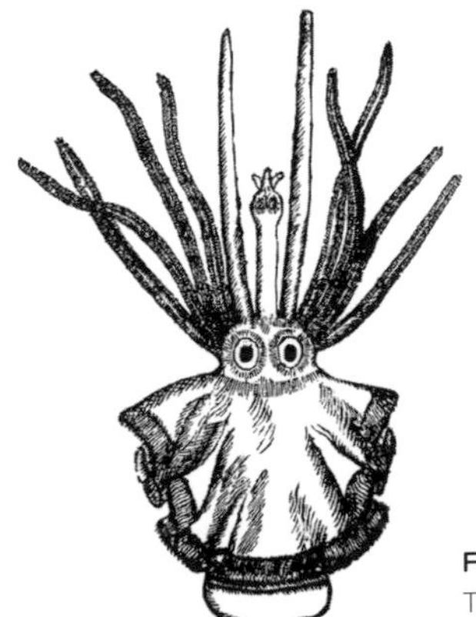

Figure 5.19
The "Strange Monster or Wonderful Fish" that washed ashore in Ireland in October 1673.

Figure 5.20 The sketch of the Stronsay Beast drawn by an eyewitness. (Redrawn from *Memoirs of the Wernerian Natural History Society* 1 [1811])

With the global publicity for Erich Pontoppidan's sea serpent, strange carcasses took on new meaning. When a huge, vaguely whale-like animal washed up on Stronsa (now called Stronsay, an island off the northern tip of Scotland) in 1808, it was immediately identified as "a sea-snake with a mane like a horse"—and, indeed, the case seemed clear. Reportedly measured at a bulky yet sinuous 55 feet in length, with a long neck and a mane, there seemed little room to doubt that this was a creature unknown to science. Multiple sworn statements and an eyewitness sketch seemed to clinch the case (figure 5.20). And then Scottish naturalist John Barclay stepped in to confirm that it "appeared to be the Soe-Ormen described above half a century ago, by Pontoppidan, in his *Natural History of Norway*," assigning it to a brand-new genus and species: *Halsydrus pontoppidani* (Pontoppidan's sea-snake).[93]

Alas, nothing was as it seemed. Today, Barclay's *Halsydrus pontoppidani* is remembered alongside *Nessiteras rhombopteryx* (proposed for the Loch Ness monster), *Hydrarchos sillimani* (an alleged fossil sea serpent), and *Cadborosaurus willsi* as a premature taxonomic misstep. The problem? The creature rotting on the Stronsay beach, scavenged by wheeling gulls, was a basking shark. The animal's skull, several vertebrae, and other samples were sent to Everard Home, a surgeon and leading scientist who had recently received the Royal Society of London's prestigious Copley Medal. With these specimens, Home was able to firmly identify the animal as a basking shark and to ascertain that the drawing and descriptions were wildly inaccurate distortions of the anatomy of the carcass. So stark was the discrepancy between the eyewitness testimony and the forensic evidence that the Stronsay Beast has lived on as a cautionary tale. "There can be little doubt that this creature was, actually, an enormous basking-shark, partly decomposed," wrote sea serpent advocate Rupert Gould, "but the original reports are so curious, and the accepted explanation so much at variance with them, that the case deserves more than a cursory mention, if only as an instance of how misleading it is possible for honest testimony to be."[94]

Figure 5.21 It is typical for a decaying basking shark to resemble a plesiosaur. (Illustration by Daniel Loxton)

But was Home's analysis accurate? Barclay did not think so. He published a rebuttal, objecting that the head of a basking shark is 5 feet across, while the skull of the Stronsay Beast was only 7 inches wide. But Barclay had fallen into a trap that has misled people ever since. The huge, gentle basking sharks—like all sharks—have skeletons made of cartilage, rather than bone. Moreover, basking sharks are filter feeders (similar to humpback and other baleen whales). As a result, the pattern of decay in basking shark carcasses can be counterintuitive (figure 5.21). The huge jaws fall off quickly, leaving a surprisingly tiny skull perched at the end of a long spine—looking for all the world like a rotten sea serpent or long-necked plesiosaur. This is what happened to the Stronsay carcass. In the two centuries since the carcass washed ashore, the vertebrae (preserved in museum collections) have been reexamined more than once. Home's findings have been confirmed each time. In 1933, for example, James Ritchie (then professor of natural history at the University of Aberdeen and later president of the Royal Society of Edinburgh) wrote in the *Times* of London about his own conclusions after reexamining the vertebrae of the Stronsay Beast at the Royal Scottish Museum:

> They tell their own tale to scientific examination. They are obviously part of the backbone of a gristly fish; by no possibility could they belong to a plesiosaur, or to any reptile or amphibian, nor could they be part of a whale or even of a bony fish. The texture of the segments of the vertebrae, their size, and curious pillared structure, agree exactly with those of the basking shark, a monster which may be 40 ft. long and which occasionally appears in British waters. . . . The sea-serpent of

> Stronsay, which a century and a quarter ago raised so great a commotion in the scientific world, has fallen from its unique estate, but it remains a not-to-be-forgotten memorial to the credulity of the inexperienced and of the scientists who built upon so shaky a foundation.[95]

The Stronsay Beast may never have been quite forgotten (at least within the niche sea serpent literature), but its lesson has never been learned. Again and again, people have fallen for the same grisly illusion. High-profile cases continue to arise, with many "sea serpent" or "plesiosaur" carcasses making headlines during the twentieth century—only to prove, time and again, to be basking sharks. A mere three months after Ritchie confirmed that the Stronsay Beast was really a shark, headlines around the world gleefully announced that a similar unidentified carcass (what is now called a "globster" in cryptozoological parlance) had washed up in Cherbourg, France. The *Los Angeles Times* declared it the "First Genuine Sea Monster Captured," while the front page of the *New York Times* trumpeted that "France Has Sea Monster."[96] After several stories on the topic, the *New York Times* made the inevitable retraction: "After a careful examination of the extensive remains of the 'sea monster' found on the shore near Cherbourg, Professor [Georges] Petit and his colleagues of the French Museum of Natural History have definitely concluded that this fish is a basking shark."[97] This was not the last time this clichéd narrative would play out; bizarrely, it was not even the last time it played out in that year. In November 1934, newspapers from Scotland to Chicago trumpeted the discovery of a 30-foot sea serpent carcass on Henry Island (in British Columbia, Canada). The *New York Times* got right back on that sea horse, reporting that this "strange sea monster" had a "head resembling that of a horse" and no bones except the vertebrae.[98] Just four days later, the *Times* revealed, predictably, that the director of the Pacific Biological Station in Nanaimo, British Columbia, had positively identified the remains as those of a basking shark.[99]

These misidentifications are among cryptozoology's silliest banalities, but they are not so passé that they have stopped. The most famous case may be that of the Japanese fishing trawler *Zuiyo Maru*, whose crew in 1977 hauled aboard a badly decomposed carcass from the Pacific Ocean off New Zealand (figure 5.22). Repulsed by the "overpowering stench and the unpleasant fatty liquids oozing onto the deck," the fishermen photographed the carcass, took tissue samples, and then dumped it overboard.[100] The loss of the specimen

Figure 5.22 The carcass of a basking shark aboard the Japanese fishing trawler *Zuiyo Maru* in 1977. (Reproduced by permission of Fortean Picture Library, Ruthin, Wales)

did not deter the *Los Angeles Times* from excitedly speculating that this reeking, slimy mess could be a relict plesiosaur: "a huge reptile thought to have died out 100 million years ago."[101] But it was not to be. The tissue samples nailed down what was already obvious to anyone with a sense of history: it was a basking shark.[102] Again.

The full-color photographs taken by a member of the crew of the *Zuiyo Maru* fired my imagination as a kid, and clearly I am not alone. The story has never completely faded. I have in my office a fantastic recent Japanese toy that lovingly re-creates the carcass in miniature—a lump of plastic painted to resemble a heap of rotting flesh. But the currency of the story goes beyond cryptozoology-themed collectibles: even today, some believers argue that the carcass truly was that of an extant plesiosaur. In a fairly typical overlap between scientific creationism (the attempt to confirm the literal truth of the account of creation in Genesis—or at least to disprove evolution—through purportedly scientific means) and cryptozoology, an article in the journal of the Creation Research Society argues that "the interpretation of the *Zuiyo Maru* cryptid as a shark is false" and concludes that a "Sauropterygia [plesiosaur or related marine reptile] identification remains viable."[103] Cryptozoologically inclined creationists seem to hope that locating a plesiosaur or another so-called living fossil would pose problems for evolutionary theory, but it is hard to see why a plesiosaur would be more disruptive than the continuing existence of crocodiles and sharks, which first appeared about 220 million and 400 million years ago, respectively. Regardless, keen interest from creationists (and from wide-eyed kids) have ensured that the survival of a relict population of plesiosaurs has remained among cryptozoology's fondest dreams. But how did plesiosaurs emerge as a dominant cryptid, and how do they relate to the hippocamp-based Great Sea Serpent?

The "Plesiosaur Hypothesis"

In their nineteenth-century heyday, sea serpents enjoyed more mainstream scientific respectability than any other cryptid has ever achieved. This high-profile scientific attention was based on a truly spectacular coincidence: just as the sea serpent was reaching the peak of its popularity, new fossil discoveries taught scientists that gigantic marine reptiles genuinely used to exist. In primeval seas, sinuous plesiosaurs and ichthyosaurs swam beside dinosaur-dominated shores (figure 5.23). As the world reflected on a recent rash of sea serpent sightings on the East Coast of North America (especially in and around Gloucester, Massachusetts, in 1817), astonishing fossil finds shone a warm light of plausibility on the modern serpent legend—a spotlight under which the sea serpent thrived.

It is worth taking a moment to picture the English scientific scene as it existed in this pioneering, pre-Darwinian period, vividly evoked by paleontologist Christopher McGowan. Reflecting on the romantic appeal of nineteenth-century geology books, McGowan wrote, "They enchanted me with their accounts of antediluvian creatures. And they spirited me away to their bygone world of Regency and Victorian England. I could picture myself—silk hat and frock coat—perambulating the foreshore of the Dorset coast, searching the cliffs for fossils. Or cabbing through London's horse-drawn traffic to attend a meeting of gentlemen geologists."[104]

Top hats. Meetings by flickering gaslight. Snifters of brandy. In the halls of Oxford, cutting-edge geological lectures still taught that the features of Earth had been carved by Noah's flood. But the world was changing fast: the Age of Steam was dawning, and the brave new science of geology was turning the understanding of the history of Earth on its head. It was becoming clear that

Figure 5.23 Many nineteenth-century artists and writers depicted epic battles between the newly discovered ichthyosaurs and plesiosaurs.

the planet was very, very old—and that life on Earth had been very different in the distant past. In France, the great anatomist Georges Cuvier convincingly argued that some creatures known from fossils (such as mammoths) had vanished entirely from Earth, but other scientists still argued that extinction was incompatible with God's perfection. Perhaps even mammoths had survived in some distant land?[105] And if remnant populations of fossil creatures had hung on in Earth's inaccessible places, they might be very much unchanged, for no convincing mechanism for the "transmutation" (or evolution) of species was yet known.

Onto that stage stepped Mary Anning, a working-class Englishwoman whose sharp eye and sheer physical stamina made her (in an era of gentleman geology) perhaps the greatest fossil collector of her time. She is a romantic figure in the history of science, although she must not have considered herself in that light. I can't help but picture her, walking along seaside cliffs near Lyme Regis, in Dorset, on damp, frigid mornings; kneeling in her long skirts; swinging her rock hammer with her calloused hands and bloodied knuckles—and all for the fossils that she would sell to survive, letting moneyed men take the credit of discovery. Yet, the large marine reptiles whose fossils she found triggered a stunning upheaval of human understanding of the past. These were creatures utterly unlike anything known. Ichthyosaurs were reptiles with shark-shaped bodies, while plesiosaurs were even weirder: long-necked reptiles that propelled themselves through the water with four paddle-like flippers. Their fossilized remains opened a window onto a lost world, paving the way for Charles Darwin—and incidentally remixing the popular legend of the sea serpent.

By nineteenth-century standards, the ink was hardly dry on newspaper reports of the sea serpent sightings around Gloucester in 1817 when ichthyosaurs were shown to be reptiles in 1821. The first nearly complete plesiosaur skeleton was described in 1824 in a presentation before the Geological Society of London—at the same meeting that announced the first dinosaur genus name: *Megalosaurus*.[106] Almost immediately, naturalists made the connection to sea serpents. In 1827, botanist Sir William Hooker employed the newly discovered plesiosaur and dinosaur fossils as a rhetorical device, asking why it should be, when "the recent discoveries of the Plesiosaurus and Megalosaurus have made demands upon our powers of credence far greater than the serpent, the descriptions of the latter animal have received very little trust, and even much ridicule and contempt." Hooker went on to argue that, in light of eyewitness testimony, sea serpents may now "be assumed as a sober fact in Natural History. . . . We cannot suppose, that the most ultra-sceptical can now continue to doubt with regard to facts attested by such highly respectable witnesses."[107]

A much more direct argument was advanced in 1833 by geologist Robert Bakewell, who matter-of-factly stated in his textbook *Introduction to Geology*, "I am inclined to believe, that the ichthyosaurus, or some species of a similar genus, is still existing in the present seas." Bakewell went on, "I remember one of the most particular descriptions of the sea serpent was given by an American captain . . . ; it had paddles somewhat like a turtle, and enormous jaws like the crocodile. This description certainly approaches to, or may be said to correspond with, the ichthyosaurus, of which animal the captain had probably never heard."[108]

In a footnote to Bakewell's speculation, chemist Benjamin Silliman, a professor at Yale, made the connection that would, a century later, give form to the Loch Ness monster: "Mr. Bakewell's ingenious conjecture, that it may be a Saurian, agrees, however, much better with the supposition that it is a Plesiosaurus than an Ichthyosaurus, as the short neck of the latter does not correspond with the ordinary appearance of the sea serpent."[109] The image of the sea serpent as a relict plesiosaur was swiftly taken up by other scientists. In 1835, John Ruggles Cotting noted, "The Sea-serpent, which has frequently visited the waters of New England, is supposed to belong to the genus Plesiosaurus. Its existence has been so often attested by thousands of competent witnesses, that its identity is no longer problematical."[110]

Living prehistoric monsters. How cool would that be? At a time when the history of life on Earth was very much up for grabs (Darwin's *On the Origin of Species* was not published until 1859), the possibility that plesiosaurs could (just maybe!) still exist had a powerful fascination for laypeople, journalists, and scientists alike. "The beliefs of the sea-serpent's supporters were inspired by the fashion of the day," noted Bernard Heuvelmans. "The discovery of bones of the great saurians in Britain in the first half of the nineteenth century, and in America in the second half, had fired everyone's imagination."[111] Of course, this popular interest had an impact on the form and frequency of reports of sightings of sea monsters. "It was not mere coincidence that as paleontologists began dredging up plesiosaurs and other relics from the past a dramatic increase in sightings of sea serpents also occurred," deadpans science writer Sherrie Lynne Lyons.[112]

Plesiosaurs were, for example, proposed as an explanation for a sighting in 1848 by officers of the British frigate HMS *Daedalus* (figure 5.24). This case had (and still has) a very high profile, thanks to the esteem granted to naval officers and to the crisply worded report to the Admiralty submitted by Captain Peter M'Quhae. (The order of events should be noted at the outset, however. Oddly enough, the log of the *Daedalus* failed to record any hint of a sea serpent.[113] When the *Times* of London nonetheless publicized an alleged sighting from almost three months earlier, the Admiralty understandably demanded details.[114] Only at that point did M'Quhae report the alleged sighting to the Admiralty, "in reply to your letter . . . requiring information as to the truth of a statement published in *The Times* newspaper, of a sea-serpent of extraordinary dimensions having been seen from Her Majesty's ship *Daedalus*.")[115]

M'Quhae described a relatively traditional "enormous serpent," complete with "something like the mane of a horse," that, with head raised, passed the ship at a moderate distance (close enough to recognize a human acquaintance, according to M'Quhae, or about 200 yards at its closest approach, according to fellow witness Lieutenant Edgar Drummond).[116] The *Daedalus* case remains unsolved, but the usual three possibilities exist: a hoax, a mis-

Figure 5.24 The creature reported by officers of the HMS *Daedalus* in 1848.

take, or a genuinely unknown animal. We may regard as naive the argument that "M'Quhae's honesty was not—and, obviously, could not be—seriously called into question," as Rupert Gould put it.[117] (While a hoax may have struck fans of the Royal Navy as "frankly unthinkable," skeptics know that hoaxes are uncomfortably common and cross all boundaries of class and occupation. Indeed, another London newspaper published a different sea serpent report less than three weeks after M'Quhae's, only to learn to its embarrassment that the story was a complete fabrication.)[118] Still, most commentators have taken the report of the officers of the *Daedalus* as sincere, in which case the question becomes: What did those men see? Swiftly the suggestion appeared in a letter to the *Times*: "[T]he enormous reptile in question was allied to the gigantic Saurians, hitherto believed only to exist in the fossil state, and, among them, to the Plesiosaurus."[119]

The *Daedalus* case became an enduring part of the cryptozoological canon after it attracted a detailed public critique from one of the world's weightiest scientific authorities: Richard Owen, of the British Museum, who had coined the term "dinosaur" just six years earlier. Owen's reputation in his field was overpowering ("[H]is opinion on a zoological question has almost the force of an axiom," as one of his contemporaries put it),[120] although he was known personally as an unpleasant character. His unrelenting campaign against Darwinian evolution and its supporters, which included a venomous anonymous review of *On the Origin of Species*,[121] led the normally genial Darwin to admit, "I used to be ashamed of hating him so much, but now I will carefully cherish my hatred & contempt to the last days of my life."[122] Nonetheless, Owen was an awfully sharp guy, and his critique of the *Daedalus* case stands as one of the clearest skeptical statements on sea serpent mythology. Owen argued, plausibly enough, that M'Quhae may have misidentified an elephant seal or a sea lion. These massive mammals would match fairly closely with M'Quhae's description of the sea serpent and the drawings he endorsed, but it remains a purely speculative explanation—parsimonious, but unproved. And that is where the mystery remains: a deadlock between Owen's argument that M'Quhae did not know what he was looking at and M'Quhae's retort that Owen did not know what he was talking about: "I now assert—neither was it a common seal nor a sea-elephant, its great length and its totally differing physiognomy precluding the possibility."[123]

Owen knew that his elephant seal explanation was speculative, but he was sure about one thing: there was no way a living plesiosaur had casually glided past a Royal Navy frigate. His caustic response remains relevant to cryptozoology:

> Now, on weighing the question, whether creatures meriting the name of "great sea-serpent" do exist, or whether any of the gigantic marine saurians of the secondary deposits may have continued to live up to the present time, it seems to me less probable that no part of the carcass of such reptiles should have ever been discovered in a recent or unfossilized state, than that men should have been deceived by a cursory view of a partly submerged and rapidly moving animal, which might only be strange to themselves. In other words, I regard the negative evidence, from the utter absence of any of the recent remains of great sea-serpents, krakens, or *Enaliosauria*, as stronger against their actual existence than the positive statements which have hitherto weighed with the public mind in favor of their existence. A larger body of evidence, from eye-witnesses, might be got together in proof of ghosts than of the sea-serpent.[124]

Other scientific superstars were more easily seduced by the possibility of surviving plesiosaurs, including the great geologist and paleontologist Louis Agassiz of Harvard University. Agassiz's scientific legacy includes the distinction of being among the first to propose a "glacial" period (ice age) in Earth's history. Less well known is his advocacy for the sea serpent as a prehistoric survivor. Agassiz touched on the topic in a lecture he gave in March 1849:

> I have asked myself . . . whether there is not such an animal as the Sea-Serpent. There are many who will doubt the existence of such a creature until it can be

> brought under the dissecting-knife; but it has been seen by so many on whom we may rely, that it is wrong to doubt any longer. The truth is, however, that if a naturalist had to sketch the outlines of an Ichthyosaurus or Plesiosaurus from the remains we have of them, he would make a drawing very similar to the Sea-Serpent as it has been described. . . . I still consider it probable that it will be the good fortune of some person on the coast of Norway or North America to find a living representative of this type of reptile, which is thought to have died out.[125]

Agassiz amplified this opinion in a letter written on June 15, 1849, concluding that while he was "not at all disposed to endorse all the reports" of sea serpent sightings, "from the evidence I have received I can no longer doubt the existence of some large marine reptile, allied to Ichthyosaurus and Plesiosaurus, yet unknown to naturalists."[126]

An example of this new type of sighting, of an overtly plesiosaur-like creature, is that reported from the HMS *Fly*, described by Edward Newman, editor of the journal *Zoologist*, in 1849:

> Captain the Hon. George Hope states, that when in H.M.S. Fly, in the gulf of California, the sea being perfectly calm and transparent, he saw at the bottom a large marine animal, with the head and general figure of the alligator, except that the neck was much longer, and that instead of legs the creature had four large flappers, somewhat like those of turtles . . . the creature was distinctly visible, and all its movements could be observed with ease: it appeared to be pursuing its prey at the bottom of the sea.[127]

In 1959, cryptozoology pioneer Willy Ley characterized the *Fly* report as "very significant, in my opinion,"[128] but it is hard to see why he would have thought so. As Ley noted, the case was "deplorably lacking in detail," but it was worse than that: the story was unsubstantiated hearsay. As Newman put it, "Captain Hope made this relation in company, and as a matter of conversation: when I *heard it from the gentleman to whom it was narrated* [emphasis added], I inquired." That is, Newman heard from an unnamed source that Hope had mentioned socially that he once saw a weird-looking animal. Not a lot to hang your hat on—but, oddly enough, Newman went on to argue that the *Fly* case "appears to me in all respects the most interesting Natural-History fact of the present century, completely overturning as it does some of the most favourite and fashionable hypotheses of geological science."[129] Despite its weaknesses, the report remained influential because it seemed to confirm a hypothesis that many people wished to be true.

Wishful thinking did not, however, stop one later supporter of the "plesiosaur hypothesis," Rupert Gould, from pointing out the flaws in the story that Newman and others had ignored: "If the matter were so important, why (one wonders) did not Newman get into touch with Captain Hope and obtain a first-hand version of his experience (preferably in writing): with, if possible, some confirmatory evidence? As it stands, the most remarkable feature of the story is the absence of such elementary data as the time, place, and date of the occurrence."[130] A century after the sighting, Gould made the effort to track down the ship's log and the details of Hope's career. When a lieutenant, Hope had indeed served on the *Fly*, ten years before Newman heard the monster tale at second hand—but there is no record of any monster sighting. As Gould summed up, "It does not appear that the creature was seen by anyone except Hope" (if it was seen at all); thus the tale "cannot be regarded as carrying very much weight." Indeed.

The *Fly* tale does show both the influence of the new fossil discoveries, introducing a second culturally available major template for the sea serpent—a competitor to the classic hippocamp-based form—and the expansion of the geographic range of alleged sea serpent habitat. No longer constrained to the North Atlantic, let alone to the coasts of Norway, by 1849 the sea serpent had successfully established itself in the congenial waters off California.

The idea that sea serpents were members of a relict population of plesiosaurs continued to gain traction, with acclaimed science writer Philip Henry Gosse making a high-profile pitch for the idea. "I express my own confident persuasion," he concluded, "that there exists some oceanic animal of immense proportions, which has not yet been received into the category of scientific zoology; and my strong opinion, that it possesses close affinities with the fossil *Enaliosauria* [extinct marine reptiles, including plesiosaurs]."[131] This was hardly the most extraordinary proposal of his career. Gosse is best remembered for his (now widely mocked) book *Omphalos: An Attempt to Untie the Geological Knot*, in which he attempted to unify the new findings of geology with his own biblical literalism by supposing that God had created the living world with built-in evidence of a past that had never existed. (Gosse was correct, in one sense: such fictive prior history is implicit in the whole concept of instantaneous creation.)[132] With this creationist argument as his unfortunate legacy, it is easy to forget that Gosse stood, in his own lifetime, among the premier writers of popular science. Stephen Jay Gould described him as "the David Attenborough of his day, Britain's finest popular narrator of nature's fascination."[133] Gosse was especially respected on topics related to marine life. He was among the first to experiment with saltwater "aquariums"—a word he popularized with his books on the topic, inspiring both the household hobby and the tourist industry of marine-animal exhibition parks.[134]

While noting the need for skepticism, especially about eyewitness testimony ("every man of science must have met with numberless cases in which statements egregiously false have been made to him in the most perfect good faith; his informant implicitly believing that he was simply telling what he had seen with his own eyes"),[135] Gosse was persuaded that sightings of sea serpents established their existence. We need not rely, he felt, purely on speculation about the possibility of surviving prehistoric animals. "On this point," he wrote, "actual testimony exists, to which I cannot but attach a very great value." What testimony? That same old bit of unsupported hearsay, the report of George Hope of the *Fly*!

Gosse addressed Richard Owen's objections to the "plesiosaur hypothesis" (in short, that there are no plesiosaur carcasses today or any plesiosaur bones in the fossil record after the end of the reign of the dinosaurs). Like modern Bigfooters, Gosse argued that the absence of modern remains is consistent with the existence of modern plesiosaurs. (Surely their carcasses would sink. And even if one of these rare animals were to wash up on a remote shore, who would recognize it?) He also maintained that the negative evidence of the fossil record was hardly a deal breaker. Those who know Gosse's name only in association with disreputable creationist arguments may be surprised that he approvingly cited Charles Darwin:[136]

> It must not be forgotten, as Mr. Darwin has ably insisted, that the specimens we possess of fossil organisms are very far indeed from being a complete series. They are rather fragments accidentally preserved, by favouring circumstances, in an almost total wreck. The *Enaliosauria*, particularly abundant in the secondary epoch, may have become sufficiently scarce in the tertiary to have no representative in these preserved fragmentary collections, and yet not have been absolutely extinct.[137]

Even more surprising, perhaps, Gosse rested his sea serpent case on an evolutionary argument!

> Not that I would identify the animals seen with the actual *Plesiosaurs* of the lias. None of them yet discovered appear to exceed thirty-five feet in length, which is scarcely half sufficient to meet the exigencies of the case. I should not look for any species, scarcely even any genus, to be perpetuated from the oolitic period to the

> present. Admitting the actual continuation of the order *Enaliosauria*, it would be, I think, quite in conformity with general analogy to find important generic modifications, probably combining some salient features of several extinct forms. Thus the little known *Pliosaur* had many of the peculiarities of the *Plesiosaur*, without its extraordinarily elongated neck, while it vastly exceeded it in dimensions. What if the existing form should be essentially a *Plesiosaur*, with the colossal magnitude of a *Pliosaur*?[138]

With popular-science writers like Gosse behind it, it is not surprising that popular entertainment also embraced the idea of the sea serpent. As do their present-day descendants, early science-fiction writers looked to dramatic scientific discoveries for storytelling possibilities—and plesiosaurs and ichthyosaurs were as dramatic as they come. In Jules Verne's novel *Journey to the Center of the Earth* (1864), for example, protagonists clinging to a raft on a secret underground sea witness a battle between "the most formidable of antediluvian reptiles, the ichthyosaurus," and "a serpent disguised by a shell like a turtle's, the terrible enemy of the first, the plesiosaurus."[139] Modern human characters likewise see or interact with surviving plesiosaurs in Arthur Conan Doyle's novel *The Lost World* (1912) and Edgar Rice Burroughs's tale *The Land That Time Forgot* (1918), in which the crew of a German U-boat finds plesiosaur steaks delicious.[140]

The theme was further advanced by Rupert Gould, who argued that—despite his criticism of Hope's report from the *Fly*—the eyewitness evidence supports as many as three types of undiscovered species. Foremost among them is a gigantic creature "much resembling in outline and structure the *Plesiosaurus* of Mesozoic times." Echoing Gosse, Gould invoked evolution to make his case: "I do not suggest that the last-named is actually a Plesiosaurus, but that it is either one of its descendants or has evolved along similar lines. In either case, I suggest that there is little doubt that it has much the same characteristics—a slender neck and tail and a comparatively large body with propelling flippers."[141]

The "plesiosaur hypothesis" remained popular throughout the twentieth century, coexisting with the hippocamp-like sea serpent. Dramatically different though these two major types of sea serpents may be, they morph easily into each other throughout the world's seas and monster lakes—with witnesses dipping liberally into both culturally available templates on a case-by-case basis. Consequently, reports and reconstructions offer no end of hybridized monsters, including plesiosaurs with manes, plesiosaur-headed serpents, and plesiosaurs with an improbable series of coil-mimicking humps on their backs. The Loch Ness monster, for example, comes in all these forms and more (including straight-up traditional sea serpents), despite Nessie's status as poster girl for the plesiosaur-survivor mythology.

Hydrarchos: 114-Foot Fossil Sea Serpent?

The discoveries of fossilized plesiosaurs and other extinct marine reptiles in the nineteenth century led to an enduring new subcategory of plesiosaur-like sea monsters and yet also boosted the profile of the traditional hippocamp-style sea serpent. Of all the dramatic fossil evidence to support the sea serpent's apparent plausibility, nothing was quite so stupendous—or quite so on the nose—as the fossil reconstruction of a creature dubbed *Hydrarchos sillimani* (figure 5.25). According to its discoverer, Albert Koch, it was nothing less than the bones of the sea serpent itself. Reconstructed in a hippocamp-like posture—massive head rearing high above fore-flippers, body arching, and tail curling—it could hardly be anything else. Or could it?

This intensely hyped skeleton of a "GIGANTIC FOSSIL REPTILE 114 Feet in Length" (as the advertisement screamed) was exhibited in New York

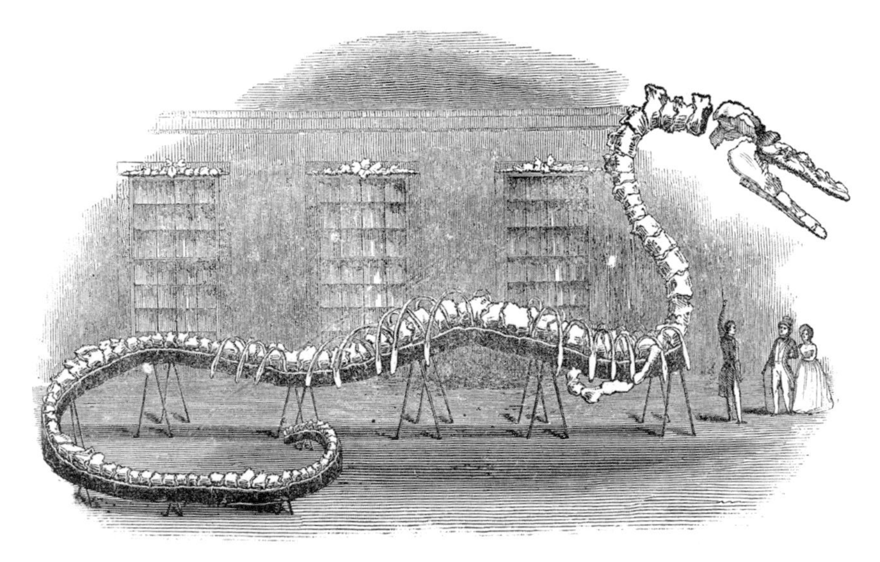

Figure 5.25 The chimera fossil assemblage that Albert Koch dubbed *Hydrarchos sillimani*.

City in 1845, in the Apollo Rooms on lower Broadway (a large saloon that had hosted the debut of the New York Philharmonic three years earlier). Crowds flocked to pay 25 cents (children half price!) to see the fossil sea serpent.[142] This amounted to a tidy fortune. Consider that New Yorkers were then paying 4 cents for a quart of milk.[143] The press was as gobsmacked as the crowds. The *New York Daily Tribune* hailed it as "indisputably the greatest wonder that ever was brought to light out of the strata which form the crust of our globe"—and a lasting monument to American greatness.[144] Alas, not all was as it seemed.

Commercial fossil collector "Doctor" Albert Koch was a colorful figure: a successful and important fossil hunter, a shameless self-promoter, and (most of all) a hustler of the first order. His previous major fossil find, likewise profitably toured and hyped, was presented to the world as *Missourium*, "the Missouri Leviathan"—a web-footed, aquatic predator that had once lurked in the rivers and lakes of Missouri. According to Koch's analysis of his newly unearthed creature, *Missourium* was remarkable for its hard armor (on which no "barbed iron, harpoon, or spear, would make any impression") and its "enormous magnitude, ferocity, and strength, as well as fleetness in swimming." Its "fleetness in swimming" may well have been overstated, however, for despite Koch's colorful assertions (and the several extra ribs and vertebrae he had added to his specimen), *Missourium* was in fact an American mastodon. (Nonetheless, it was a very nice American mastodon. The British Museum knew exactly what it was buying when it snapped up the skeleton in 1843 for a whopping £1,300.[145] Anatomist and sea serpent skeptic Richard Owen promptly restored the skeleton to its rightful anatomy. It still stands in the Natural History Museum in London, properly identified as a mastodon.)[146]

When Koch revealed as his next big trick that he had personally discovered definitive proof of a 114-foot sea serpent, it drew crowds—but also scrutiny. *Hydrarchos* seemed too good to be true. And, indeed, it was.

Jeffries Wyman, a professor of anatomy at Harvard, examined the massive skeleton during its New York exhibition and saw immediately that something was wrong. He communicated his findings to the Boston Society of Natural History. First, it was clear from the double-rooted teeth that the jaws of *Hydrarchos* belonged to a mammal, not a reptile. This identification slammed the scientific door on the sea serpent pretensions, but it wasn't the worst of it. It was clear to Wyman that the vertebral column was cobbled together from "a series of bones which could never have belonged to the same individual, as is obvious from the fact that they manifest different degrees of ossification, and must, therefore, have belonged to individuals of different ages."[147] As a final humiliation, Wyman discovered that the hippocamp-like paddles were at least partly constructed "not of bones, but of casts of the cavities of a camerated shell, a species of Nautilus. . . . These could not fail to strike the eye at a glance, when examined by any one acquainted with the forms of fossil nautiloid shells." Ouch. *Hydrachos* was not an animal, but a chimera—a sculptural creation that combined bones from multiple fossil animals. It was assembled from several specimens of an extinct whale (now called *Basilosaurus*).

If the problems were this obvious to any trained eye, does this expose Koch as merely incompetent or as a deliberate fraud? The jury remains somewhat split on the question of Koch's competence. One modern historian cautions, "To dismiss Albert Koch . . . as simply one of the more ingenious showmen of his day would be unfair. Koch appears to have been a keen observer of geologic phenomena and a competent natural historian, although he was not professionally trained."[148] Others have been less kind. In particular, geologist James Dwight Dana of Yale was ferociously critical of Koch, writing that his papers collectively "sustain the conclusion that Dr. Koch knew almost nothing of geology, and that what he gradually picked up from intercourse with geologists he generally made much use of, but seldom was able to use rightly. In zoological knowledge he was equally deficient."[149]

Regardless of Koch's level of competence, two truths stand out: his fossil finds were genuinely spectacular, and he was intentionally deceptive about those finds. "Dr. Koch was a man of large pretensions," said Dana, concluding that "Koch appreciated the absurdity" of his claims, but made them anyway "merely to get a full house" of paying customers for his exhibits.[150] Dana singled out as evidence Koch's willingness to improbably assert that *Missourium* was the inspiration for the biblical Leviathan—and then turn around with a straight face and publicize the notion that Leviathan was based on *Hydrachos*!

Following up on his original exposé of *Hydrachos*, Wyman explained in a letter to a colleague that "Dr. Koch, who by the way is no doctor, is a shrewd man [and] knows very well that few are sufficiently acquainted with bones to give an opinion as to the nature of the beast,"[151] and asked for further information to help him determine the true origin of the fossil whale bones from which Koch's sea serpent had been constructed. Koch had asserted that all the bones had been found together in a life-like assembly, but Wyman's discovery that they came from multiple individuals showed Koch's yarn to be "a mere fabrication" (as was the judgment of the Boston Society of Natural History). Sure enough, Wyman's findings were confirmed by Dr. Lister, a resident of the area where Koch had found the bones. Lister investigated, informing the Boston scientists,

> Dr. Koch found a considerable portion of the bones now constituting the Hydrarchos, lying upon, or near, the surface of the earth. They were not lying in their natural position, so as to constitute an unbroken series, but were scattered here and there. Some days after finding some of the bones of the anterior extremities, and some ribs and vertebra . . . [h]e also procured others of the bones in Clark county, at a place twenty miles distant, and some wagonloads of them at another place seven miles distant from the spot where he got the most interesting part of them.[152]

Interestingly, much of this story is corroborated by no less an inside source than Koch himself. His travel diaries describe his discovery of the largest cache of incomplete bones that would be used to create *Hydrachos*: "After the softer parts had decomposed, the bones were lying for some time unpetrified . . . individual bones of the skeleton suffered displacement; indeed, a part of the ribs, the foot bones and several of the tail vertebrae were lost completely." Although loosely arranged in a "sort of half-circle," the remaining bones were damaged and scattered, with the surviving portions of the head "completely turned around"; nonetheless, Koch was confident that "of the whole so much existed that the missing parts could be replaced artificially."[153] And, as we now know, that is what he did—artificially, and very creatively, using bones from other locations. (*Basilosaurus* vertebrae were so common in the area, Koch wrote, that he found them built into fireplaces, used to support a garden gate, and even serving "a Negro for a pillow.")

Even more striking, Koch's diaries refer to these fossils as belonging to "Zygodon"—a bastardization of *Zeuglodon*, which was another name for *Basilosaurus*. Koch clearly understood from the get-go that *Hydrachos* was not a startling new form of life, but a genus that had been discovered, described, and mounted by other naturalists. Koch's diaries reveal that he deliberately chose to do his fossil hunting in "the region in Alabama where the big Zygodon is found."[154] Moreover, Koch's use of the term "Zygodon" reveals that he knew in advance that his sea serpent was a mammal! The genus *Zeuglodon* was proposed by Richard Owen, whose examination of fossil bones from the inaccurately named *Basilosaurus* (king lizard) in 1839 revealed that "the fossil was a Mammifer of the cetaceous order"—a whale. Just as Wyman found when he examined Koch's specimen six years later, Owen concluded that the creature's double-rooted teeth were a dead giveaway. Thus Owen renamed it *Zeuglodon* (yoked tooth).[155]

Nonetheless, the bones comprising Koch's so-called sea serpent were rare and valuable. Despite insightful criticisms from European scientists, the assemblage was snapped up by King Friedrich Wilhelm IV of Prussia, who gave Koch a yearly pension for life in order to secure *Hydrarchos* for the Royal Anatomical Museum in Berlin.[156] (Sadly, it was, like so many of Europe's natural and cultural treasures, destroyed during World War II.)[157] Once again, a generous reward for the dishonest "Dr." Albert Koch—the man who, as Robert Silverberg put it, made a "career of creating supermonsters."[158]

False-Positive Sightings: Fuel for the Global Legend

Running a case-sensitive Google Ngram search on "sea serpent" and "Sea Serpent," I see without surprise that the frequency of the term in English-language books spiked dramatically in 1849 (in the wake of the *Daedalus* case of 1848), soaring to heights of fame that it would never again attain.[159] This tidal wave of celebrity subsided, but the sea serpent never did recede into obscurity. Having become a truly global legend, sea serpents continued to appear wherever ships sailed and wherever eyes looked out to sea. (Nor was it just saltwater: sea serpents appeared inland as well, slithering into the folklore of freshwater bodies from Lake Erie to Lake Okanagan.)

The sea serpent thrived not only in books and popular culture, but, according to eyewitnesses the world over, also in nature. What, if anything, were all those people seeing? To answer that question, we might stop first to consider an insight or two from the critical literature on other paranormal claims—for, to repeat Richard Owen's insight: "A larger body of evidence, from eyewitnesses, might be got together in proof of ghosts than of the sea-serpent."[160]

When considering UFOs, ghosts, Bigfoot, telekinesis, faith healing, and similar elusive, paranormal phenomena, advocates must grapple with the two fundamental vulnerabilities of eyewitness-dependent cases: people lie, and people make mistakes. Traditionally, advocates for these phenomena have responded with a "where there's smoke, there's fire" argument:

- There are many eyewitness encounters.
- It is unreasonable to suppose that lies or mistakes are so extremely common as to account for all encounters.
- Therefore, some of the paranormal encounters record genuine paranormal events.

But as population geneticist George Price explained in 1955 in a widely read article, this argument is backward. Price's criticisms focused on fraud in research on extrasensory perception (ESP), but his arguments apply just as well to sightings of sea serpents:

> When we consider the possibility of fraud, almost invariably we think of particular individuals and ask ourselves whether it is possible that this particular man, this Professor X, could be dishonest. The probability seems small, but the procedure is incorrect. The correct procedure is to consider that we very likely would not have heard of Professor X at all except for his psychic findings. Accordingly, the probability of interest to us is the probability of there having been anywhere in the world, among its more than 2 billion inhabitants, a few people with the desire and ability to produce false evidence for the supernatural.[161]

We have seen that deliberate fraud by both witnesses and investigators is a serious (perhaps crippling) problem for cryptozoology, but let's set aside fraud for the moment. Price's argument also goes to the issue of false-positive misidentification errors. As with fraud, cryptozoologists often argue, in essence, that a reasonable observer in a given eyewitness case would not have mistaken an ordinary phenomenon or animal for a cryptid. This retort to skeptics is typical: "[T]he men who see sea-serpents are familiar with seals, and . . . are not likely to make such mistakes."[162] By extension, if many reasonable observers report having seen cryptids, a significant percentage of those sightings should be genuine. Right? Not so fast.

As Price pointed out in regard to ESP researchers, this argument approaches the evidence from the wrong end. Those who claim to have seen cryptids are not randomly selected average observers. They are among the few people who, out of the billions of people on Earth, have already filtered themselves into the minuscule outlier population of those *who claim to have seen cryptids*. Even ignoring the issue of fraud, we run into the Law of Large Numbers: given large enough numbers, very unlikely things become inevitable. (As Michael Shermer has often observed, "The Law of Large Numbers guarantees that one-in-a-million miracles happen 295 times a day in America.")[163] We could grant the dubious assertion that people in general are decent observers—we could even unrealistically suppose that almost everyone is accurate almost all the time—and we would still expect some small percentage of the population to mistake ordinary animals, objects, or events for cryptozoological creatures. This becomes an inevitability simply because so many billions of people see so many billions of things.

As it happens, however, there is no need to rely on theoretical arguments and certainly no need to speculate. As with other cryptids, there are many records of false-positive sightings of sea serpents. Let's look at a few of these documented misidentifications and their variety of causes.

- *Smaller animals*: It is important to recognize that eyewitnesses do not usually report having seen animals shaped like serpents. Instead, they describe a series of discrete coils, humps, or dark rounded objects ("like a string of buoys" is typical)[164] and infer that they are connected beneath the water's surface. The problem, of course, is that such sightings are by their nature ambiguous: a humungous serpentine animal might resemble a string of buoys, but a group of smaller individual objects (say, an actual string of buoys) also might resemble a string of buoys. For this reason, seals, sea lions, dolphins, porpoises, and waterfowl have always been obvious sources for false-positive sea serpent sightings. Edward Newman noted in 1864 that the large groups of barrel-like harp seals common to Norwegian waters could easily generate reports of sea serpent sightings:

> These were very fond of swimming in line, their heads alone above water, engaged in a game of "follow-my-leader"; for on the first seal making a roll over, or a spring into the air, each seal of the whole, procession, on arriving at the same spot, did the like, and exactly in the same manner. While viewing this singular proceeding (and I had many opportunities of doing so), I could not but be struck with the plausibility of one of the suggested explanations of the appearance which has obtained so widespread a notoriety under the name of the "great sea serpent." . . . I could quite understand any person, not an unromantic naturalist, on witnessing for the first time such a sight as I have tried to describe, honestly believing that the mythical monster was actually before his eyes.[165]

Newman was correct: pods of seals are mistaken for sea serpents. In one example cited by Cadborosaurus advocates Paul LeBlond and Edward Bousfield, two British Columbia police officers observed a "huge sea serpent with a horse-like head"—only to discover, using binoculars, that the "serpent" was actually a group of seven sea lions. "Their undulations as they swam appeared to form a continuous body," reported the *Province*, "with parts showing at intervals as they surfaced and dived. To the naked eye, the sight perfectly impersonated a sea monster."[166]

Likewise, groups of dolphins or porpoises may also impersonate sea serpents. Proponents of monsters are often scornful of this explanation

Figure 5.26 The sea serpent sighted from the *St. Olaf* near Galveston, Texas—"about two minutes" after a school of sharks passed the ship.

(which Willy Ley called "one of the favorite 'explanations' of many people who do not take the time to acquaint themselves" with the evidence),[167] but I have seen this illusion. Sitting in a window seat of the tiny ferry that carries foot passengers between Seattle, Washington, and Victoria, British Columbia, I saw Cadborosaurus in broad daylight. When I noticed its rolling coils, my heart leaped into my throat. There I was, a "professional skeptic," and I wanted to shout out loud. It was only with sustained observation that I was able to break down the illusion—and even then, my brain kept *trying* to interpret those porpoises as a sea serpent. The school of porpoises in Puget Sound looked exceedingly similar to a sketch made in 1872 after a sighting in the Gulf of Mexico near Galveston, Texas (figure 5.26). According to Captain Hassel of the three-masted *St. Olaf*, a school of sharks passed under the stern of his ship, and about two minutes later the crew spotted a 70-foot sea serpent. However, the sketch of the "serpent" shows three or four distinct, rounded objects—each with its own prominent triangular dorsal fin! That positively screams porpoise, dolphin, or shark—all of which are found in the region. (Assuming, of course, that it happened at all. A. C. Oudemans wrote of this case: "I don't know whether the following . . . is a true hoax or an optical illusion, but I think it is a hoax.")[168]

From the perspective of many cryptozoologists, these "group of *X* swimming in a line" explanations seem forced. "If so many otters around the world are 'swimming in a line' to fool eyewitnesses into thinking they are the loops of Sea Serpents or Lake Monsters, where are all of the photographs of such visual demonstrations?" asked Loren Coleman.[169] Going for the pun, Coleman called the theory "otterly ridiculous"—only to post a photograph suggested by fellow cryptozoologist John Kirk a week later, acknowledging that "now I've been sent a clear instance where this behavior has been observed and photographed."[170] Naturalists observe such behavior in many species, but it is largely beside the point: the illusion does not depend on the animals moving in a line, but merely moving in any clustered group. This is one of the least appreciated important facts relevant to the sea serpent literature: thanks to perspectival effects, *any distant cluster of objects at sea appears as a line when viewed from near sea level*, as from a small boat or the shoreline. It is an effect that you can observe on your kitchen table. Just plunk down some small objects in a random-looking cluster at one end of the table, and then bend down to view the scene from the other end in edge-on perspective. I just did this with some blobs of modeling clay: presto, a sea serpent (figure 5.27). Many animals swim on the water's surface in groups, from otters to ducks to dolphins. Given perspective, many of them will appear sea serpent–like; therefore, it's predictable that some witnesses will believe that they have seen sea serpents

Figure 5.27 When viewed from a low angle (as from a boat or beach), any loose cluster of distant objects on the water's surface appears serpent-like. The effect is demonstrated with blobs of modeling clay on a kitchen table, photographed first from a high angle (*top*) and then from a low, edge-on angle (*bottom*). (Photographs by Daniel Loxton)

when in fact they have not. And, again, we need not rely on theory; this effect is well documented within cryptozoology. Consider, for example, Nessie witness Alex Campbell's misperception of a flock of birds:

> I discovered that what I took to be the Monster was nothing more than a few cormorants, and what seemed to be the head was a cormorant standing in the water and flapping its wings as they often do. The other cormorants, which were strung out in a line behind the leading bird, looked in the poor light and at first glance just like the body or humps of the Monster, as it has been described by various witnesses.[171]

It is tempting to suppose that while groups of sea lions or other animals may be mistaken for sea serpents, the illusion cannot last more for than an instant. After all, in the examples just cited, weren't the witnesses able to pierce the illusion themselves? But mistakes of this kind are not always so ephemeral, according to Ian McTaggart-Cowan. In 1950, while a professor in the Department of Zoology at the University of British Columbia,[172] McTaggart-Cowan explained to *Maclean's* that he had seen this illusion convince people, persist, and then make press as genuine sightings of the Pacific Northwest sea serpent, Cadborosaurus: "'On two occasions I saw what was reported to be Caddy,' says Cowan. 'In each case Caddy consisted of a bull sea lion with two others following it. I must admit that they looked convincingly like a sea serpent and you could hardly blame an untrained person for not recognizing them as sea lions. Each time the papers came out the next day with reports that Caddy had been seen.'"[173]

- *Seaweed*: The idea that anyone could mistake ordinary, inert seaweed for a living monster has been widely mocked. "It will not be necessary to point out that this hypothesis is not deserving of any notice on our part," huffed Oudemans.[174] The argument that seaweed cannot explain all sea serpent reports is true and yet trivial. No one on any side of cryptozoology should be looking for a single explanation. A cryptid is not a single animal to be identified—either *this*, or else *that*—but an umbrella under which diverse reports are grouped. It's just barely possible that genuine sightings of new creatures may be among the evidence collected under those umbrellas; if so, they must huddle alongside many false positives, generated by a very wide range of phenomena.

  One of the known sources of erroneous reports of sea serpents is seaweed, of which heroically proportioned, serpentine pieces can be found drifting throughout the world's oceans. Nor is it only landlubbers who make the elementary-seeming error of mistaking algae for monsters. Groups of experienced sailors have made exactly this mistake (and in broad daylight, at that). In 1849, for example, Captain J. A. Herriman of the *Brazilian* sighted what was clearly a sea serpent through his telescope. This huge and lively creature swam with its head held high. The mane running down its neck was as clear as its forked tail. Herriman called over an officer and several passengers—all of whom, "after surveying the object for some time, came to the unanimous conclusion that it must be the sea-serpent" that had been widely publicized in recent news. Herriman grabbed a harpoon and ordered a boat into the water in pursuit:

> The combat, however, was not attended with the danger which those on board apprehended, for on coming close to the object it was found to be nothing more than an immense piece of sea-weed, evidently detached from a coral reef, and drifting with the current, which sets constantly to the westward in this latitude, and which, together with the swell . . . gave it the sinuous snakelike motion.[175]

It's important to reflect how history would have recorded this account if Herriman had not ordered the pursuit. Without the opportunity to "capture" the seaweed, this would stand as an important, sustained sighting in broad daylight by multiple witnesses—including experienced sailors.

The case of the *Brazilian* is in no sense unique. In 1858, Captain Smith of the *Pekin* shared a nearly identical experience (said to have occurred in 1848).[176] "With the telescope," Smith recalled, "we could plainly discern a huge head and neck, covered with a long shaggy-looking kind of mane, which it kept lifting at intervals out of the water. This was seen by all hands, and declared to be the great sea-serpent." Smith dispatched a boat to hunt the monster. When the sailors hauled the creature on board, they were shocked to discover that it was (of course) just seaweed. Once again, seasoned mariners had mistaken ordinary seaweed for a monster in broad daylight. "So like a living monster did this appear," Smith warned, "that, had circumstances prevented my sending a boat to it, I should certainly have believed I had seen the great sea-serpent."[177]

There is even a dramatic case in which armed forces apparently opened fire on drifting seaweed. According to the recollection of Captain J. H. Taylor, a government official responsible for India's important Madras Harbor,

> About fifteen years ago . . . an enormous monster, as it appeared, was seen drifting, or advancing itself . . . into the Harbour. It was more than one hundred feet

> in length, and moved with an undulating snake-like motion. Its head was crowned with what appeared to be long hair, and the keen-sighted among the affrighted observers declared they could see its eyes and distinguish its features. The military were called out, and a brisk fire poured into it at a distance of about five hundred yards. It was hit several times, and portions of it knocked off. So serious were its evident injuries, that on its rounding the point it became quite still, and boats went off to examine it and complete its destruction. It was found to be a specimen of the sea-weed above mentioned, and its stillness after the grievous injuries inflicted was due to its having left the ground swell and entered the quiet waters of the Bay.[178]

- *Boat wakes and waves*: It is extremely telling that sea serpents are so often spotted in calm water. As Erich Pontoppidan put it in 1755, "It is never seen on the surface of the water, but in the greatest calm"[179]—an observation that many other writers and witnesses have repeated since that time. "Evidently the animals feel comfortable in fine weather and when there is no wind," concluded Oudemans, citing dozens of cases in which these conditions were described.[180] This circumstance is so common that some have even argued that sea serpents may be too fragile to withstand waves or wind.

  Of course, calm water also provides perfect conditions for mistaking boat wakes for sea serpents. Many skeptical writers fall prey to the temptation to champion one or another prosaic explanation for monster sightings. Naturalist Henry Lee interpreted virtually all important sea serpent cases as misidentifications of giant squid. I strongly recommend against such silver-bullet explanations—false sea serpent sightings clearly emerge from a wide variety of phenomena—but I am convinced that boat wakes are an underappreciated cause. Given calm water and the right light, the characteristic V-shaped waves left by passing boats or ships can create an extremely compelling illusion. As the two wake lines diverge, each can appear as a series of sinuously animated, dark-colored humps that travel with apparent purpose. Few people realize how far boat wakes can travel. Witnesses may spot wakes miles from their point of origin, with no boats nearby or even anywhere in sight, thus leading to the conclusion that they have seen a sea serpent. Nor is it just boat wakes that can resemble sea serpents; natural waves can also create this effect. Because waves are, well, waves, they interact—some canceling each other, some reinforcing each other (a bit like the famous "double bounce" familiar to kids with trampolines). Even in relatively calm water, natural waves can sometimes stand out, serpent-like, from their surroundings. And, of course, natural waves also interact with boat wakes.

I am persuaded that many popular monster books include photographs or descriptions of boat wakes. Indeed, it would be surprising if they did not. Next time you are near a marina or beach (or reading a monster book), keep your eye out for this illusion. It does not take long to spot waves that could convince a reasonable observer under some circumstances that he or she has just seen a sea serpent. Remember, a given source of false positives need not seem persuasive to most people under most circumstances. Imagine that boat wakes fooled only 1 in 100,000 people, 1 in 1 million, or even 1 in 100 million. Wakes and waves need convince only a handful of people a year in order to become a major part of monster lore!

CADBOROSAURUS: THE GREAT SEA SERPENT OF THE NORTHWEST

The Great Sea Serpent may seem like a mystery from a bygone era—a legend that faded out of fashion alongside top hats and buggy whips. But while it may have enjoyed its greatest surge of scientific interest and public belief during the Age of Steam, the traditional sea serpent is still spotted. Even today, cryptozoologists seek proof of the existence of these "aquatic mega-serpents."

The best-known modern iteration of the hippocamp-inspired Norwegian serpent is seen not in the North Atlantic off Europe, but in the Pacific Northwest of North America (figure 5.28). This Victorian legend swam into the twentieth century, suitably enough, off the Canadian city of Victoria (at the southernmost tip of Vancouver Island). The provincial capital of British Columbia, Victoria is a bustling tourist destination with a busy cruise-ship port. Today it bills itself as the City of Gardens, but in 1933 it enjoyed worldwide fame for something altogether more mysterious—nothing less than an 80-foot sea monster. According to legend, it was an awesome, primeval monster: a huge serpent with fore-flippers, a lobed tail, a mane, and a head something like that of a camel or horse. But how did that legend get started? Why would anyone suppose that a monster out of time would slide undetected through the frigid waters off British Columbia and Washington State? That question hinges on a moment in history.

Figure 5.28 The horse- or camel-like head of the maned *Cadborosaurus willsi*. (Illustration by Daniel Loxton)

Imagine yourself in 1933. The Great Depression was causing tremendous hardship at home. As Archie Wills, editor of the *Victoria Daily Times* recalled, a trip to Ottawa

> let me see first hand the devastating effect of the Depression across the country. In Ottawa; close to Parliament Hill, men slept in the open and panhandled for money. As we travelled by train we saw thousands of youths riding the rails, crammed into box cars; seeking work when there was none anywhere. I saw . . . homeless families. I saw youths in Kamloops searching for any kind of food under [a railway] bridge. I was shocked at the conditions.[181]

The news from overseas was just as grim. Daily newspaper headlines carried ever more bad news about Adolf Hitler. As tensions continued to mount between the new Nazi government and the rest of Europe, war seemed increasingly likely.

"After my return to Victoria during this troublesome period," Wills continued, "it occurred to me that we should try to inject a bit of humor in the newspaper. Rumors were abroad that a sea serpent was disporting itself in our waters and I felt if the story was handled circumspectly we might have a little fun."[182] Wills was forthright about his motivations for publicizing the serpent (people needed a pick-me-up to balance all the bad news, and, not incidentally, the newspaper needed the money),[183] but he was hardly the first newspaper editor to turn to the sea serpent in an hour of need. As the *Critic* put it in 1885:

> There is a difference between a policeman and a sea-serpent. It is a familiar saying that you can never find a policeman when you want him. But whenever the sea-serpent is needed he comes up smiling, at the very moment when the editor, short of a topic, has sent his Sister Anne into the look-out to see if she can descry a "sensation" in the distance. No stronger proof could be asked of the high intelligence and self-sacrificing good-nature of these aquatic ophidians than the precision with which they inform themselves as to the exact date when the Silly Season has set in.[184]

In a 1950 interview, Wills went so far as to say, "Caddy's a psychologist." The too-perfect timing was "significant," hinted Wills, adding, "We certainly needed distraction then."[185]

With the *Victoria Daily Times* backing the 1933 serpent story, Cadborosaurus, as the paper dubbed the creature (or Caddy, for Cadboro Bay, allegedly a favorite haunt for the creature), became a media darling, making headlines from London to Los Angeles, from the *New York Times* to *Time*.[186] But how did the traditional Atlantic sea serpent take up residence in the Pacific? Assuming that Wills did not invent it from whole cloth, how did the "rumors" about the serpent get started? And why, then, in 1933—the same year that Nessie was born?

My strong suspicion is that Cadborosaurus may have been inspired by the film *King Kong* and, even more so, by the global press surrounding the recent monster innovation at Loch Ness in Scotland—which was itself based on *King Kong*. Let's look at how events unfolded.

According to Wills, it literally began on a slow news day. "One morning things were dull in the newsroom of The Times," Wills recalled. "The police court reporter had returned from his beat without record of even a drunk. In his despair he blurted out: 'A couple of guys say they've seen a sea serpent off Cadboro Bay. What about that for a headline?'"[187] Wills dispatched reporter Ted Fox to talk first to witness W. H. Langley and then to another man, Fred W. Kemp.[188] The next day's front-page headline proclaimed, "Yachtsmen Tell of Huge Serpent Seen off Victoria."[189] This seized the imagination of the city, inspiring a rash of copycat sightings and launching an enduring legend.

Both Langley and Kemp claimed that they had seen huge sea monsters, allegedly in independent sightings over a year apart. According to Langley, he and his wife were sailing when they heard "a grunt and a snort accompanied by a huge hiss," and then "saw a huge object about 90 to 100 feet off," of which "the only part of it that we saw was a huge dome of what was apparently a portion of its back." It was, he said, visible for only a few seconds before diving. Contemporary critics were quick to point out that it swam like a whale, sounded like a whale, and looked like a whale. Langley disputed this interpretation,[190] but whales do abound in the area: humpback whales, gray whales, sperm whales, and others. Given that no other evidence corroborates this momentary, undeniably whale-like anecdote, the Langleys' sighting seems to be a completely trivial case.

Kemp's sustained daylight sighting was more interesting. According to Kemp, he and his family were picnicking one afternoon in 1932, on one of a group of tiny islands just off Victoria, when they saw something extraordinary. A huge creature swam up the channel between Chatham and Strongtide islands, leaving an impressive wake. In a signed statement, Kemp recalled,

> The channel at this point is about 500 yards wide. Swimming to the steep rocks of the island opposite, the creature shot its head out of the water on to the rock, and moving its head from side to side, appeared to be taking its bearings. Then fold after fold of its body came to the surface. Towards the tail it appeared serrated, like the cutting edge of a saw, with something moving flail-like at the extreme end. The movements were like those of a crocodile. Around the head appeared a sort of mane, which drifted round the body like kelp.[191]

Kemp estimated that the animal was more than 60 feet long. Although it was indistinct with distance—it was at least 1,200 feet away, maybe 1,500—this was no fleeting sighting. According to Kemp, he and his family watched the monster for several minutes before it slid off the rocks and swam away.

What was it? It seems to me like a group of sea lions among the distant kelp, viewed at too great a distance and interpreted with too great a dollop of imagination. The interesting question is: Whose imagination?

Kemp's description may provide a clue. Despite the copycat sightings and editorial assumptions that gave Cadborosaurus its canonical appearance as a hippocamp-type sea serpent, and despite Kemp's inclusion of the traditional "mane . . . like kelp," his description was as much like that of a dinosaur as a sea serpent. The wash thrown up by the monster, he said, gave the impression that it was "more reptile than serpent to make so much displacement." He added, "It did not seem to belong to the present scheme of things, but rather to the Long Ago when the world was young."[192]

Responding to the Kemps' sighting, one letter to the editor of the *Victoria Daily Times* offered an opinion that Caddy might be a sauropod dinosaur called *Diplodocus*. The writer noted Caddy's long neck and long tail, and called it "probable that it has legs with webbed feet with which it propels itself."[193] Kemp seized on the dinosaur idea with enthusiasm: "*Diplodocus* describes better what we saw than anything else. My first feelings on viewing the creature were of being transferred to a prehistoric period when all sorts of hideous creatures abounded." He said that the creature's movements "were not fishlike, but rather more like the movement of a huge lizard."[194] Kemp drew a sketch of the half-submerged monster that was consistent with the appearance of a sauropod: long neck, rounded body, and long tail (figure 5.29). He explained that "the body must have been very bulky. From the movements I firmly believe the creature had flippers or some kind of legs."[195]

This combination of elements—a swimming sauropod dinosaur and the notion of being transported to a prehistoric world full of terrible monsters—is very familiar. It is reminiscent of another sauropod swimming in another

Figure 5.29 The "huge lizard" reported by Fred Kemp. (This drawing, done in 1933, is identified in Archie Wills's notes as having been "made by an artist," but it closely resembles a rougher sketch identified as "Mr. Kemp's original sketch," in Bruce S. Ingram, "The Loch Ness Monster Paralleled in Canada, 'Cadborosaurus,'" *Illustrated London News*, January 6, 1934.) (Image courtesy of University of Victoria Archives, Archie H. Wills fonds [AR394, 4.3]; see also "Caddy," *Victoria Daily Times*, October 20, 1933)

primal environment teeming with hideous creatures: the *Diplodocus* or *Apatosaurus* (formerly *Brontosaurus*) that attacks the crew of the *Venture* on King Kong's perilous Skull Island!

We have seen that the legend of the Loch Ness monster was created immediately after the release of *King Kong* and that George Spicer's key Nessie sighting seems to have been lifted from that film. Could the movie also have inspired Kemp's Cadborosaurus story? The time line certainly works: *King Kong,* it happens, opened in Victoria on May 20, 1933—less than five months before the birth of Cadborosaurus.[196] It blew away moviegoers, scared the socks off viewers, and stuck in the memories of all who saw it. The similarity between Caddy and *Kong* may not be a smoking gun, but it is certainly suspicious.

If Kemp's sighting story was sincere, it may well have been influenced by the film. After all, he told no one about his astonishing brush with a gigantic super-monster at the time that it allegedly occurred—speaking up only the following year, months after the release of *King Kong*. His sighting, he admitted, was extremely uncertain because of the tremendous distance involved. He could not make out the key details. For example, the creature's head was just a blob. If the sighting occurred at all, it is likely that he saw a distant group of marine mammals swimming and climbing on the rocks (as is entirely typical in the area) and plausible that Hollywood may have contaminated his already vague impressions of what he had seen.

If, however, the story was completely invented at a date close to its first publication, the influence of the recently screened *King Kong* would be even easier to explain. (There are thin but tantalizing hints that the case could be a hoax. In 1950, for example, *Maclean's* quoted Kemp as claiming that Cadborosaurus had super-speed: "'I wouldn't like to say how fast he travels, people might laugh.' But he added under persuasion, 'Only a bullet would travel anywhere near as fast!'"[197] There's also Wills's recollection, "We convinced several noted citizens to lend their names to a story stating they had seen the marine creature."[198] This probably refers to the reluctance of sincere witnesses to go on record, although Wills did make this remark in the context of generating a media sensation on purpose. For now, I will assume that Kemp's story was probably sincere, although, of course, completely anecdotal.)

Where does this leave Cadborosaurus? Could it be that the entire legendary edifice—the subsequent sightings, the books and television programs, the public sculpture, and the place in popular culture—rests on a foundation of smoke and the flickering screen of a cinema?

Caddy as a Going Concern

Questionable as Caddy's origin story may be, enthusiasts continue to advocate for and seek the creature, which means that the hippocamp-type Great Sea Serpent of the North Atlantic remains a going concern in cryptozoology. This may come as a surprise to some readers, but there it is.

An effort called CaddyScan, for example, uses digital-camera traps to record passing wildlife in the hope of catching evidence of Cadborosaurus. The British Columbia Scientific Cryptozoology Club (of which I am a member) records and reports on new sightings as they come in. Honestly, I dig all that. As a lifelong Cadborosaurus fan, I find myself pleased that people are still searching. I also find it easy to appreciate the reasons—good and bad—that people find the mystery compelling.

To start with one of the weakest, it is occasionally pointed out in the monster's defense that Caddy was included in the *Guide to Marine Life of British Columbia*, published by the Royal British Columbia Museum. Written by G. Clifford Carl, the guide does indeed include Cadborosaurus, but the minuscule listing is clearly tongue in cheek. (Caddy's total length is described as "variable, depending upon circumstances or condition of observer," while its status is listed as "Questionable.")[199] This humorous approach is not surprising. As Archie Wills said, recalling his deliberate manufacture of a popular-culture monster legend, "Even Dr. Clifford Carl, director of the Provincial Museum, went along with the idea; knowing that world-wide coverage would prove beneficial to Victoria's tourist trade."[200] As *Maclean's* described him, Carl was a fan of the Caddy character (call him a willful agnostic), but hardly a believer in its biological reality: "'I'm all for Caddy myself,' says Dr. Carl. 'I don't want to see him (or her, or it, or them) exposed. And if Caddy by some strange chance does actually exist it would be a pity to capture him, stuff him and put him on view in some museum.'"[201] (In fact, Carl created one light-hearted Caddy hoax himself. When the *Victoria Daily Times* offered a cash reward for a photograph of Caddy in 1951, the first, comically fake entry came from Carl.)[202]

A much better argument is that sea monsters are inherently plausible in light of both the available habitat and the fossil record. There's an awful lot of ocean out there—room to hide any number of large aquatic creatures and *perhaps* even air-breathing vertebrates. The rates of discovery for new species can be plausibly extrapolated to estimate the number of discoveries remaining to be made, although not with any great certainty. For example, when applying various means of extrapolation to the question, "How many extant pinniped species [seals and relatives] remain to be described?" Michael Woodley, Darren Naish, and Hugh Shanahan arrived at estimates as high as fifteen and as low as zero.[203] (The number of pinnipeds left to be discovered is especially relevant because the mainstream of cryptozoological thought now favors the hypothesis that undiscovered, long-necked seals lie behind many sea serpent sightings.) Finally, the coincidence that mosasaurs, plesiosaurs, and other large marine reptiles existed in the geologic past remains as seductive today as it was in 1848.

Another reason that the argument for the existence of Cadborosaurus appears compelling is that sightings continue. In 2010, for example, cryptozoologist John Kirk reported a personal sighting of a Cadborosaurus with a cow-like head swimming in the wide Fraser River (within a few miles of the point where the Fraser empties into the Strait of Georgia, near Vancouver).[204] Also in 2010, reports emerged of a tightly guarded video of Cadborosaurus that became part of the reality-television series *Alaskan Monster Hunt*.[205] As flimsy as it may appear to skeptics, the continuing accumulation of eyewitness accounts builds up to a "mountain of evidence" that proponents find unparsimonious to reject. (Skeptics may bend over backward to explain all these cases, runs the argument, but c'mon! Surely the simplest explanation is that these people saw what they said they saw!) Here we must note once again that many people do mistake ordinary phenomena for sea serpents and repeat Richard Owen's observation that a "larger body of evidence, from eye-witnesses, might be got together in proof of ghosts than of the sea-serpent."[206] Mountains of anecdotal testimony exist for many paranormal beliefs that cryptozoologists would reject—but, of course, this does nothing to make anecdotes in favor of their own favorite mysteries seem less compelling.

For those who pursue Cadborosaurus, however, the most powerful piece of evidence is something known as the Naden Harbour carcass (figure 5.30). Found in 1937 in the stomach of a sperm whale processed at the Naden Harbour Whaling Station in Haida Gwaii (a large cluster of Canadian islands near the Alaska Panhandle that were known as the Queen Charlotte Islands until 2010), this long, unidentified biological specimen seemed odd to the whalers. They photographed the specimen. As a result of those photographs, generations of cryptozoologists and critics alike have been compelled to share the whalers' impression: it looks really weird. Moreover—and this has sailed, if not a thousand ships, at least some flights of fancy—it looks somewhat like

Figure 5.30 The unidentified carcass recovered from the stomach of a sperm whale at the Naden Harbour Whaling Station in Haida Gwaii in 1937. (Image I-61404 courtesy of Royal BC Museum, BC Archives)

Cadborosaurus. Or, rather, it looks superficially similar to the features that the hypothetical Cadborosaurus is alleged to possess—the features of a hippocamp.

In 1992, that resemblance provoked leading Cadborosaurus proponents Edward Bousfield and Paul LeBlond to formally propose in a scientific paper that the Naden Harbour carcass be considered the type specimen for a new species to be recognized as *Cadborosaurus willsi*.[207] Although this naming followed the lead of Clifford Carl, whose *Guide to Marine Life of British Columbia* had jokingly dubbed the creature "Cadborosaurus (Wills)," Bousfield and LeBlond were widely criticized for having taken this step. Despite being friendly to cryptozoology in principle, paleontologist and science writer Darren Naish pointed out that this taxonomic leap put the cart before the sea horse:

> Firstly, establishing a new species on the basis of a photo is just not acceptable: article 72(c)(v) of the International Code of Zoological Nomenclature states that the actual specimen figured or described, and not the illustration or description, must serve as the holotype. The name *Cadborosaurus willsi*, based on a photograph and not the specimen it depicts, therefore has no official standing and should be ignored.[208]

Worse, according to Naish, was that "the number of speculations that Bousfield and LeBlond included within the paper are inappropriate for a work masquerading as a technical description, never mind the fact that those speculations were fantastic and logically flawed." Naish noted some oddities that also had leaped out at me. According to Bousfield and LeBlond, Caddy's very long body (which should have a lot of drag as it moves through the water and, therefore, be slow) can be reconciled with the extremely high speeds attributed to it by witnesses if it bunches itself up into a series of loops that approximate the torpedo shape of a tuna. They also speculated that the beast's famous (hippocamp-based) mane may be "some gill-like, gas-exchange organ" able to extract oxygen from the water.[209] A neat trick for an animal that Bousfield and LeBlond believe is a reptile—and an outright guess. "If these proposals sound nonsensical, or just extremely speculative and lacking in justification," Naish concluded, "that's because they are."

Many cryptozoologists, though, applaud Bousfield and LeBlond for their forward-looking vision; even skeptics acknowledge that the risks they have taken require some serious guts.[210] But why speculate at all? If the Naden Harbour carcass can settle the case, so much the better. So . . . where is it now? Well, that's the problem. As LeBlond and Bousfield made clear in 1995, "Exactly how the carcass was disposed of or whether any part of it was preserved is not known."[211] One lead suggests that it was sent to the Pacific Biological Station at Nanaimo, British Columbia. However, LeBlond and Bousfield found that "nobody there knows anything about it." Another possibility is that it was sent to the Royal British Columbia Museum in Victoria, as the *Los Angeles Times* reported in 1937.[212] When I contacted the museum to follow up on this lead, Jim Cosgrove had to disappoint me:

> The fact is, we have no record of ever having received the specimen in the photograph from Naden Harbour nor do we have any record of the examination of the specimen or who would have done the necropsy. Lastly, we have no record of how the specimen was disposed of. Nothing appears in any of the Museum's annual reports which were often remarkable documents of the previous year's activities. In short we have no records of this specimen at all.[213]

And that's it. That's the beginning and end of the case of the Naden Habour carcass. Without the carcass (or material from it), there is no way to identify it.[214] Cryptozoologists do not know what it was, as much as they would like to; I do not know what it was, for all that I wish I did. Could the photograph really show the carcass of a genuine Cadborosaurus? Sure. Why not? Unfortunately, it is lost to history. If, indeed, it was a Cadborosaurus, we can only mourn for the lost opportunity and hope that it comes again.

The Nature of Cadborosaurus

Why did Cadborosaurus appear in 1933? Local factors had to align to make it happen (Archie Wills's personal creativity and sense of humor, economic hardship, and, of course, proximity to the water), but the release of *King Kong* and the resulting emergence of the Loch Ness monster earlier that year no doubt primed the pump.

But why did Caddy take the form that it did? On the simplest level, the nature of *Cadborosaurus willsi* is obvious: it is fakelore, a manufactured legend, an entertainment product created by regional media. (There is even a bogus Indian legend associated with the story.)[215] On a deeper level, Caddy became a genuine cryptid—genuine in that people "see" it, making it a living part of twenty-first-century folk belief. And, at the deepest level, it is a modern descendent of the classical Greek artistic motif: the hippocamp.

Caddy's namesake, Archie Wills (dubbed "Victoria newspaperman and fish story expert" by the Associated Press),[216] conceived a popular mystery, and then he made it happen. Although he maintained a winking ambiguity in his writing, there is good reason to suspect that Wills did not believe in the serpent's literal reality. When the *New York Sun* put the question to Wills directly, for example, he dodged—and yet also said it all. "It is not my function to dispute these eyewitnesses or to erect barriers impugning them," Wills answered. "News is news. The story attaching to this particular sea serpent belongs to *The* [*Victoria Daily*] *Times* wherever it turns up. When Caddy goes out of his way to call at this port, we shall continue to treat him as front page material for all he is worth."[217] When Paul LeBlond and John Sibert contacted Wills in 1970 to solicit his help in building a scientific case for the creature, Wills politely responded that "the best success is found in writing stories with 'tongue in cheek.' I have always found that 'Caddy' appeared here when morale was low, for instance, in 1933, when the Depression was bad." Wills wished the men "good luck in your quest for knowledge," but warned, "In view of the fact that you wish to be factual and actually prove the existence of the species, maybe my collaboration would be a handicap."[218]

Be that as it may, Wills deserves his reputation as Caddy's godfather. As a result of the publicity he gave it, many other people certainly do believe in Cadborosaurus. And as skeptics know all too well, believing is seeing.

"Expectant attention" has been known by psychologists since the nineteenth century, and its role in the paranormal has been discussed since at least 1852 (when William Carpenter invoked the combination of expectant attention and what he coined as "the ideo-motor principle" in order to explain the action of dowsing rods).[219] Cryptozoologists acknowledge this problem. As Loren Coleman and Patrick Huyghe explained, the human propensity for simple mistakes of perception is amplified by "psychological contagion—if people are told to look for monsters, they will see them."[220] The Caddy database contains fairly obvious hoaxes (such as the yarn that Jack Nord told about shooting at a 110-foot serpent with 8-inch fangs at a range of a mere 35 yards),[221] but also many sincere sighting reports. We have looked at several sources of false-positive misidentification errors—such as sea lions, waterfowl, seaweed, and boat wakes—and I may add just one more: elephant seals. Sometimes found in the waters (and on the beaches) off Victoria,[222] these 16-foot giants are rarely seen and unfamiliar to most locals, and they bear an intriguing resemblance to the hypothetical Cadborosaurus. Depending on ambient conditions, they may appear black, brown, gray, or even green, and (most interestingly) the bulbous noses of male elephant seals give them an exaggerated, camel-like profile.

But neither elephant seals nor boat wakes explain why the canonical Caddy takes the form it does. Eyewitness reports of Cadborosaurus are highly variable (up to and including the claim that "'Caddy' has three heads"),[223] which is to be expected from a mixed population of witnesses interpreting a wide variety of phenomena through the multiple lenses of modern sea monster mythology, Hollywood, and their own imaginations. Nonetheless, the influence of the classical hippocamp on Cadborosaurus is very clear.

Anatomically, the canonical Caddy is extremely similar to the hippocamp: snaky and horse-headed, with a lobed tail like a whale, vertical tail arches, and two fore-flippers below the base of the neck. Caddy's improbable mane so obviously resembles that of the hippocamp that I find it hard to regard these features as anything other than cultural homologues—which neatly explains something that has puzzled me since I was a child. (The hippocamp template is even clearer when looking at a Caddy-inspired children's book about a sea serpent named Serendipity, which was very popular among children of my generation. As author Stephen Cosgrove told me, Serendipity's design borrows "a whole lot of Caddy. I think she is related on her mother's side.")[224]

But what of Caddy's famous camel-shaped head? Wills argued on a television program in 1950 that Caddy "is not a horse-face[d] sea serpent like some inventions of the past, but has a camel-like head."[225] Doesn't this distance Caddy from the hippocamp? Not really. "Camel-like" is a minor variation on the traditional "horse-headed" characteristic, and the shape of sea serpents' heads have always been said to resemble those of a variety of large terrestrial mammals: the hippocamp-descended "horse-like" head dominates, but "cow" and "sheep" are common variants, and even "giraffe" shows up from time to time. For their part, those who claim to have seen Cadborosaurus also use a range of animal comparisons to describe their monster. But contrary to the creature's camel-headed reputation, "horse-like" is by far the most common description for Caddy's head! When Darren Naish, Michael Woodley, and Cameron McCormick broke down the 178 eyewitness reports recorded by LeBlond and Bousfield in their book *Cadborosaurus*, they found an astonishing range of descriptions. The witnesses mentioned body lengths of anywhere from 5 to 300 feet, for example, and practically any natural-seeming color from tans to blacks to greens. But the descriptions that the witnesses gave of Caddy's head are much more uniform: about 75 percent compared it with that of a large animal; of those, a whopping 42 percent said that Caddy's head looked like that of a horse, and about 13 percent chose the next most popular description: "camel-like." Combined, the closely similar head-shape descriptions of "horse," "camel," "giraffe," "cow," and "sheep" comfortably dominated, with a majority of almost 70 percent.[226]

The continuing influence of the horse-headed hippocamp template seems clear. Yet, it is also true that sea monster reports are as wildly unpredictable as the sea—and as kaleidoscopically varied as the human imagination.

AN EMBARRASSMENT OF SERPENTS

Writing in 1893, the great English biologist Thomas Henry Huxley (known to history as "Darwin's Bulldog") paused to consider the mystery of the Great Sea Serpent. Huxley was a major thought leader in his day (he coined the word "agnostic"), so it pleases me to credit him with one of the most important critical comments ever offered in regard to sea serpents—and, indeed, to paranormal claims in general. Like skeptic Richard Owen before him, Huxley granted that there was "no *a priori* reason that I know of why snake-bodied reptiles, from 50 ft. long and upwards, should not disport themselves in our seas, as they did in those of the cretaceous epoch, which, geologically speaking, is a mere yesterday." But Huxley saw a serious problem in the way that eyewitness testimony was used to support belief in the serpent. After reading one compelling eyewitness case, "I was almost convinced," he wrote—until he compared the testimony with that of a second witness. It became clear "beyond doubt that the circumstance under which the first deponent saw the apparition were such as to make it impossible that he could have properly assured himself of the facts to which he testified. He had done what we are all tempted to do—mixed up observations and conclusions from them, as if they rested on the same foundation."[227] This is a subtle but powerful point. When people say that they saw a sea serpent or any other cryptid (or a ghost, Elvis, or an Ivory-Billed Woodpecker), what they mean—or *should* mean if they were to take Huxley's caution to heart—is that they saw a phenomenon that looked a certain way, *and also* they think it may have been a sea serpent.

Cryptozoology thrives on the failure to distinguish observations from conclusions. The inappropriate certainty that people feel for their conclusions is itself treated as primary evidence. Few things in cryptozoology feel as hair-raisingly persuasive as "I was there! *I know what I saw.*" The problem is that different people "know" that they have seen many very different things. Taken at face value, then, the eyewitness evidence implies the existence of many very different types of undiscovered sea monsters—indeed, many competing monsters in every cryptozoological subcategory. (For example, cryptozoologists Loren Coleman and Patrick Huyghe report more than fifty kinds of just bipedal, humanoid cryptids as described by different witnesses at different times.)[228]

This exuberant proliferation of monsters posed a challenge for the "author zero" of the modern sea serpent mythology, Bishop Erich Pontoppidan. Discussing an earlier sighting of a sea monster described by clerics Hans and Poul Egede (figure 5.31),[229] Pontoppidan was forced to "conclude (what by other accounts I have thought probable) that there are sea snakes, like other fish, of different sorts."[230] This was honest of him. Rather than annexing Egede's quite different monster in service of his own maned Norwegian serpent, Pontoppidan itemized the many points of difference and dubbed it a new species. But he could not stop there. Pontoppidan was likewise forced to admit the existence of other sea monsters: the vast kraken 1½ miles in diameter—and even mermen! How could he not? In his own diocese were "several hundreds of persons of credit and reputation, who affirm, with the strongest assurance, that they have seen this kind of creature," and he personally spoke with many of these highly certain merman eyewitnesses.

As the sea serpent concept spread beyond Scandinavia, new yarns covered the range from unlikely to implausible to flat-out ridiculous—and each story

Figure 5.31 A detail from a map, showing the widely reproduced (and frequently re-drawn) illustration of the somewhat whale-like sea monster described by Hans Egede and his son Poul.

jarred against others. In 1786, for example, a newspaper reported that two sailors had sworn before a magistrate and a justice of the peace that they had seen a monster "at least three English miles" in length.[231] (Even in those days, this was a bit much to swallow. A rival newspaper soon mocked that this "droll fish" had been killed and that a colossal lead kettle was being constructed to boil it.)[232] A 3-mile monster is clearly an outlier in the sighting database, but the report did involve sworn testimony from multiple witnesses—cryptozoology's gold standard, such as it is.

This situation would not improve with time. As Charles Gould noted in 1886, "The narratives of different observers disagree so much in detail that we have a difficulty in reconciling them, except upon the supposition that they relate to several distinct creatures . . . the various creatures collectively so designated being neither serpents nor, indeed, always mutually related."[233]

So variable are eyewitness accounts of sea serpents that Bernard Heuvelmans found it necessary to propose several distinct, undiscovered species of marine mega-fauna: "The legend of the Great Sea-Serpent, then, has arisen by degrees from chance sightings of a series of large sea-animals that are serpentiform in some respect," including "three Archaeoceti or very primitive Cetecea . . . then two pinnipeds . . . and several eel-like fishes," and "finally a pelagic saurian shaped like a big crocodile."[234] Yet, not even this expanded scheme is sufficient to encompass the diversity of eyewitness accounts. Discussing Heuvelmans's "Merhorse" (hippocampus)-type serpent, Coleman and Huyghe struggled to explain why "observers report several sizes and forms," invoking sexual variation and regional differences as "probably responsible for the appearance of two distinct categories within the Waterhorse type; one larger and hairier, the other smaller with smoother skin and what appears to be a longer neck in proportion to a slighter body."[235] Even this level of subdivision glosses over enormous variation. For example, Coleman and Huyghe tentatively folded the Scottish traditions of freshwater water-horses and water-bulls into their two mer-horse subcategories, even though these distinct folkloric creatures are typically described as supernatural, amphibious quadrupeds. (All European water-horse traditions possibly owe something to cultural common descent from the hippocamp, but that does not mean that the multiple traditions are not clearly distinct from one another. The Cadborosaurus-type sea serpent bears extremely little similarity to the shape-shifting kelpies described in the discussion of the Loch Ness monster.)

"Sea-serpents themselves seem to be as variable as fashions," despaired Heuvelmans.[236] He struggled to make eyewitness reports fit within the classification scheme of his several hypothetical species, but many refused to fit. Many reports described characteristics from more than one of his categories or creatures that fit into none. He wrote, "The problem seems to get more and more confused, as more and more different types of monsters appear, until at times one can almost bear it no longer. I have often felt, during the last seven years, that it was beyond my powers to solve and that I should have to confess myself beaten."[237] The truth is, he probably should have. Efforts to classify sea serpents (or other cryptids) on the basis of sighting reports are deeply problematic, as Huxley understood in 1893. An insightful article by Charles Paxton compared such efforts to obsessing over the *Monster Manual* from the role-playing game Dungeons and Dragons. Subjectively "shoe-horning" varied reports into preconceived categories is "misleading and can lead to invalid conclusions," Paxton explained. The monsters of folklore and of eyewitness reports "are not easily classified. They are conceptually amorphous and difficult to define. Fireside tales are not associated with morphological or zoological precision. Names and characteristics of monsters do not conform to the rigid minds of modern-day cryptozoologists. Snarks overlap with boojums."[238] Disconnecting cryptozoology from the fundamental fuzziness of eyewitness reports destroys the foundation on which monster hunting rests. According to Paxton's very good advice, it is best to avoid the temptation even to assign names to cryptids (unless used simply to designate a geographical location, such as "Loch Ness monster"). "The raw data of cryptozoology should not be 'cryptids'—as many cryptozoologists seem to believe—but *reports*," Paxton emphasizes. "So when cryptozoologists refer to a particular 'cryptid' they really should be referring to a number of reports which may or may not have a common source."

Does Paxton's warning about assuming a common source for divergent reports pose a problem for the identification of sea serpents with the hippocamp of classical Greek art? It would if the argument were that witnesses are "really" describing hippocamps, but that is not my position. Instead, I argue, as does Paxton, that myths and legends are amorphous and fluid, with individual storytellers (including eyewitnesses; sincere or not, cryptid reports are ultimately stories) drawing freely from any and all culturally available templates. When witnesses attempt to explain ambiguous phenomena, the hippocamp-derived Great Sea Serpent is one of those available templates. Plesiosaurs are another.

The characteristics of the hippocamp, including the mane and horse's head, were built into the sea serpent legend at its inception—with the hippocamp inspiring the *hrosshvalr* and *havhest*, these Nordic mer-horses engendering the Scandinavian sea serpent, and the Scandinavian sea serpent informing all that has come since. In this evolution of ideas—"memetic" evolution, to employ biologist Richard Dawkins's metaphor[239]—the hippocamp's cluster of connected characteristics have allowed it to remain a successful and long-lived creature in the ecology of the imagination. It has evolved slightly since classical times and has come to perform new functions and occupy new habitats, but the hippocamp has remained such a coherent and competitive form that it clearly is echoed in the canonical Cadborosaurus. In parallel with the hippocamp's line of descent, the discovery of fossils of extinct marine reptiles in the nineteenth century introduced a successful new cluster of dominant traits: those of the small-headed, flippered, long-necked plesiosaur. In the memetic competition taking place in the imaginations of individuals, in popular culture, and in scientific (and pseudoscientific) speculation, ideas may not only change vertically over time, but also trade characteristics horizontally with their competitors. This horizontal meme transfer results in no end of hybridized creatures, as witnesses and storytellers mix and match characteristics from the hippocamp-style Great Sea Serpent, from plesiosaurs, from dragons, from whales and other known animals, and (in memetic evolution's engine of mutation) from mistakes and imagination. In most alleged water monster habitat, serpent types are reported alongside plesiosaurs. In some niches, the plesiosaur becomes dominant (as is the Loch Ness monster); in others, the hippocamp-type serpent emerges as the apex monster (as is Caddy). But the truth is that sea serpents are shape-shifters. How could they not be? They are, after all, creatures of culture, not of nature. In all environments—in fiction, in the cryptozoological literature, and in the oceans of the mind—sea monsters teem and vary and return to type, as unpredictable, as unique, and yet as familiar as the waves themselves.

CLOSING THOUGHTS

Richard Owen's famous critique of the report of a sighting of a sea serpent by Captain Peter M'Quhae and other officers of the *Daedalus* contains a sentiment that speaks to my own values and approach to cryptids and other paranormal mysteries:

> I am usually asked, after each endeavour to explain Captain M'Quhae's sea-serpent, "Why there should not be a great sea-serpent?"—often, too, in a tone which seems to imply, "Do you think, then, there are not more marvels in the deep than are dreamt of in your philosophy?" And freely conceding that point, I have felt bound to give a reason for scepticism as well as faith.[240]

Arguing about the burden of proof is one of skepticism's great clichés. It is common for skeptics to have to state the obvious: the world is not obligated to accept anyone's personal claims or speculations. If someone wants others to believe that there is a worldwide population of aquatic mega-serpents, he or she must present evidence that this is the case; others do not have to prove the claimant wrong. Yet when skeptics emphasize this "burden of proof," we may forget a deep truth: *Doubt is cheap. Finding out is hard.*

Incredulity requires no training, no knowledge, no investigation. "Any fool can disbelieve in sea serpents," as Archie Wills put it.[241] By itself, doubt does nothing to advance knowledge. Even Occam's celebrated razor is just a betting strategy. Parsimony tells us which proposed explanations to investigate *first*; it does not tell us which explanations are *true*. Often nature is simple and elegant, but sometimes it is not.

For this reason, Owen should be applauded for voluntarily taking up his own burden of proof. After all, the goal of science is not to stonewall weird ideas, but to find out what is true. He rested his case on the negative evidence that no bone or carcass or geologically recent fossil of a sea serpent had yet emerged, on all the coasts and in all the museums of the world. It is often said that "absence of evidence is not evidence of absence," but this is only sometimes true. If a given hypothesis ("The world has giant sea monsters") predicts that we should see certain evidence ("Therefore, the world has sea monster carcasses") and we do not see that evidence (for the record, we do not), then that actually *is* evidence of absence. It may or may not be strong evidence, depending on the particulars of the case—but it is relevant, and it does count.

In this chapter, I have tried to present not only the negative evidence, but also a positive historical case for the sea serpent. The sea serpent can be shown, in my opinion, to be a cultural creation: a concept rather than an animal. It arose out of art, and then evolved and spread with changing fashions, popular media, and the happenstance of fossil discoveries. Yet even now, knowing all that, I must admit: that does not mean that there could not *also* be a sea serpent.

In 1933, a letter writer to the *Times* of London argued—somewhat oddly—that the existence of the sea serpent or Nessie would be disturbing to skeptics:

> If there is one thing the sceptics doubt more than religion, it is the sea serpent, and now the sea serpent is in a fair way to becoming an established fact. The sceptics by their teachings and writings have made this life of ours drab and dreary enough. It is time the faithful rebelled, and if we can prove that the sea serpent exists there is no limit to the discomfiture that can be inflicted upon those who doubt.[242]

If only this correspondent had been able to enjoy the chance to gloat! If a sea serpent were ever to be discovered, it would be among the happiest days of my life. But one cannot prove a negative—or at least not easily, and not often. Do we really know that sea serpents do not exist? If there is one lesson to be drawn from what we do know, it is to avoid arbitrarily shoehorning diverse events into our favorite explanatory categories. Each individual sighting is its own mystery, resting on its own particular set of facts, posing its own challenge, requiring its own investigation. Most remain unsolved.

And what of my parents' sighting of Cadborosaurus—the event that propelled me into a career of skeptical paranormal investigation? I'll leave you with my father's closing thoughts:

> Then it occurred to me: a half-dozen sea lions swimming nose-to-tail would look just like what we were seeing. . . . We stared and stared, trying to figure it out, debating the possibilities until it turned and swam out to sea. Even at that moment, looking right at it, we couldn't be sure. Now, 30 years later, I'm almost certain that I saw six or seven sea lions swimming together in a line, one after the other.
>
> But, my wife is still 100 percent certain that she saw Cadborosaurus.
>
> Maybe she's right.[243]

MOKELE MBEMBE

THE CONGO DINOSAUR

THE *MONSTERQUEST* AFFAIR

IN LATE 2008, I was asked to participate in an episode of a cable-television show called *MonsterQuest*. Like many of the newer "reality" shows on formerly scientific cable stations, it was pseudoscientific tabloid journalism: lots of moody music and dark, foreboding camera shots promoting one kind of legendary monster or another, with dubious "eyewitness" testimony and sketchy "evidence"—nothing concrete like an actual body or bones. Normally, I ignore such programs as a waste of time and focus on trying to improve real science documentaries. However, this episode concerned the alleged dinosaur in the Congo: Mokele Mbembe. Thought by cryptozoologists to represent a surviving population of sauropod dinosaurs, Mokele Mbembe has often been identified with *Brontosaurus*—the old name for the gigantic, long-necked, long-tailed *Apatosaurus*. As a vertebrate paleontologist with considerable experience with dinosaur fossils, I was qualified to speak to at least some of the usual claims made about the creature. I knew that I would be the token skeptic on the program, but I decided that at least one skeptic should appear on the show—otherwise, it would be entirely pseudoscientific fluff.

In January 2009, a two-man crew arrived to film my segments. I set up a quiet classroom with controlled lighting, so they had an undisturbed setting in which to film. I selected a number of real dinosaur specimens and casts to use as props or to fill the background of the shots. The crew set up my "talking-head" interviews in front of a cast of a large duckbill dinosaur skull and filmed me moving teaching fossils around on tables and rolling the cast through hallways on a cart. Most of the questions were relatively straightforward, and I gave them the answers that are found in this chapter.[1]

The one surprising moment came when I was handed a wrapped package and asked to unwrap it and interpret the contents as the camera rolled. In the package, I found a fist-size shapeless lump of plaster that looked like absolutely nothing. Hoping for a "gotcha" moment, the crew tried it again, showing me photographs of where the plaster cast had been taken and trying to get me to admit that it looked like a dinosaur footprint. They filmed the shot again and again, each time prompting me with more details about where the cast had been made, but it changed nothing. The cast was simply a lump of plaster that had been formed when it was poured into a random hole in the ground. It was clearly *not* a dinosaur track or, indeed, any animal track. All experienced vertebrate paleontologists have seen many photographs and casts

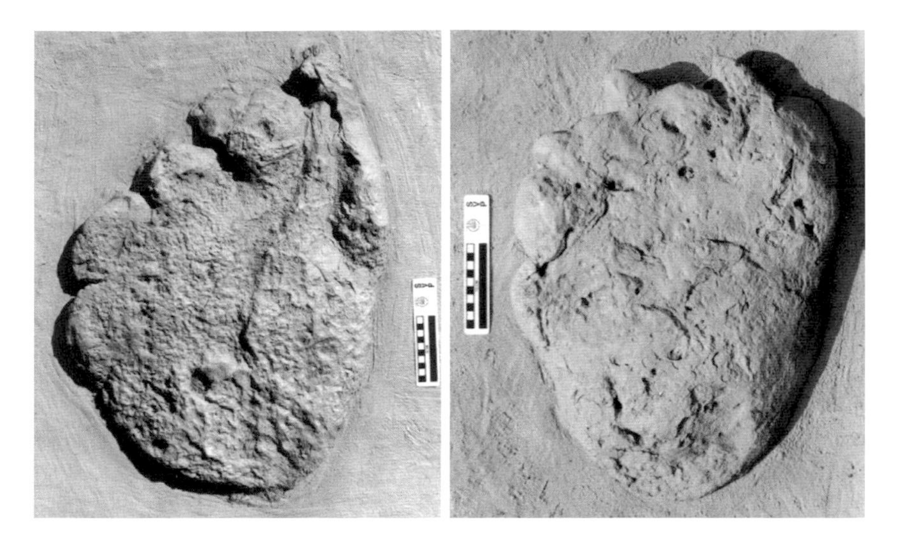

Figure 6.1 Typical sauropod footprints from the Morrison Formation of the western United States, dating to the late Jurassic period (ca. 156–147 million years ago), showing their characteristic shape and evidence of front claw marks: (*left*) from the Tidwell Member; (*right*) from the Salt Wash Member. (Photographs courtesy of John Foster)

of dinosaur footprints (figure 6.1), and most have visited a number of the important track sites, such as the Paluxy River site in Texas. Martin Lockley, a paleontologist who has spent his career documenting trackways, has written several excellent books on dinosaur tracks.[2] Thus paleontologists know what dinosaur tracks actually look like—the tracks of not only sauropods, like the alleged Mokele Mbembe, but also three-toed theropods and many others. And the lump of plaster bore no resemblance to the track of any animal: no symmetry, no flat footpad impression, no distinct toe impressions. It was a fizzle, and the filmmakers were disappointed.

The episode on Mokele Mbembe first aired in June 2009. It was generally as I had expected from other episodes of *MonsterQuest*: a soundtrack of spooky music and shots of "explorers" trying to find their way up the Congo Basin through the jungles of Cameroon—but no concrete evidence. No footage of the animal, not even old photographs from past films; no carcasses, bones, or other physical remains. The "explorers" tried to interview local people, but immediately biased their efforts by showing them pictures of a sauropod, thus "leading the witnesses." If they had let the natives do the drawing, they would have been somewhat more believable, although the Western conception of an "African *Brontosaurus*" has been widely disseminated across

Africa for more than a century. The specter of contaminated testimony is raised by anecdotes told by Mokele Mbembe proponent William Gibbons (one of the *MonsterQuest* explorers), who relates that when an expedition he led in 2001 pulled into a random "non-descript, one-horse town" in rural Cameroon, a young man turned to a friend and said, "These must be the people looking for the dinosaur."[3] Even worse was an exchange that took place in 2003, while Gibbons was visiting Langoue, Cameroon (the same site to which he took the *MonsterQuest* crew in 2009). Cryptozoologists find it highly significant when African villagers pick out sauropod pictures from a set of animal flashcards or draw a sauropod-like profile, because (as the *MonsterQuest* narrator pronounced) they have "virtually no contact with the outside world."[4] But consider the witness who presented himself to Gibbons, and then "immediately picked out a picture of the *Diplodocus*. 'Brontosaurus,' he said, without hesitation. Brian [Sass] and I looked at each other. A brontosaurus? 'Why did he call it that?'" It turned out that the witness simply recognized the dinosaur from television.[5] Finally, it is difficult to know when and if the local people make the same distinctions among myth, legend, and reality that Westerners do.[6]

The *MonsterQuest* episode showed a few fuzzy images on underwater sonar, but nothing definitive—especially in waters that support crocodiles, hippos, huge fish, and many other large aquatic animals. The investigators made a big fuss about the holes in the ground from which the "footprint" casts had been made. Finally, they reached the peak of absurdity when they concluded the episode by poking around a large hole in the bank of a river and claiming that a Mokele Mbembe had dug the burrow and was inside. The burrow—if that's what it was (there are many random holes along riverbanks that are not true burrows)—was definitely not large enough for any creature matching the size of the dinosaur they were hunting. If it was a burrow, it may have been big enough for a snake, a crocodile, or possibly one of the monitor lizards that live in the area. But the investigators claimed that their huge hypothetical dinosaurs squeeze into riverbank caves or burrows, and then seal up the entrances behind themselves, leaving only relatively narrow "air vents." (These supposed hibernation burrows are vaguely similar in concept to the nesting chambers created by hornbill birds.) Although the sheer size of sauropods renders the burrowing idea silly, the larger problem is that *this whole scenario is made up from whole cloth.*[7] There is absolutely no reason to guess that small holes in the riverbanks may in fact lead to hibernating dinosaurs—nor do the dinosaur hunters seem to have made any effort to support their hunch with any actual digging. It is utterly baseless speculation, pulled as completely from thin air as if they announced that the holes led to pirate treasure or the remains of Jimmy Hoffa. Such breathtaking leaps showed how little actual training in field zoology they have had.

It turns out that none of the "explorers" seems to have had any relevant training in biology. The "chief scientist" and most experienced Mokele Mbembe hunter in the episode, William Gibbons, is a creationist with degrees in religious education.[8] His fellow "explorer," Robert Mullin, is a creationist as well.[9] Gibbons has published books on cryptids from a creationist perspective and with no peer review, the most recent of which includes the *MonsterQuest* expedition.[10] Neither their lack of appropriate training, including the simplest field biology, nor their creationist bias was mentioned in the film or in the narration. Both Gibbons and Mullin were treated as legitimate scientists. To anyone, such as myself, who has done a *lot* of field research in both biology and geology, their ignorance of basic procedures was painfully obvious by their approach and conclusions. Adding to the play-acting feel of the affair was Mullin's later declaration that "we were aware that the animal had moved on long ago, we knew when we set out on this particular expedition that it was really more of an opportunity to make a television episode raising awareness of the animal itself than it was a full-fledged expedition."[11]

My own part in the film was chopped down considerably, but I was pleased that most of my statements were left intact and not edited to contradict what I had really said (except for the final segment, in which a key phrase had been cut). Oddly, the filmmakers gave me more screen time pushing the cart with the duckbill skull and sorting fossils on the table during the voice-over, compared with the time I actually talked into the camera about the facts of the case. This is consistent with how the rest of the episode was filled with fluff about "explorers" in the Congo, rather than with evidence to support the claims for the existence of Mokele Mbembe.

THE SEARCH FOR MOKELE MBEMBE

Although not as well publicized as those of Bigfoot, Nessie, and the Yeti, the legends of Mokele Mbembe, the alleged dinosaur of the Congo, have a long history. The name is said to come from the Lingala language and is usually translated as "one who stops the flow of rivers." The creature is most often reported in the upper reaches of the Congo Basin in the Democratic Republic of the Congo, Equatorial Guinea, the Republic of the Congo, Gabon, Cameroon, and the Central African Republic, with the most intense interest focused around Lake Tele and the surrounding regions in the Republic of the Congo.[12] Like those of other cryptids, the lore and eyewitness descriptions of Mokele Mbembe are very discrepant, but a canonical image has emerged in the cryptozoological literature and in popular culture: it is envisioned as a sauropod the size of an elephant (or larger), with a long neck, no hair, and a long tail. Its skin is reddish brown, brown, or gray, depending on the report. It is said to live in the deeper water of the lakes in the Congo Basin and in the deep channels in the cut banks of the rivers. Some descriptions suggest that it has pillar-like legs and leaves tracks with a three-clawed foot impression, although other accounts differ about its trackways.

Early Testimony

Rumors of enormous beasts hidden in the Congo region date back to at least the sixteenth century. In 1776, French missionary Abbé Lievain Bonaventure Proyart's *History of Loango, Kakonga, and Other Kingdoms in Africa* alleged, "The missionaries have observed in passing along a forest, the track of an animal which they have never seen; but it must be monstrous, the prints of its claws are seen on the earth, and formed an impression on it of about three feet in circumference. In observing the posture and disposition of the footsteps, they concluded that it did not run this part of its way, and that it carried its claws at a distance of seven or eight feet one from the other."[13] Of course, stories about giant footprints are as widespread as human storytellers, and there seems no reason beyond geographic coincidence to infer any connection between the Proyart anecdote (the two sentences quoted from his book are the entirety of the story) and the modern Mokele Mbembe cryptid. While cryptozoologists have sought confirmation of their "living dinosaur" story in both African rock painting[14] and Middle Eastern art and literature of antiquity—citing as evidence even the *sirrush* dragons on the Ishtar Gate in Babylon, which was 2,800 miles away from Lake Tele in present-day Iraq (figure 6.2)[15]—the idea of an elusive African dinosaur-like animal seems to have developed only *after* the discovery in the nineteenth century of fossil dinosaurs and other reptiles from the Mesozoic era (250–65 million years ago). This is not to say that there weren't monster yarns in Africa; of course there were. It's

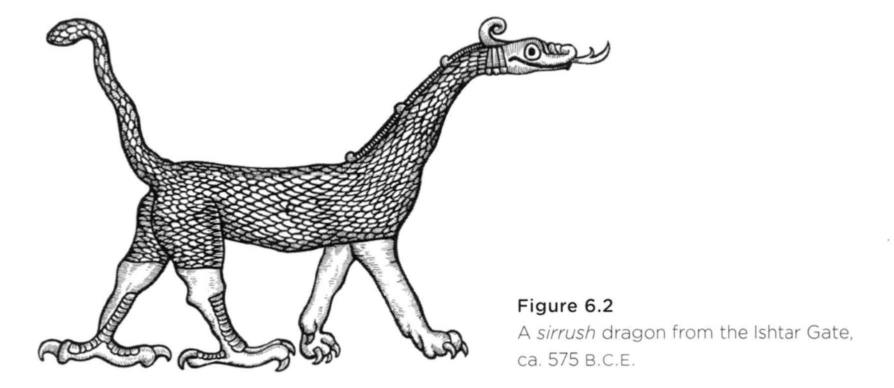

Figure 6.2
A *sirrush* dragon from the Ishtar Gate, ca. 575 B.C.E.

just that Africa's teeming menagerie of folkloric monsters did not in any clear way describe dinosaurs—not, that is, until the twentieth century.

This is a familiar story. As seen in the discussion of sea serpents and the Loch Ness monster, dinosaur discoveries influenced popular fiction—especially pulp and science fiction—and the folklore (and fakelore) of monsters as well. As early as 1833, arguments were advanced that ichthyosaurs and plesiosaurs might survive in the oceans; by 1864, Jules Verne and other authors had made encounters between modern humans and relict prehistoric beasts into a familiar literary trope.[16] The golden age of dinosaur paleontology reached a frenzied peak at the dawn of the twentieth century, igniting the imagination of millions. In particular, the world's first exhibits of mounted sauropod skeletons opened to teeming crowds and buzzing press in 1905. At the American Museum of Natural History in New York, *Brontosaurus* was unveiled at a gala luncheon attended by celebrities, titans of industry, and scientific superstars (including J. P. Morgan and Nikola Tesla) (figure 6.3).[17] When the doors opened to the public, "crowds poured by thousands" into the museum for a peek.[18] Meanwhile in London, the Natural History Museum revealed with similar fanfare its mounted, cast replica of another enormous sauropod, *Diplodocus*—which was a personal gift from Scottish American industrialist Andrew Carnegie to King Edward VII.[19] Other heads of state clamored for *Diplodocus* replicas for their own national museums, and Carnegie was happy to oblige. In 1907, American crowds stood in awe of Carnegie's original *Diplodocus* in its mounting in the newly constructed Carnegie Museums in Pittsburgh. Further replicas of *Diplodocus* were installed in Berlin in 1908[20] and Paris in 1910 (figure 6.4),[21] with still others going to the great museums of Austria, Italy, Russia, Spain, Argentina, and Mexico.[22]

Figure 6.3 The mounted skeleton of *Brontosaurus* (now called *Apatosaurus*) in the American Museum of Natural History, New York. (From W. D. Matthew, *Dinosaurs: With Special Reference to the American Museum Collections* [New York: American Museum of Natural History, 1915], fig. 19)

Anyone who has ever marveled at a dinosaur skeleton—perhaps even one of the original mountings of *Apatosaurus* or *Diplodocus*—will agree that these fossilized animals are an impressive and humbling sight. Yet it may not be possible for us to truly appreciate what it was like for the crowds who flooded into museums in the years 1905 to 1910 to see sauropod dinosaur skeletons *for the very first time*. They had never before seen a toy or a movie or a video-game depiction of these titanic creatures. To behold these reptiles' impossible-looking necks stretching into the rafters was a shock to the imagination. The wondering question echoed from country to country: *What would it have been like to encounter dinosaurs in the flesh?* Fiction writers leaped to answer that question, with Sherlock Holmes creator Arthur Conan Doyle's novel *The Lost World* (1912) stranding its bold adventurers on a remote plateau teeming with prehistoric animals. Edgar Rice Burroughs's books explored several variations on this "lost world" genre, with his heroes confronting dinosaurs and primeval beasts in a hollow Earth (*At the Earth's Core* [1914] and its sequels), on a mysterious island (*The Land That Time Forgot* [1918] and its sequels), and in a hidden African valley (*Tarzan the Terrible* [1921]).

Figure 6.4 Museum workers installing a replica of Andrew Carnegie's *Diplodocus* in the Musée d'histoire naturelle in Paris. (George Grantham Bain Collection, Prints and Photographs Division, Library of Congress, Washington, D.C.)

As much a promotional coup as a marvel of natural history, Carnegie's *Diplodocus* was an international sensation. Admiring accounts appeared in the African press, just as they did elsewhere. In Africa, however, mentions of *Diplodocus* and *Brontosaurus* took on a competitive edge after 1907, when mining engineer Bernhard Sattler and paleontologist Eberhard Fraas discovered the fossilized bones of a massive new sauropod in German East Africa (a colonial territory that encompassed most of present-day Rwanda, Burundi, and Tanzania).[23] It was against this backdrop of sensational sauropod skeleton exhibits in Europe and dramatic sauropod fossil discoveries in Africa that famous exotic-animal dealer, animal and ethnographic entertainment showman, and zoo pioneer Carl Hagenbeck[24] stepped forward to present a seductive possibility: What if sauropods like *Brontosaurus* were not truly extinct, but instead *lived on in the remote swamps of Africa*? According to Hagenbeck's book *Beasts and Men* (1909) there was reason to think that was the case:

> Some years ago I received reports . . . of the existence of an immense and wholly unknown animal, said to inhabit the interior of Rhodesia. Almost identical stories reached me, firstly, through one of my own travellers, and, secondly, through an English gentleman, who had been shooting big-game in Central Africa. . . . The natives, it seemed, had told both my informants that in the depth of the great swamps there dwelt a huge monster, half elephant, half dragon. . . . [I]t seems to me that it can only be some kind of dinosaur, seemingly akin to the brontosaurus. . . . At great expense, therefore, I sent out an expedition to find the monster, but unfortunately they were compelled to return home without having proved anything, either one way or the other. . . . Notwithstanding this failure, I have not relinquished the hope of being able to present science with indisputable evidence of the existence of the monster. And perhaps if I succeed in this enterprise naturalists all the world over will be roused to hunt vigorously for other unknown animals; for if this prodigious dinosaur, which is supposed to have been extinct for hundreds of thousands of years, be still in existence, what other wonders may not be brought to light?[25]

This startling suggestion was merely an aside in Hagenbeck's book, but his celebrity as an expert on novel foreign animals (he was the first person to introduce many now-familiar species to Europe—and even to science)[26] ensured that his claim made headlines from New York to New Dehli.[27] "Brontosaurus Still Lives," proclaimed the *Washington Post*.[28] With those headlines, Hagenbeck's book launched what would become the modern cryptozoological legend of Mokele Mbembe. This press no doubt enhanced the appeal

Figure 6.5 The 66-foot cement *Diplodocus* was the largest replica of the life-size prehistoric animals that Carl Hagenbeck installed on the grounds of his zoo. (Photograph from *Travel*, January 1917, courtesy of David Goldman)

not only of the book, but of the cement-dinosaur attraction that Hagenbeck added to his zoo in Hamburg the following year—featuring as its centerpiece a 66-foot *Diplodocus* ("an exact copy of the skeleton of the same animal . . . with the addition of having the flesh on") (figure 6.5).[29]

Hagenbeck's alleged African "dinosaur, seemingly akin to the brontosaurus," made news around the globe, but it was nothing more than third-hand rumor, presented without details—except for that of location. According to the case that started the cryptozoological legend of Mokele Mbembe, the creature was "said to inhabit the interior of Rhodesia" (a British colonial territory that comprised present-day Zambia and Zimbabwe), an area 1,200 miles from the Lake Tele region of the Congo Basin, where Mokele Mbembe is currently supposed to reside. And while newspaper readers in Europe and America found it perfectly plausible that the primitive giants newly mounted as skeletons in their museums could still lurk in the heart of the "dark continent" alongside the "bloodthirsty savages," it is important to understand that reactions in the African colonial press were less enthusiastic.[30] "There may be no truth whatever in the story, and probably there is not," opined the *Uganda Herald*.[31] The resident zoologist at the Rhodesia Museum, E. C. Chubb, had nothing but polite scorn for Hagenbeck's claim. Having recently spent a month doing field research in the region where Hagenbeck's dinosaur was said to reside, reported the African press, Chubb "regretted that he did not meet with the creature on his travels, nor did he hear any rumours of its existence. In fact not only was this the first time that this land edition of the Great Sea Serpent has been sprung upon him, but with due deference to Hagenbeck he begged to doubt the existence of such a monster." After all, if it were true, "it is curious that nothing has been heard of it before through local native sources"—and, besides, the half-elephant, half-dragon description suggested a biologically impossible chimera, rather than a dinosaur.[32] While equally skeptical, another man wrote in response to Chubb that he had heard tales of a monster when in the same region of Rhodesia—and had, he said, talked to "two natives who said they had actually seen it." The creature he described was not a sauropod, however, but something more like a plesiosaur—or even more like an imaginary chimera. It had "paddles or flippers" that it "used to propel itself with. The general description was the head of a crocodile, with rhino horns, neck like a python, body of a hippo and a crocodile tail, all of tremendous size."[33] In a private letter to Chubb that he evidently shared with the *Buluwayo Chronicle* (which ran it under the headline "The 'Brontosaurus': More Hearsay"), another writer reported rumors of a "three-horned water rhino" that "kills the hippo in Lake Bangweolo" and lives in swamps in that region.[34]

In 1911, another famous German adventurer, Lieutenant Paul Graetz, collected further tales—and, he said, some physical evidence—about a creature that was known to the locals of the Lake Bangweulu region of Zambia as *nsanga*. Graetz was a celebrity, having become the first person to cross Africa in an automobile (a two-year journey), and in June 1911 he set out to cross the continent again—this time by motorboat. Widely circulated press hype about his aim to reach "parts of central Africa, which are at present practically unexplored" and "in particular Lake Banguelo" attracted some backlash from people familiar with those regions.[35] One correspondent wrote "ridiculing the pretentions [*sic*] of Lieut. Graetz" to explore "the mysterious Lake Bangweolo. The correspondent says he lived there for years, shot over it, fished it, and says hundreds of white men have thoroughly explored this hunters' paradise." He added, "The stories of the Clupekwe (the mythical Brontosaurus) are just legends" and explained that regional folklore describes dozens of equally fabulous monsters, including "a wonderful animal in shape like the lion, but of monstrous size, which even frightens the lordly elephant."[36] Despite this warning, Graetz reached Lake Bangweulu at the end of October and found people willing to tell tales of monsters. He wrote, "The crocodile is found only in very isolated specimens in Lake Bangweulu . . . but in the swamp lives the nsanga, much feared by the natives, a degenerate saurian which one might well confuse with the crocodile, were it not that its skin has no scales and its toes are armed with claws. I did not succeed in shooting a nsanga, but . . . I came by some strips of its skin."[37]

How did the notion of a dinosaur in Rhodesia in 1909 morph into the legend of a dinosaur half a continent away in the Congo today? Whatever the description of this Rhodesian monster (or monsters), sauropods were all the rage in Europe and America. The international press latched onto the idea of a living African sauropod and never let go. We can infer that this African dinosaur was a cultural creation, a creature from the imagination projected onto the African landscape because it was projected *everywhere*—eventually including the Congo (figure 6.6). "Rumours about creatures of this kind have been reported in a ring of country 2,000 miles in diameter," explained Bernard Heuvelmans.[38] This is not an exaggeration. In 1910, for example, a character named Charles Brookes came forward to say that "pygmies" from the southern rim of the Sahara Desert had told him that colossal dinosaurs lived around "great lakes existing in the heart of the desert itself"—along with giant humans.[39] Notably, he cited direct inspiration from Carnegie's sauropod skeleton: "[T]he 80-foot diplodocus, now in the Paris Museum of Natural History . . . brings back to my mind many stories I heard from natives and pigmies . . . regarding monsters that are still in existence."[40] (Perhaps also notably, he was pursuing government funds to mount an expedition.) Brookes's claimed dinosaurs, living 1,000 miles or more from Lake Tele, the home territory of Mokele Mbembe, were still close neighbors compared with the *Brontosaurus* reported in 1921 in South Africa. Allegedly spotted lurking in the Orange River by multiple witnesses on multiple occasions, this "strange, gigantic beast . . . swims in the rapids and is so tall that he stands upon his feet and stretches his neck into the trees, where he devours the topmost branches."[41] These Orange River sightings took place almost as far from Lake Tele as New York is from Los Angeles.

Rather than recording genuine ethnozoological knowledge of a population of animals in the Congo Basin, the Mokele Mbembe lore is a distillation of many creatively varied stories from widely separated regions. Nonetheless, there *are* signs that a recognizable version of the modern Mokele Mbembe tradition existed in the Congo by 1913—even if most of the dinosaur sightings in Africa were occurring elsewhere across the continent. That year, according

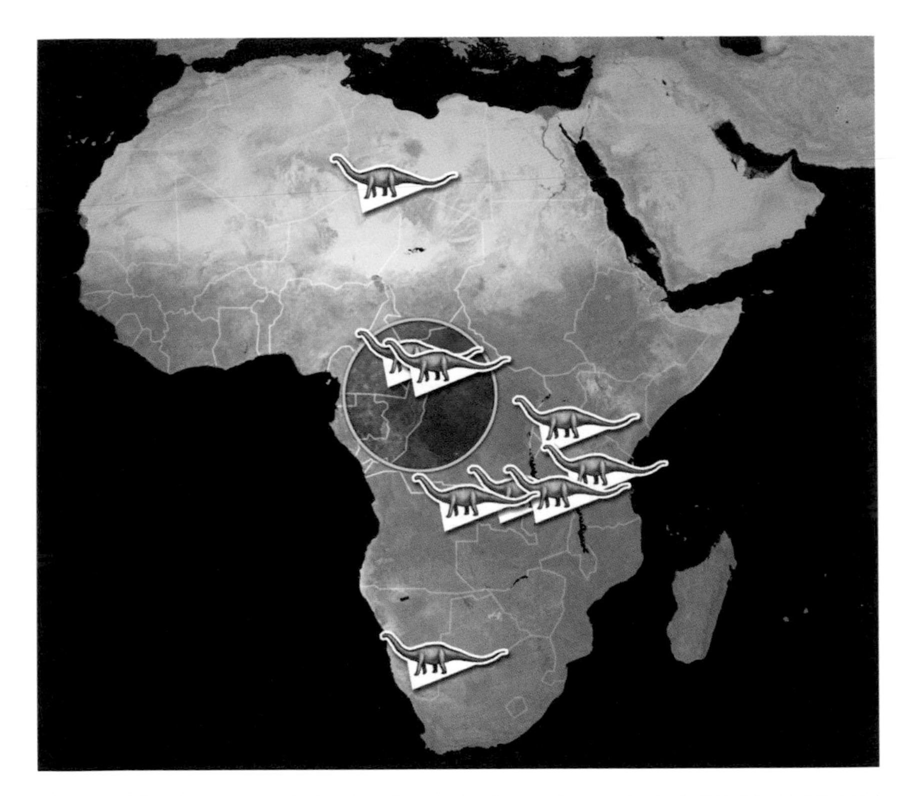

Figure 6.6 The canonical Mokele Mbembe's alleged Congo Basin habitat is highlighted in this map of Africa (Lake Tele is near the center), but reports have placed relict African dinosaurs in many other regions of the vast continent, thousands of miles apart. This map indicates just a few of the early reports of *Brontosaurus*-like animals in Africa. (Illustration by Daniel Loxton)

to Willy Ley, a German officer named Ludwig Freiherr von Stein zu Lausnitz led an expedition into the interior of what was then the German colony of Kamerun. He is reported by Ley to have written a detailed manuscript that deals in part with "narratives of the natives" concerning a creature they called *mokéle-mbêmbe*:

> The animal is said to be of a brownish-gray color with a smooth skin, its size is approximately that of an elephant; at least that of a hippopotamus. It is said to have a long and very flexible neck and only one tooth but a very long one; *some say it is a horn.* A few spoke about a long, muscular tail like that of an alligator. Canoes coming near it are said to be doomed; the animal is said to attack the vessels at once and to kill the crews but without eating the bodies. The creature is said to live in the caves that have been washed out by the river in the clay of its shores at sharp bends.

> It is said to climb the shores even at daytime in search of food; its diet is said to be entirely vegetable. This feature disagrees with a possible explanation as a myth. The preferred plant was shown to me, it is a kind of liana with large white blossoms, with a milky sap and applelike fruits. At the Ssômbo River I was shown a path said to have been made by this animal in order to get at its food. The path was fresh and there were plants of the described type nearby. But since there were too many tracks of elephants, hippos, and other large mammals it was impossible to make out a particular spoor with any amount of certainty.[42]

This account is obviously extremely important. If correct, it establishes a reasonably early date of development for the Mokele Mbembe legend in form and name, and even (arguably) suggests some now-canonical details. However, important questions of provenance have yet to be resolved. As Heuvelmans notes, this "report was never published, because Germany lost interest in the Cameroons when she lost the colony itself after the 1914–18 war, but it still exists in manuscript."[43] Stein's important account is known to the cryptozoological literature *only* from the fragment translated by Ley in the 1940s—and the background for it is known only from Ley's brief discussion. When was the manuscript written, and does it still exist? Is Ley's translation accurate? What were the circumstances of Stein's expedition? What other relevant information does the manuscript contain? Reliance on Ley regarding these questions is a major gap in scholarship about Mokele Mbembe. (There is possible partial corroboration for this account—or possible refutation—in a recollection from Chalmers Mitchell, secretary of the Zoological Society of London. Commenting in 1919 on a current yarn about a Congo dinosaur, Mitchell recalled a remark made by former Kaiser Wilhelm: "When visiting the London Zoo he described the existence of a similar pre-historic monster in German East Africa."[44] German East Africa was on the opposite side of the continent from Cameroon.)

Whatever the provenance of Stein's manuscript, it was in any event unknown to the public. So how did the "African *Brontosaurus*" legend—originally focused on Rhodesia—become entrenched as a belief about the Congo? In some ways, either location is equally suitable for a Western media legend about an African dinosaur: both regions feature vast, lush swamps and river systems, making them seem to early-twentieth-century audiences like the perfect habitat for sauropods. As one 1909 response to Hagenbeck's dinosaur claims conceded, "Those creatures lived under conditions probably not very dissimilar from those that obtain to-day in the swamps around Lake Bangweolo and Lake Mweru."[45] As we now know, sauropods were not swamp dwellers and did not spend their time immersed in lakes or rivers—but in those days, people incorrectly imagined just that. In the public imagination, all sauropods were essentially Mokele Mbembe.

Nonetheless, it was a specific case that brought the Rhodesian dinosaur to "the Congo"—though only just. In 1919, London newspapers rang with astonishing dinosaur news from "the native village of Fungurume" in the Belgian Congo (the southeastern corner of the present-day Democratic Republic of the Congo), a mere 250 miles from Lake Bangweulu in Rhodesia:

> The head of the local museum here has received information from a Mr. Lepage, who was in charge of railway construction in the Belgian Congo, of an exciting adventure last month. While Lepage was hunting one day in October he came upon an extraordinary monster, which charged at him. Lepage fired but was forced to flee, with the monster in chase. The animal before long gave up the chase and Lepage was able to examine it through his binoculars. The animal, he says, was about 24 ft. in length, with a long pointed snout adorned with tusks like horns and a short horn above the nostrils. The front feet were like those of a horse and the hind hoofs were cloven. There was a scaly hump on the monster's shoulders. The animal later charged through the native village of Fungurume, destroying the huts and killing some of the native dwellers.[46]

As the horror of World War I unfolded, Africa's dinosaurs seem to have vanished—but with the dawn of peace, the world was once again ready for an entertaining monster mystery. Headlines buzzed with the news of Lepage's sighting, with some newspapers explicitly likening this encounter to those in Conan Doyle's *The Lost World*.[47] It was swiftly followed by a report from a second man:

> A Belgian prospector and big game hunter, named Mr. Gapelle, who has returned here from the interior of the Congo, states that he followed up a strange spoor for 12 miles and at length sighted a beast certainly of the rhinoceros order with large scales reaching far down its body. The animal, he says, has a very thick kangaroo-like tail, a horn on its snout, and a hump on its back. Mr. Gapelle fired some shots at the beast, which threw up its head and disappeared into a swamp. The American Smithsonian expedition was in search of the mysterious monster referred to above when it met a serious railway accident, in which several persons were killed.[48]

It is important to note that Lepage's and Gapelle's accounts do not describe sauropods and that they were set 1,000 miles from the regions where cryptozoologists now seek Mokele Mbembe. But these were not the only flaws. Fearing that the tales "may grievously harm scientific research," a correspondent knowledgeable in the details of the cases wrote to explain how they had unfolded as a practical joke. "I have absolute authority to say that the whole report is fabulous," he wrote, "and I trust that the same publicity will be given to the correction that was given to the rumour."[49] The stories swiftly came apart under a crossfire of critiques from people who knew the joker, David Le Page. "It is all a yarn," one wrote to a relative in England. "A missionary arrived at Fungwrame mine . . . on his way south and stayed at the mine to mess, waiting for a train. Dave le Page, whom I know well, pulled the missionary's leg by reciting the yarn. . . . The missionary gave an account to the Press just as it appeared. The story caused a lot of amusement up here."[50] This exposé was confirmed in detail two years later by a Mrs. Simon, who had been in Fungurume when the lark took off. She explained to the papers that Le Page, "well-known in Southern Rhodesia for his elastic imagination," fooled a Mr. Raymer, who then communicated the bogus yarn to the head of the Port Elizabeth Museum. The museum director, in turn, passed it to the press.[51] Speaking to the remaining detail of the story, Wentworth D. Gray wrote as the representative of the Smithsonian African Expedition to say that the explorers were not, in fact, hunting dinosaurs. Moreover, he pointed out, Gapelle "does not exist except in the imagination of a second practical joker, who ingeniously coined the name" as an anagram of Le Page.[52] And yet, predictably, the unraveling of Le Page's story did little to dampen its inspirational effect. Copycat sightings soon emerged—one of a monster "with a head like a lion, with fangs resembling a walrus, 18 feet in length, and its body covered with scales and spottings like a leopard"[53]—and brave souls set off in search of dinosaurs: "Amongst the British big game shooters who are leaving in quest of the alleged Brontosaurus is Capt. Lester Stevens, who is taking with him the dog 'Laddie,' who is half a sheep dog and half a wolf," the press announced, adding, "Capt. Stevens believes that the giant reptile lives in a subterranean lake."[54]

Modern Testimony

The modern cryptozoological legend of the Mokele Mbembe of Lake Tele grew slowly. Through the 1940s and 1950s, early cryptozoological authors

Willy Ley, Ivan Sanderson, and Bernard Heuvelmans picked up and discussed the turn-of-the-century "African *Brontosaurus*" idea.[55] It was a question, Sanderson wrote, "doubtless born of wishful thinking, that all of us have probably at one time or another asked ourselves—namely, could there be a few dinosaurs still living in the remoter corners of the world?" In the 1960s, independent explorer James H. Powell Jr. took up this very question, inspired by the material collected by Heuvelmans (and by the correspondence he struck up with Heuvelmans, Sanderson, and Ley). "Of these early accounts," Powell wrote, "none was more tantalizing than that of the Baron [Ludwig Freiherr] von Stein zu Lausnitz from northern Congo." As with so many popular ideas, the person who first seizes the reins sets the course. By zeroing in on Stein's report, and resolving to pursue it, Powell did just that. In 1972, he received a grant from the Exploration Fund of the Explorer's Club to study crocodiles in the region. Unable to get an entry visa for what was then the People's Republic of the Congo, Powell used the Explorers Club grant to pursue the African dinosaur (or at least to ask questions about it) in neighboring Gabon and Cameroon in 1976 and in Gabon again in 1979.[56] Meanwhile, he had introduced himself to Roy Mackal, who had been investigating reports of the Loch Ness monster. Together, they ventured to the Republic of the Congo in 1980—and in doing so began the fully modern cryptozoological legend of Mokele Mbembe.

As detailed by Mackal, numerous sighting reports, traditional descriptions, and dubious yarns have accumulated over the years, although they differ greatly in details and are highly inconsistent about many important physical features.[57] Most accounts come from a variety of local peoples, often in the form of hearsay. It is hard to determine how much Moklele Mbembe testimony is based on actual events and how much is based on legends that have been passed down and distorted through retelling. Worse, it is impossible to know how much of the lore of this cryptid has been invented to meet the expectations of foreign guests. As one of Carl Hagenbeck's colonial-era animal wilderness scouts explained, it was next to impossible to get reliable information on the alleged Rhodesian dinosaur because "the natives, wishing to please the white visitor and hoping for a valuable gift at the same time, are only too ready to assert that they know of an animal in their territory with blue skin, six legs, one eye, and four tusks. The size is entirely up to the questioner; the native will tell him what he thinks the white man wants to hear."[58] Neither this flexibility nor this conflict of interest vanished with the end of colonialism. For an example of the first, consider that one of Powell's informants in Gabon told him in 1976 that he had never seen the beast, only to dramatically change his story three years later. By 1979, the man was claiming to have, decades earlier, built a hut from which he staked out the monster for many days and many nights and finally watched it emerge from the water. He was now able to obligingly take Powell to the exact spot where he had seen the creature climb from the river, but objected in apparent fear when Powell attempted to take a depth sounding. "I have never seen a man more truly terrified," wrote Powell, seemingly unconcerned that the same man had previously denied seeing the creature at all. "If he were acting, then he ought to be in Hollywood picking up Academy Awards." (Nor did Powell seem troubled that most of the people in the area did not recognize a picture of a *Diplodocus*—or that one villager explicitly told Powell that the regional name *n'yamala* described an imaginary animal.)[59]

Although Mokele Mbembe seeker William Gibbons has insisted that "these people have nothing whatsoever to gain from telling stories because we don't pay them, they get no reward from us for doing this,"[60] he also has described both financial transactions and monetary disputes between foreign monster hunters and, for example, the residents of the village of Boha near Lake Tele. Gibbons reported that the villagers were angry that the expedition led by Herman and Kia Regusters in 1981 had failed to fulfill "promises of gifts and money" and that one expedition had been left stranded at the lake in 1987 by the local guides after refusing to pay more than the agreed price. Gibbons's own 1985 expedition paid the Boha elders for access to Lake Tele.[61] Similarly, a member of the 1988 Japanese expedition to Lake Tele described meeting with "Boha village elders, who demanded a substantial fee for our expedition to gain access to Lake Tele."[62] Gibbons even asserts that during negotiations with a Japanese expedition in 1992, the Boha village elders "promptly held the expedition hostage while the sum of $12,000 (USD) was sent . . . to secure their release" (though we have been unable to confirm this anecdote).[63] Clearly, there are plausible financial incentives for local informants to wish to entice further investigators and television crews to visit the region. Moreover, virtually all the eyewitness testimony in the Mokele Mbembe literature has been solicited and translated by a small number of guides—often paid guides who lead multiple cryptozoological tours to the same villages and draw on existing relationships year after year.[64] Notably, in countries with a guided tour industry, such as Cameroon (and, indeed, in

Figure 6.7 According to one poorly corroborated but frequently repeated anecdote, hunters killed and ate a Mokele Mbembe in the 1950s. As the story goes, those who ate the animal mysteriously sickened or died. (Illustration by Daniel Loxton and Jim W. W. Smith)

any service industry anywhere in the world), it is not uncommon for operators to quietly distribute funds to others in the service chain (and to receive commissions when bringing their tour groups to preferred vendors). Finally, Mokele Mbembe proponents themselves benefit when wealthy patrons and television companies provide tens of thousands of dollars in funding for their expeditions.

For a study of the problems with soliciting local testimony, consider the famous tale that a Mokele Mbembe was killed and eaten at Lake Tele, which is difficult to access, in 1959 (figure 6.7). According to this dramatic and mysterious story, all who ate the meat of the animal sickened or died. This has become such a canonical part of this cryptid's lore that it was featured as part of the action in Hollywood's Mokele Mbembe movie, *Baby: Secret of the Lost Legend* (1985). But where did the tale come from? During a one-month fact-finding mission to the Congo in 1980, Mackal and Powell heard the story of a Mokele Mbembe killed (but not eaten) at Lake Tele—first as "vague rumors," then as an anecdote that an unnamed soldier heard from his wife,[65] and finally as an unsourced account related by a local official who was variously identified as Antoine Meombe or Miobe Antoine.[66] Mackal and Powell followed these rumors to the district of Epena, where they put the question to President Kolonga (or Kolango) of the Epena District of the Likouala Region. To their surprise, Kolonga "smiled at the words Mokele-mbembe, declaring that the word only meant 'rainbow.'"[67] Mackal argued with him—and fed the story to the informant: "We have heard several times over that a Mokele-mbembe was killed some time in the past in Lake Tele. We have heard, too, that this Mokele-mbembe is very dangerous, although its food is strictly vegetable material; the malombo is its favorite food. If your people, or rather the pygmies at Lake Tele, are able to kill a rainbow with spears, and the rainbow eats malombo fruit, we are very interested."[68] It was a ready-made template—essentially, "This is the story we want to hear." (Note as well that this is merely the extent of the leading that Mackal *tells* us they did.) Sure enough, Kolonga announced the next day that he was willing to "provide us with the truth about the Mokele-mbembe." Powell, seemingly unaware of how suspicious this sounds, explained how this went down:

> Gradually, as we gained his confidence, and he came to realize we were serious, and not laughing at the traditions of his people, he became cooperative, and promised to collect for us informants who had lived near Lake Tele, as well as others who could give information on "deep places where the animals are seen."
>
> When he had done this, and we were all gathered together at his house, Mr. Kolango made a little speech to the assembled informants, explaining that we had come from far away to get information on the mokele-mbembe, that we took the existence of the animal seriously and were not laughing at their traditions, and exhorting them to tell only the truth, neither inventing nor holding back.[69]

Sure enough, two of the informants thus provided and primed by Kolango obligingly recited the tale that the Americans had *already announced that they wanted to hear*. Both the first storyteller (identified as Mateka Pascal) and the second ("a fisherman") acknowledged that they were repeating hearsay:

> He (Mateka) did not personally see the animal, as he was only a small child at the time. According to his account, the mokele-mbembe had been entering Lake Tele from the moliba in which it lived via one of the waterways which enter the lake

> on its western side. After the animal had entered the lake, the pygmies blocked off its waterway by constructing a barricade of large stakes across it. When the mokele-mbembe tried to return to its moliba, it was trapped by the barricade and killed with spears. Some of the stakes used to construct the trap were large tree trunks, and are still there. The pygmies cut up the animal and ate it. All who ate of it died. The animal killed was said to be one of two. The other one—possibly a mate—is said to still be there, but has become wary and difficult to approach.[70]

The second, unnamed man merely confirmed the hearsay yarn that *he had just heard*, adding only the flourish of a sole survivor: "Those who ate of the meat died. The one survivor, who did not eat of the meat, had died about ten years ago." Even Mackal seems to have been uneasy about the possibility of informants influencing or inspiring each other's stories. "If there was anything negative about this informative meeting," he reflected, "it is that all or most of the eyewitnesses were in the same room, hearing everything that was being reported."

What are we to make of the "all who ate of it died" story? It is hearsay at best, and the possibility that the story was concocted to order looms large. Only one independent telling of the "all who ate of it died" yarn was recorded—and that by someone admittedly not yet born when the event allegedly took place. Nor is there any corroborating evidence, despite the tantalizing claim that "because Pascal still goes to fish in the very molibo where the killing occurred, he could affirm that the stakes are there to this day, and, therefore, the story is true." Because Mackal's and Powell's visas were about the expire, they left without visiting Lake Tele. "Roy and I now wanted to go on to Lake Tele and the site of the alleged killing of the mokele-mbembe in 1959," Powell explained. "There we hoped to find bones or other physical remains, perhaps even to sight and photograph the surviving animal. But this was not to be."[71] (Mackal returned to the region the following year, but seems not to have verified this one straightforwardly testable part of the hearsay "all who ate of it died" tale.)

That's pretty thin stuff, but it gets worse. Mackal reported that on several occasions local Africans either denied any knowledge of Mokele Mbembe or *asserted that the creature did not exist*—and that he refused to accept such negative testimony! At the village of Moungouma Bai, for example, the locals explained that they had "heard of the Mokele-mbembe from [their] fathers" but "never saw it" themselves. "I was astonished," Mackal wrote. "Here we were, only a few kilometers from Lake Tele, where a Mokele-mbembe had been killed, yet these villagers claimed to know nothing about it." Mackal's confrontational response to this testimony is jaw-dropping (especially when one considers that Mackal's party was dispersing beer with their pressure to provide congenial testimony and included Congolese security guards armed with AK-47s):

> Through Gene and Marcellin as interpreters I responded, demonstrating our rather extensive knowledge of the episode at Lake Tele, including descriptions of the appearance and habits of the animal, what it ate, where it had been seen and by whom. When confronted with such a barrage of information, they were visibly disturbed, and some, in their confusion, admitted to a great deal more knowledge. . . . It became clear that the people of Moungouma Bai were hiding information and knew a great deal about the Mokele-mbembe but were not going to share it with us. . . . Georges made an impassioned plea for cooperation, first conciliatory and then threatening.[72]

Given this pressure, it is not surprising that some of the villagers began to offer testimony that complied with the story the Americans had provided. Still, Mackal remained displeased with their lack of enthusiasm for his Mokele Mbembe narrative. The party left some desperately needed medical supplies with a Red Cross medic who was visiting the village, but Mackal stated his opinion "that these people did not deserve our medical largess."

Local testimony about Mokele Mbembe is badly burdened by hearsay and by obvious leading (and pressuring) of alleged witnesses. Moreover, regional informants have also supplied descriptions of a whole menagerie of additional, distinct monsters, including "a giant turtle, a giant crocodile, a giant-snake-like creature, a water elephant with a great horn but no trunk, an animal with plank-like structures growing out of its back, and of course, the Mokele-mbembe proper."[73]

A few Westerners (mostly missionaries) claim to have caught brief glimpses of Mokele Mbembe, but their accounts are also highly inconsistent and difficult to interpret. Every attempt to obtain reliable photographs, films, or footprint casts of this alleged creature has failed. The film shot by zoologist Marcellin Agnagna in 1983 is also useless. Agnagna claimed first that he had failed to remove the lens cap and then that the camera had been set at macro-focus rather than telephoto. (The circumstances are also suspicious. Visiting the same Boha village three years later, Gibbons reported that "we could not find any of the witnesses from 1983 who could confirm Marcellin's story."[74] Even more damaging to Agnagna's claim is this exchange between British travel writer Redmond O'Hanlon and the son of a Boha village elder, which took place in 1989: "'So, Doubla,' I said softly, 'why did Marcellin swear he saw the dinosaur?' 'Don't you know?' said Doubla, giving me his first real smile. 'It's to bring idiots like you here. And make a lot of money.'")[75] The photos taken in 1985 by Rory Nugent cannot be usefully interpreted. As one critic described them, "One is a very distant snapshot of what appears to be a log floating in a lake; the other might as well be a flying, out-of-focus wedding bouquet in transit past a bed sheet."[76] (Even cryptozoologists are unimpressed with these images. "Rory Nugent's alleged Mokele-mbembe photos could be anything," according to Gibbons.)[77]

More revealing is how many expeditions have traveled through the Congo Basin in search of Mokele Mbembe without obtaining any convincing evidence of the creature (or in some cases, even managing to find locals willing to affirm that it exists). Between 1980 and 2000, "almost twenty expeditions . . . searched unsuccessfully to find *mokele-mbembe*," Gibbons reflected.[78] Let's briefly consider a sample. In 1981, the husband-and-wife Regusters expedition traveled to Lake Tele in competition with an investigation led by Mackal. (These expeditions were initially one, but split as a result of a dispute between Mackal and Regusters.)[79] After their return, the couple held a press conference in which they swore to have seen Mokele Mbembe multiple times—and to have taken photographs, not yet developed. "Even if we had the best photographs in the world," said Regusters, "there would still be people who do not believe it."[80] This turned out not to be a problem , since none of their thousands of photos showed evidence of a dinosaur.[81] In 1985/1986, creationist and cryptozoologist Gibbons spent time around Lake Tele under the guidance of Agnagna, but "did not find any tangible evidence for the existence or otherwise of *mokele-mbembe*."[82] With the end of that expedition, Gibbons and his team fell out with Agnagna, who was alleged to have deliberately concealed the location of Mokele Mbembe (among other matters of dispute).[83] In 1988, a Japanese expedition and film crew searched the Lake Tele region for thirty-five days, but "found no evidence of the existence of Mokele-Mbembe in the area."[84] O'Hanlon toured the area in 1989 and interviewed many local people, who told him that the creature was a spirit and not a physical being.[85] In 1992, a Japanese expedition captured aerial footage of something large and blurry in Lake Tele.[86] Although ultimately useless—the object was too distant to allow any positive identification—it has been suggested that this Japanese team may have filmed a canoe. As cryptozoologist John Kirk explained, "What some have taken to be the head and neck of the creature could also just as easily be a man standing at the front of the boat while oaring his way across the lake, while behind him sits another man who might be mistaken for a hump."[87] Gibbons's second trip, in 1992 (apparently financed, oddly enough, by Mick Jagger, Ringo Starr, and other rock musicians),[88] was actually a missionary and sightseeing trip, with the team's short stay in the region occupied with dispensing medical treatments and spreading the Gospel.[89] As a highlight, Gibbons describes testifying in a thatch-roofed village church: "I began to speak, telling them my story of salvation and finding Christ in the jungle. . . . After I had finished speaking, Sarah gave an altar call, and three people . . . came forward and accepted Christ. . . . The power of Satan had been broken. It was a glorious day indeed!"[90] Notwithstanding this spiritual victory, no tangible sign of Mokele Mbembe was located. In 1999, J. Michael Fay led a 2,000-mile, 456-day biological transect through the region on foot and reported nothing relevant to Mokele Mbembe.[91] Gibbons spent most of November 2000 scouting for Mokele Mbembe in Cameroon[92] and led a well-funded, all-Christian search into Cameroon in 2001, accompanied by a film crew for the BBC and Discovery Channel.[93] They heard stories of several monsters, including the claim to have seen a gray-haired Bigfoot-like creature carried out of the jungle, bound to a pole, by a group of European hunters. Although the members of the expedition allegedly spotted a series of UFOs (one of which "took off into space at a 45-degree angle at a speed that no known earthly craft could possibly match"), they found no tangible evidence for Mokele Mbembe.[94] (Disappointed by the entire affair, the Discovery Channel dumped the documentary "due to insufficient film material"; the Christian businessman who had helped them secure the contracts and funding for their creationist expedition declared them "amateurs" and parted ways.)[95]

Despite this embarrassment, an American insurance broker and creationist named Milt Marcy offered to foot the bill for Gibbons's next expedition, which visited Langoue, Cameroon, in 2003.[96] This trip was notable for introducing yet another type of monster: giant spiders. Having heard from an English woman who claimed that her parents once had spotted a spider "at least four or five feet in length," Gibbons asked the villagers "if they knew of any such giant arachnid, and indeed they did!" In a remarkable coincidence, these monsters—"strong enough to overpower and kill a human being"—lived right around the village! No one had mentioned the giant spiders to Gibbons on his previous two visits because he had not asked, which was a shame because a giant spider had been living right behind the camp of one of his informants at the time. "At that moment," wrote Gibbons, "I felt like drowning myself in the river. A golden opportunity to capture a rare and completely unclassified species of giant arachnid had eluded us."[97] (Such a B movie–style giant spider probably is not possible, physically: limited by an exoskeleton, it could neither support its own weight nor extract enough oxygen from the air to survive.) The group poked around for a few days and delivered some sermons, with the group's Cameroonian Pastor Nini "even casting out demonic spirits."[98] Then, on their last day on the river, "Pastor Nini boldly proclaimed that this would be the day when we would encounter the la'kila-bembe [allegedly a synonym for Mokele Mbembe]. The Lord had assured him that this was truly the day"—and right on schedule, it was! As they floated over the water, their guide stood up in the canoe, declaring that he could see "a very big animal crossing the river just ahead of us." The cryptozoologists saw nothing and captured nothing on film, but they were impressed with the guide's detailed description of a reddish-brown creature with "typically reptilian" eyes.[99] And then they went home.

In 2004, Marcy funded a return expedition to Langoue, headed this time by creationists Peter Beach and Brian Sass.[100] They used plaster to take casts of some marks that they intuited, based on nothing in particular, to be claw prints from a dinosaur. I examined these casts during the *MonsterQuest* production and determined that they were not sauropod tracks. But Beach and Sass also conceived a notion that now constitutes the cutting edge in Mokele Mbembe belief: the legendary dinosaurs seal themselves up in riverbanks for long periods. Supposedly, the animals first climb into caves or burrows and then wall up the caves from the inside, with small air vents near the top. As no part of this behavior has ever been observed—even allegedly—this wild speculation can only be described as baseless and bizarre. While examining the small riverbank holes that they supposed to be air vents, they heard "a distinctive scraping sound, as though something was attempting to claw its way out of the sealed chamber." Could it bc that Mokele Mbembe was only *feet away from them*—and about to emerge? We will never know because Beach and Sass got spooked, and left immediately.[101] In 2006, Marcy himself ventured to the same region with Robert Mullin and Beach. They located what they imagined to be several sealed caves containing Mokele Mbembes—which is to say, some small burrows or holes in the riverbank—

and that's it. Mullin described it "a dry run as far as Mokele-mbembe was concerned."[102]

In 2008, an episode of *Destination Truth* filmed possible Mokele Mbembe footage on location at Lake Bangweulu in Zambia, the original Rhodesian home of the "African *Brontosaurus*," but this long-distance sequence proved to be of two partially submerged hippos. "I think the people in this lake, they know this myth, they see something like this, which certainly we couldn't identify, and that's I think how this legend gets more and more heat on it until people are really just, you know, believing it," said the host, who concluded, "I think we can lay the mokele to rest."[103]

This brings us back to the *MonsterQuest* expedition of 2009, with all its flaws. Perhaps the silliest sequence features the small riverbank holes taken to be air vents for concealed, hibernating Mokele Mbembes. On camera, Gibbons pokes ineffectually at a bank with a small shovel, declaring, "Once they're sealed in there, it's very difficult to get them out. It's just a pity we didn't have any other way of finding out what's on the other side of this mud wall." Not so much "a pity" as absurd: this was the *third* expedition to travel all the way to Africa to examine the riverside burrows, only to (according to the leaders' confident assertions) stand idly chatting *within a few feet of hibernating Mokele Mbembes*—and not once did they bother to come prepared to dig out the dinosaurs they were there to find? Or even to send a camera inside the alleged air vent? The investigation of the "cave" for *MonsterQuest* literally consisted of poking a stick into the hole.[104] Then, Gibbons explained, "By 3:00 P.M., we all had quite enough of the baking heat and headed back to Langoue for an early dinner."[105] The weakness of this investigation is so obvious that Mullin has acknowledged, "Some have wondered why better equipment hasn't been taken on some of these trips." His explanation is that "we go with what we can afford; not every expeditioneer is rich, and often we do what we can at great cost to ourselves and our personal lives."[106] Perhaps, but many expeditions do little more than arrive and turn around, each burning several thousand dollars in the process. The Gibbons-headed BBC expedition alone enjoyed a budget of at least $65,000. In 2006, Marcy shipped an entire inflatable zodiac boat and outboard motor to Africa for his trip.[107]

As more and more expeditions have been mounted in the past thirty years, with access to increasingly sophisticated technology, much *less* evidence has been recovered. It is also striking that the descriptions of Mokele Mbembe are so inconsistent that one group of Westerners, having heard about its long neck, thought that it was a sauropod dinosaur, while another, having heard about its horns, assumed that it was a ceratopsian dinosaur. Finally, many of the leads have turned out to be hippos or crocodiles in the water. (Rhinoceroses sometimes have been suggested as culprits. Since they usually live in savannas, scrublands, and grasslands, not in the Congolese jungle, out-of-place rhinoceroses presumably would appear to be as exotic as dinosaurs.)

REALITY CHECKS

Sauropod Biology

What is even more revealing about the descriptions of Mokele Mbembe is that they are based on the antiquated image of sauropod dinosaurs as lumbering beasts that lived in swamps, ate aquatic plants, and dragged their tails in the muck (figure 6.8). This was still the standard depiction of sauropods when early-twentieth-century explorers in the Congo Basin interpreted eyewitness accounts—and even as late as the 1970s, when proponents such as Roy Mackal formed their impressions. But more recent scientific research has shown that sauropods were nothing like this century-old image. Numerous trackways of actual sauropods that date from the Jurassic period (208–144 million years ago) show that they were not slow-moving, tail-dragging dinosaurs. On the contrary, they walked fairly quickly and efficiently on dry land, had a relatively upright straight-legged posture, and held their tails out behind them (figure 6.9), as portrayed in movies like *Jurassic Park* (1993). The reinterpretation of the famous sauropod bone beds in the Morrison Formation in the western United States and Canada, which dates to the Late Jurassic (150–144 million years ago), has demonstrated that most sauropod fossils do not come from swampy deposits, for the dinosaurs lived in seasonally dry woodlands with very little standing water.[108] Evidence from their teeth shows that sauropods did not subsist on aquatic plants, as Mokele Mbembe is alleged to do, but fed on tough conifers and cycads, as well as on ferns.[109] Thus a century-old image of sauropods still influences the ideas of the cryptozoologists who believe in the existence of Mokele Mbembe and try to fit the inconsistent accounts of local peoples into an outdated concept.

Surviving Dinosaurs?

The biggest obstacle to proving the existence of Mokele Mbembe, however, is not the absence of hard evidence. It is that cryptozoologists seldom or never address the basic principles of biology and paleontology:

Figure 6.8 The Mokele Mbembe of legend resembles Charles R. Knight's painting of *Brontosaurus* (now called *Apatosaurus*), shown dragging its tail and supporting its bulk in the water. Knight's reconstruction was cutting edge in 1897, but the modern understanding of sauropods is very different.

Figure 6.9 Modern reconstructions of sauropods feature upright legs for life on land, a horizontal tail posture—and, in some species, an adornment of iguana-like dermal spines. (Illustration by Daniel Loxton, with Julie Roberts)

- *Population constraints*: As has been pointed out in relation to Bigfoot and Nessie, Mokele Mbembe cannot be a singleton or a handful of individuals. For dinosaurs to have survived in the Congo for the past 65 million years, there would have to be a sizable population of them. If there were, there would not be just a handful of inconsistent accounts and no direct evidence, but hundreds of carcasses and thousands of skeletal parts found throughout the Congo Basin over the 150 years of Western exploration. Although the swampy regions of the Congo are not ideal habitats in which to preserve skeletons, skeletal remains of elephants, hippos, and many other animals are found all the time. For a population of animals as large as Mokele Mbembe is believed to be, some hard evidence of its existence would surely have been discovered by now.

- *Aerial surveillance*: A large population of air-breathing sauropods would not be able to hide underwater indefinitely, especially since they preferred dry, open habitats and fed on conifers. They would have been seen by members of the many zoological expeditions that cross the Congo Basin every year, especially by the aerial surveys that are frequently undertaken to count large animal populations. For example, if you type the coordinates 10.903497,19.93229 into Google Earth and zoom in, you can see clear images of an entire elephant herd, with the details of their trunks, ears, tusks, and tails. Indeed, with spy satellites able to spot an object that is less than 3 feet across, and Google Earth able to resolve the details of your own backyard, a population of sauropod dinosaurs should have been caught by such surveillance.

- *The fossil record*: Although the jungles of the Congo Basin are not the ideal habitat in which to preserve or find fossils, there is an excellent fossil record of the past 200 million years from many parts of Africa. In South Africa, fossil beds from the Permian (286–250 million years ago) and Triassic (250–208 million years ago) periods have yielded excellent specimens of synapsids (formerly but incorrectly called "mammal-like reptiles") and some of the earliest dinosaurs. In the Tendaguru beds in Tanzania, which date to the late Jurassic, have been found some of the best-known sauropods, including the huge *Brachiosaurus* in the Museum für Naturkunde in Berlin, which is the largest nearly complete sauropod skeleton known. There are also fossils of sauropods (mostly titanosaurs) from the Cretaceous

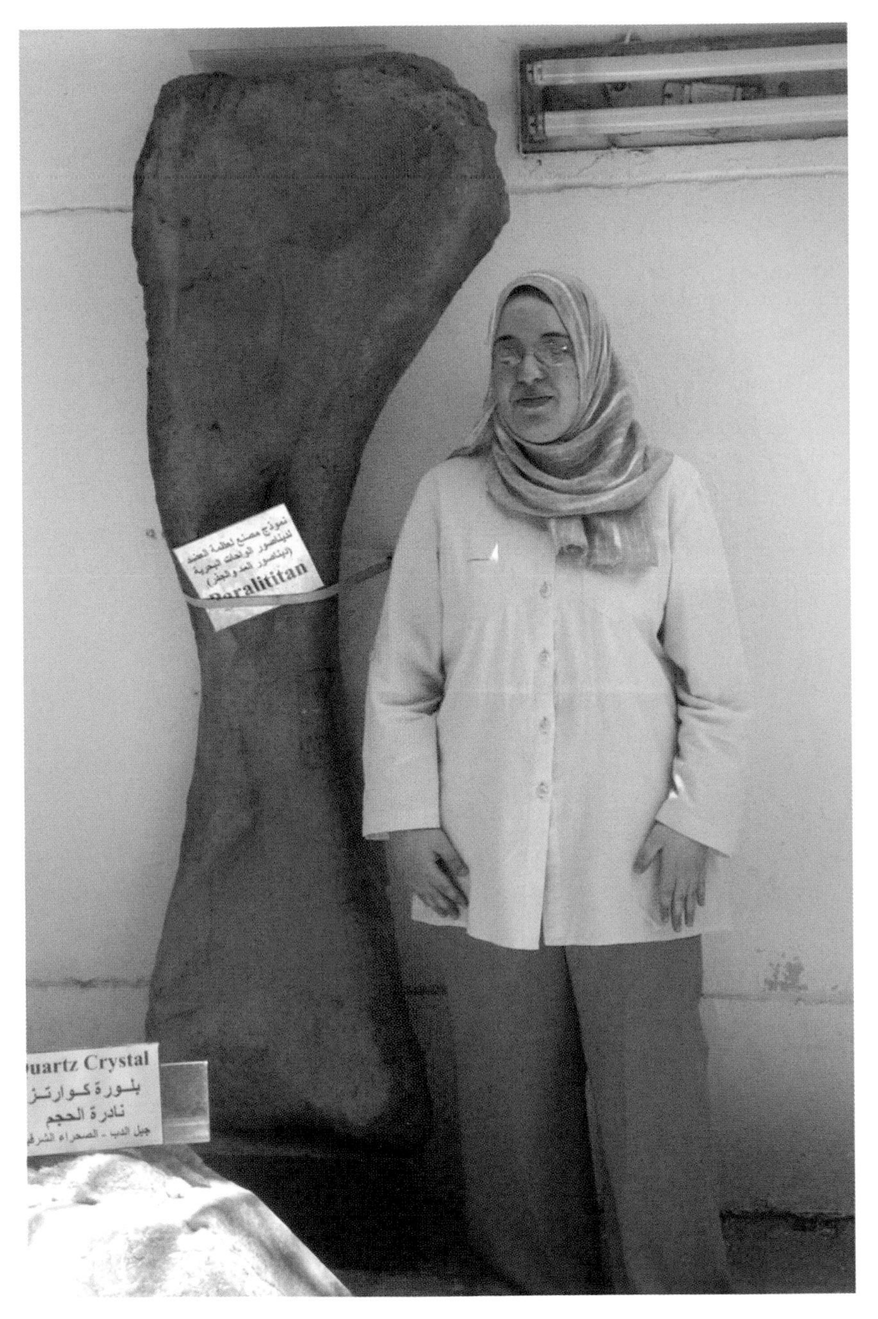

Figure 6.10 Paleontologist Enas Ahmed poses with a fossil humerus from the North African titanosaurian sauropod *Paralititan*, dating from the Cretaceous period, at the Egyptian Geological Museum, Cairo. (Photograph courtesy of Jason Loxton)

period (98–65 million years ago) in a variety of regions in Africa (figure 6.10), including the remarkable discoveries of diplodocoid sauropods by Paul Sereno in the western Sahara.[110] And then, in Africa just as in every other locality around the world, the nonavian dinosaurs vanished 65 million years ago, and not one bone has been found to show that they survived the mass extinction at the end of the Cretaceous. The African fossil record of the past 65 million years, the so-called Age of Mammals, is excellent, with many localities that have preserved large-bodied animals like mastodonts and rhino-like arsinoitheres—but not one sliver of a dinosaur.[111]

THE HIDDEN AGENDA: CREATIONISM

Most of the active explorers seeking Mokele Mbembe have a nonscientific agenda: Young Earth creationism (the evangelical Christian belief that Earth was created about 6,000 years ago by God, as described in the book of Genesis). For example, Roy Mackal's expeditions were shaped by their interpreter and guide, Pastor Eugene Thomas, an American missionary whom Mackal matter-of-factly described as "with us not only to interpret, but also to spread Christianity."[112] Thomas went on to personally convert and baptize William Gibbons in the Congo in 1986,[113] following a harrowing supernatural attack that Gibbons experienced while staying at Thomas's Impfondo mission station. Gibbons described this experience in terms that exactly match the classic symptoms of sleep paralysis (a presence in the room, physical paralysis, terror, a feeling of pressure on the upper body), a well-understood and common sleep disruption that many victims interpret in paranormal or supernatural terms:

> While I was slowly drifting toward sleep, something caused me to suddenly snap awake. There was something else in the room with me. . . . Within a minute, perhaps less, a dark and thoroughly evil entity began to fill the room. I tried to call out . . . but my voice died in my throat. I tried to sit up . . . but I was completely paralysed. Steadily the evil presence approached my bed. I was fully awake, yet completely immobile, soaked with sweat and beginning to go out of my mind with terror. . . . [T]he most unbearable pressure suddenly began to squeeze down on my chest, shoulders and arms.

In the days before this experience, Gibbons had become uneasy about his past "dabbling in the occult" (he mentions spiritualism and tarot cards) and had confided this behavior to Thomas. When experiencing his episode of paralysis in bed, Gibbons writes, "the words of Gene Thomas sprang into my mind. 'Only Christ can set you free.' With every last vestige of conscious will, I cried out in my mind to the saviour that the Christians worshipped. 'Jesus, help me!' . . . The suffocating, evil presence in the room recoiled, and then vanished." Perhaps unsurprisingly, this experience triggered a dramatic and sudden religious conversion.[114]

Gibbons went on to become the leading proponent of the Mokele Mbembe legend—a project he took on with overt missionary zeal. All the expeditions to the Congo led by Gibbons were undertaken with the objective of proving the truth of Young Earth creationism, and most of those who promote Mokele Mbembe are also creationist in their approach.[115]

In a book on cryptozoology, Gibbons and his coauthor (and flamboyant creationism activist) Kent Hovind shared their "hope and prayer" that their cryptozoological work would inspire "a new generation of Godly, Christian explorers, who will endeavor to venture forth to find and present these amazing mysteries of creation to an unbelieving world."[116] But why would creationists look to legendary monsters to support their biblical literalism?

Mokele Mbembe, in particular, is an idea that Young Earth creationists like Gibbons find natural and attractive: "To the Bible believing Christian, the idea of dinosaurs living with man in the past or even some still living today, is scientifically possible. Christians know that God made all the animals, including dinosaurs, about 6000 years ago."[117] But the creationists' fascination with Mokele Mbembe is not merely that its existence seems plausible within a creationist worldview, but that its existence has important ideological or theological ramifications. For some reason, creationists believe that the discovery of a dinosaur in Africa will overthrow the entire theory of evolution. This belief is poorly founded. The reality of evolution is based on a gigantic amount of evidence from the fossil record,[118] and a single find of a relict species does not overturn this mountain of data.

Creationists point to the discovery of the coelacanth in 1938 as upsetting the evolutionary story, but all it really did was extend the range of a species known rarely from fossil beds of the Early Cenozoic era (65 million years ago–present) into the Late Cenozoic. Gibbons shows their typical thinking:

> The Coelacanth was discovered alive and well in 1938 after having been dismissed as extinct for as long as 200 million years. The embarrassing thing about that for

> evolutionists was that it was thought to be a foundational species for the transformation of fish into amphibians and[,] as such, should be extinct, since foundation species are not supposed to continue surviving past their progeny. The Coelacanth paid no attention to those scientists and just kept right on "keeping on" to the present day.[119]

The many errors in this statement show the creationists' ignorance of the fossil record. Coelacanths were not thought to have been extinct for 200 million years, since there are fossils as young as 20 to 5 million years in age.[120] Contrary to the creationists' notions of evolution, coelacanths were not "foundational species for the transformation of fish into amphibians," but are an order of primitive lobe-finned fishes that appeared in the Late Devonian period (400 million years ago) alongside the earliest relatives of amphibians, as well as the earliest lungfish (also lobe-finned). Coelacanths were *never* thought to be ancestral to amphibians by any evolutionary biologist or paleontologist. Even more revealing is that Gibbons is following the outdated "ladder of creation" notion of evolution, according to which ancestors must die off to give rise to their descendants. This idea is completely erroneous and is akin to saying that your grandfather must have died when your father was born, and your father died when you were born.[121] Evolution is bushy and branching, and ancestral groups often survive and live alongside their descendant groups.

If it were not clear enough from their general arguments, the creationists' motives are clearly spelled out by Gibbons:

> In case there is any doubt about our motivation for this work I should tell you that we feel that the discovery of any of these creatures will be an earthshaking event. It is our belief that eliminating common objections regarding why the Bible can't be trusted, and demonstrating the historical and scientific accuracy of Scripture[,] will naturally lead people to the next logical step in thinking: If the Bible is true in other respects, what does that tell us about its spiritual ramifications?
>
> When the evolution hypothesis was proposed 150 years ago, it was with the expressed intent of destroying the church and Christianity along with it. If a wrench of this kind could be thrown into the machinery of evolution it would go a long way toward turning people back to the only real truth, the Word of God.[122]

There are many distortions in this statement, but the one that really stands out is the claim that evolution was proposed expressly to destroy Christianity. Evolution was proposed only to explain the observed pattern of life. Many evolutionary biologists are devoutly religious and have no conflict with the theory of evolution.[123] Evolution undermines the religious beliefs of only fundamentalists, who insist that Genesis must be true in every detail, despite a large amount of evidence to the contrary.[124]

Thus the quest for Mokele Mbembe is not just an idle search for a cryptid, but part of the effort by creationists to overthrow the theory of evolution and undermine the teaching of science by any means possible. As such, it cannot be dismissed or treated lightly, but must be subjected to the full scrutiny of the scientific community.

WHY DO PEOPLE BELIEVE IN MONSTERS?

THE COMPLEXITY OF CRYPTOZOOLOGY

"I want to believe."

THE X-FILES

THE CRYPTOZOOLOGY SUBCULTURE

WHO ARE THE PEOPLE who spend a significant amount of time, money, and other resources thinking about and searching for Bigfoot, Nessie, and other cryptids? What motivates them? Are they scientists or pseudoscientists? If they are not practicing real science, is it possible to transform cryptozoology into a real science? Is cryptozoology just a hobby of no particular consequence, or does it make genuinely useful contributions—or can it, perhaps, lead to harm?

Whatever the case, the cryptozoological impulse is not new. Monsters have always stirred in the shadows of the imagination, as have sincere efforts to understand them. Projects to describe, categorize, and probe claims of legendary beasts extend back through the modern era and medieval period, to such classical authors as Pliny the Elder.

Journalist Elizabeth Landau put it this way:

> Across human societies, variations on mythical creature stories like that of Bigfoot have persisted for thousands of years, and accounts of seeing or hearing them still abound. There may be some basic culture-based need for these fantastical tales, said Todd Disotell, professor of anthropology at New York University.
>
> Monsters represent dark aspects of our subconscious worlds and can be metaphors for the challenges of life, said Karen Sharf, a psychotherapist in New York.
>
> "Some monsters are scary. Some monsters are friendly. Sometimes in movies or myths, we befriend the monster, and it's just like in our inner world: There are monsters; there are dark aspects that we have to face," she said.
>
> Humans also have a fascination with the divide between their species and animals, and Bigfoot bridges that gap, said John Hawks, anthropologist at the University of Wisconsin, Madison.
>
> Believing in these creatures and following their trails in the forest is somewhat akin to an amusement park ride: They are safe ways of experiencing fear, said Jacqueline Woolley, professor of psychology at the University of Texas at Austin.[1]

As we have explored in this book, reports of mysterious creatures have drawn interest for centuries from natural historians, popular writers, and explorers—and have long fascinated the public as well. That attraction continues to this day. The Baylor Religion Survey found in 2005 that one in five Americans have read a book, consulted a Web site, or otherwise researched "mysterious animals, such as Bigfoot or the Loch Ness Monster."[2] Many of those people are interested because they are persuaded that cryptids exist, but around 13 percent of those who do *not* believe in cryptids have made the effort to seek out information about these hypothetical creatures.[3] Co-author Daniel Loxton can relate to both sides of that interest, having been first motivated to devour cryptozoology books by the belief that cryptids exist, only to sustain an enduring fascination with monster mysteries in his later role as a skeptic and critic of the topic. Loxton is not unique in this; many people feel the attractiveness of cryptozoological ideas, even if they are skeptical about the literal existence of cryptids. Disbelievers are not immune to a feeling that Joshua Blu Buhs described as a transcendent interest "shared by Bigfooters, a feeling that united them and underlay all of their activities, their expeditions, their reading, their correspondence.... That transcendent interest was love. Bigfooters *loved* the monster."[4] Linguist Karen Stollznow, co-host of the skeptical cryptozoology podcast *MonsterTalk*, described a source of that love: "When I was a kid I thought my neighbor had a jungle in her backyard, the tall stories my brother told me about strange creatures were true, the monstrous koalas and wombats of local legend really existed, and the monsters in the books were real. The kid in me now thinks that still." Her co-host Blake Smith, an incisive skeptical investigator, speaks as well of the tantalizing sense of plausibility that gives cryptozoology its spice—even for doubters: "*What if* it turns out that there really is a primate living in the Pacific Northwest? Or what if some strange tide brings in a frilled sea-serpent? What if a presumed-extinct bird or mammal is found in some remote, exotic locale? I doubt most of these monsters or cryptids are real—but it's the *what if* that keeps me excited."[5]

Like other subcultures, cryptozoology is stratified by level of knowledge and involvement. Whether believers or disbelievers, only a subset of those who form an interest in cryptozoological topics go on to pursue cryptozoology as a hobby or sustained area of investigation. Buhs provided an interesting portrait of the subset of the early Bigfoot community that was motivated to actively research Bigfoot—an inner circle of dedicated enthusiasts, but not necessarily the innermost hard core whom we might think of now as "professional" cryptozoological content producers and investigators. Buhs met with George Haas, publisher of the *Bigfoot Bulletin* since 1969. This early cryptozoological zine had a "magical effect: formalizing what had been a private, haphazard exchange of information and welding the disparate Bigfoot enthusiasts into a community." The zine placed Haas at the center of a lively analogue social network of hundreds of active correspondents who sent in sightings and documents:

> Mostly what Haas received, though, and what the newspaper reported on, were references to books and articles about Bigfoot. Sasquatch enthusiasts were readers first and foremost, cataloguers and archivists. The focus owed something to the Fortean background of many Bigfooters. Subscribers to the *Bigfoot Bulletin* combed through old newspapers, compiled bibliographies—such as one listing all the wildman articles in *Fate*—and assigned reading. "Homework," Haas called it. Swapping citations and articles helped to bind the community; it was also another way to demonstrate skill, in the library rather than the woods. Enthusiasts were to gird their loins and seek out musty tomes, plunge into newspaper morgues, battle with microfilm readers, just as Charles Fort had.[6]

Today, the "cataloguers and archivists" level of activity is pursued online, in cryptozoological discussion forums and blogs. Of these active participants, a subset takes the next step to attend cryptozoological events, such as Bigfoot conferences, and only those in a further, smaller group pursue their interest so far as to undertake fieldwork. Buhs found the hard core of amateur cryptozoologists hunting for Bigfoot in the forests of the Pacific Northwest to be mostly white, working-class men for whom Bigfoot is an icon of untamed masculinity; a populist rebel against scientific elites; the last champion of authenticity against a plastic, image-conscious, effeminate consumer society. (Ironically, Bigfoot has a career as advertising mascot and tabloid fodder, making him a major purveyor of consumerism.) Buhs shows that many Bigfoot stalkers participate at this fieldwork level of the subculture because it has the same attractions as other types of hunting: getting back to nature, tramping through the woods in search of elusive prey, and testing their manhood against the wilderness. He quotes Thom Powell: "I think I became interested in the Bigfoot thing because it gave me an excuse to get out and use my wilderness skills. My life-long love of the wilderness exploration has a purpose beyond just getting there and back." It is an allure familiar not only to hunters, but also to countless thousands of hikers, canoers, campers, birders, and photographers. Buhs quotes a similar sentiment from contractor Tom Morris, who reflected, "Maybe I'm only trying to justify all my trips to the mountains by calling them research. I like wildlife; I like to see anything I can. The more I go, the more I'm amazed at how elusive wildlife can be. I'm happy just to be up there, watching the animals move around. I want to come back with the best pictures I can. The ultimate would be a shot of Bigfoot."[7]

Few activities have more broad, populist appeal than getting out in the woods. That folksy populism also feeds a common theme across cryptozoology (and especially Bigfoot research): the conflict between amateur cryptid hunters and professional scientists. The amateurs usually resent their treatment by the academics. They believe that if they can find the elusive creatures that science rejects, they will be able to triumph over those who have ignored and ridiculed them for decades. Famously impatient with even pro-Bigfoot academics, Sasquatch pioneer René Dahinden imagined a moment of comeuppance for ivory tower scoffers. "I'd take the scientists by the scruff of their collective necks and rub their goddamn faces in—actually, I would like to see all the people—the scientists—who have opened their mouths and made their stupid, ignorant statements, fired from their jobs," Dahinden felt. "They should totally, absolutely, right then and there, without pension, without anything, just be taken and thrown out the front door. Then and there."[8] If the dream of the discovery of a cryptid should one day be realized, Buhs explained, "those who had always known the truth, those who had come to the right conclusion by the dint of hard work and the application of skill, would receive the dignity that the world had otherwise denied them."[9]

Historian of science Brian Regal has discussed the genuine conflicts between professional scientists and the largely amateur cryptozoologists, but has shown that the narrative of conflict can also be complicated by the presence of some academics among the cryptozoologists and by the willingness of some scientists to review the evidence presented by cryptid hunters.[10] A number of disincentives—disdain from colleagues, damage to reputation, possible denial of tenure, and so on—can prevent academics from voicing sympathy for unconventional views or pursuing long-shot possibilities. But there is also the problem of the attitudes of amateurs who have no idea how science works and cannot understand why their obsession with flimsy anecdotal and trace evidence, such as sightings and tracks, does not convince professional scientists. There is not only a large communication gap between the two communities, but also a culture gap, exacerbated by the contrast between the less educated Bigfoot fans and the "academic elite" whom they feel scorns and disrespects them.

Monster proponents have long complained that academics do not give their ideas a fair shake and never look into their evidence. Consider this complaint as it was articulated by science popularizer Gerald Durrell in his introduction to Bernard Heuvelmans's cryptozoology classic, *On the Track of Unknown Animals*. "Of course, scientists want positive proof of the existence of these 'mythical' creatures in the shape of skins and skulls," Durrell acknowledged, "but they would make much better use of their energies in looking for them, rather than wasting their time desperately trying to disprove the few bits of evidence we have."[11] This time-worn cliché echoes throughout the whole of the literature of the paranormal; at least in relation to cryptozoology, however, it is a barb that lands wide of the mark. After all, Heuvelmans and Durrell were themselves respectable authorities in zoology, and they were hardly alone.

Monsters have always attracted a trickle of advocates from within the academy, and they continue to do so.[12] When scientific figures as prominent in their times as geologist Louis Agassiz and rocketry pioneer Willy Ley have promoted cryptids, it is clear that the "scientists won't look at the evidence" argument has been hollow for well over a century.[13] In more recent decades, Grover Krantz and Jeff Meldrum, both professors of anthropology at major universities, retained their posts despite their quests to find Bigfoot. More important, this complaint discounts the fairly large number of scholars who have spent significant time evaluating cryptid evidence, only to conclude in good faith that it was inadequate. For example, when Ivan Sanderson and Heuvelmans (founders of cryptozoology as a modern movement) were taken in by the Minnesota Iceman, a supposed Bigfoot carcass encased in ice and displayed as a sideshow attraction,[14] scientists at the Smithsonian Institution, such as paleoanthropologist John Napier and director S. Dillon Ripley (who were old hands in the search for Yeti), were eager rather than reluctant to examine the discovery. The problem in that case was not institutional bias against the evidence, but the suspicious slipperiness of the evidence offered—or, rather, not offered. When the scientists asked for permission to see the carcass, the mysterious exhibitor suddenly backpedaled and claimed that the original specimen had been replaced by a model. The Smithsonian scientists learned that the Minnesota Iceman was probably a hoax created by a Hollywood model maker; only at that point, after they were denied the opportunity to examine the specimen, did they withdraw their interest.[15]

Nor were scientists unwilling to consider the film shot by Roger Patterson and Bob Gimlin, which supposedly shows Bigfoot walking across a riverbed and looking back toward the camera. When cryptozoologists brought the film to the American Museum of Natural History in New York in 1967 to get the museum's experienced anthropologists and mammalogists to give it their seal of approval, busy scientists took time out of their schedules of serious research to view it. They watched the film, quietly thanked the cryptozoologists for coming, and showed no further interest in viewing it a second time—because in their expert understanding of mammals and especially apes, the creature was clearly a hoax. Scientists were similarly willing to view the footage at the University of British Columbia, at the Smithsonian, and at other institutions as well.[16] Despite the widespread belief among monster hunters that cryptozoological evidence is frozen out of consideration by scholars, this pattern is quite typical: when plausible evidence is brought forward, scientists are generally willing to examine it and give their good faith assessment. But cryptozoologists do not always like what the scientists have to say. When the assessments of the scientists differ from the hopes of the cryptozoologists, that disagreement becomes "evidence" of the closed-mindedness of the very scientists who were most willing to review the evidence. Broad expert consensus that the creature in the Patterson–Gimlin film was a hoax did not, for example, convince Bigfoot proponents that the film was an unreliable plank in their case for the Sasquatch, but instead convinced them that scientists were unreasonable or lacked integrity. "My God," responded Dahinden, "what *do* you have to show them before they'll take it seriously?"[17]

Such distrustful, dismissive responses to expert opinion effectively punish open-mindedness from scientists, who quickly learn that their views will be disregarded and their time wasted. Some, like mammalogist Sydney Anderson of the American Museum of Natural History, eventually become fed up with not only obvious hoaxes, but also the stream of mail they receive from cryptozoologists demanding that they spend their precious time investigating their claims or requesting that the institutions for which they work fund amateur expeditions. That Anderson and his colleagues at the museum had even given a few minutes to cryptozoologists made them the target of endless cryptid-related media requests in the years that followed—so many requests that Anderson eventually resorted to a terse brush-off: "I don't think [cryptids] exist. My fee for writing is 10 cents per word."[18] It is even common for scientists who share their time to be rewarded with accusations of conspiracy. When Napier and the Smithsonian withdrew their initial interest in the Minnesota Iceman, for example, this triggered accusations of a cover-up. As Napier recalled, one story even "suggested that the Mafia, who were credited with having an unspecified interest in the matter, were bringing their considerable influence to bear to stop any further scientific investigations being carried out by myself on behalf of the Smithsonian."[19]

SIGNIFICANT FIGURES IN CRYPTOZOOLOGY

Before we look further into the psychology of cryptozoology, we should pause to consider a few of the figures who played significant roles in forming the modern ideas, rhetoric, and movement of cryptozoology—especially a few of those who started their careers as mainstream scientists and then became cryptozoologists later in life. The classic example is Roy Mackal (b. 1925), who spent his early career working in microbiology, in which he earned a good reputation and got tenure at the University of Chicago. Based on his background, he is qualified to do research only as a microbiologist; he has no formal training in systematics, field biology, or ecology, which might qualify him to hunt for exotic creatures. On a visit to London in 1965, he took a side trip to Scotland, visited Loch Ness, and met with the local Nessie researchers, who had founded the Loch Ness Investigation Bureau. Mackal soon was spending a lot of time at Loch Ness, helping with the monitoring, trying to develop new means of detection, and eventually becoming the head of the project. One evening in 1970, according to an interview he gave in 1981, he finally had his reward. He saw, some 30 yards away, "the back of the animal, rising eight feet out of the water, rolling, twisting. If that's a fish, I thought, it's a mighty fish indeed! To this day, when someone asks me, 'Do you believe there is a monster in Loch Ness?' my stomach does a somersault. I know what I saw."[20] Although there are hints that his sighting may not have been as persuasive in the moment as he later recalled—as he said, "For days I refused to admit that I had seen, with my own eyes, one of the strange animals in Loch Ness"[21]—this sighting led to his writing a book on the Loch Ness monster.[22] It also solidified his commitment to cryptozoology in general. "From then on, I knew the Loch Ness monsters existed. This state of mind has remained with me and most certainly has predisposed me to be receptive to reports of other unidentified animals," Mackal reflected. When confronted with reports of Mokele Mbembe, "I was ready to listen."[23] Mackal mounted two expeditions to the Congo (where he found nothing) and published a book on his search for the alleged dinosaur. Since then, he has largely been retired and not very active in cryptozoology, although old footage of him talking about Nessie or Mokele Mbembe when he was younger often reappears on television shows that promote cryptozoology.

Cryptozoologists have often promoted "Professor Roy Mackal, Ph.D." as one of their leading figures and one of the few with a legitimate doctorate in biology. What is rarely mentioned, however, is that he had no training that would qualify him to undertake competent research on exotic animals. This raises the specter of "credential mongering," by which an individual or organization flaunts a person's graduate degree as proof of expertise, even though his or her training is not specifically relevant to the field under consideration. This strategy is employed dishonestly by creationists, who flaunt their degrees in hydraulics or biochemistry as proof that they are legitimate scientists, but then make arguments about paleontology or evolutionary biology—fields in which they have had no training—and quickly show that they are incompetent. Mackal's knowledge of microscopic organisms does not necessarily translate into expertise about stalking large animals, any more than it translates into expertise about repairing cars or taming lions or playing the harpsichord. Nor does it equate to domain knowledge of the literature of cryptozoology itself. There may not be degrees in monsterology, but we can assure you from painful experience that there is, nonetheless, a good deal to know. No less a cryptozoological heavyweight than Bernard Heuvelmans made this exact point in a scathing review of Mackal's book *The Monsters of Loch Ness*. Correctly noting the "extensive literature" pertaining to the Loch Ness monster, Heuvelmans described Mackal as "a person who presents himself as a specialist in this problem (in fact the most qualified specialist), but who is unaware, or acts as if he were unaware, of this previous work." In Heuvelmans's view, Mackal used his academic credentials to present "himself from the very beginning as the long-awaited Messiah"—as the first person qualified to bring scientific rigor to the investigation of the Loch Ness monster. (Heuvelmans added without apparent irony that he previously felt as though Mackal "were one of my disciples.")[24]

The founder of the modern cryptozoology movement, the French-born, Belgian-raised zoologist Bernard Heuvelmans (1916–2001), earned a doctorate in zoology in France in 1939, studying the teeth of the aardvark. He was a student of the eccentric zoologist Serge Frechkop, whose greatest claim to fame was his support of the discredited theory of initial bipedalism (which argued that all mammals were initially bipedal).[25] Heuvelmans apparently never had formal training in field biology or ecology, disciplines that were in their infancy when he was trained and that could have qualified him to find cryptids. He performed as a jazz musician and a comedian as well as worked as a writer. During World War II, he managed to escape after being captured by the Nazis, and he spent the rest of war evading the Gestapo. (Ironically, he was a close friend of and a consultant to the Belgian artist and comics writer Georges Remi, known as Hergé, the author of the series *The Adventures of Tintin*—and a Nazi sympathizer.) Heuvelmans was interested in cryptids from an early age and credited a magazine article by Ivan Sanderson for having stimulated his interest in unknown animals.[26] For most of his career, he worked as a freelance researcher and writer, earning his living and financing his expeditions from the sale of his books and publication of his articles. His most famous work, *On the Track of Unknown Animals*, established him as the "father of cryptozoology,"[27] and he (along with Roy Mackal and Richard Greenwell) founded the International Society of Cryptozoology (ISC) in 1982.

Richard Greenwell (1942–2005) did not have a doctorate in any field relevant to cryptozoology (but received an honorary doctorate late in his life), although he did research on living mammals and published papers on mammalogy in peer-reviewed journals. He held the position of research coordinator for the Office of Arid Land Studies at the University of Arizona. Most of his time, however, was spent writing about cryptids and joining expeditions to find them. As one of the co-founders of the ISC, he was largely responsible for maintaining the *ISC Newsletter* and the journal *Cryptozoology* from 1982 until 1996, when they ceased publication. More than any other prominent cryptozoologist, Greenwell tried to rein in the rampant speculation and unscientific aspects of cryptozoology. In several articles, he proposed the establishment of rigorous standards for cryptid research. Despite his efforts, the ISC had financial problems and folded in 1998.

The final major figure among the founders of cryptozoology was the Scottish zoologist Ivan Sanderson (1911–1973), who is often credited with coining the term "cryptozoology." He earned a bachelor's degree in zoology and master's degrees in botany and geology at Cambridge University, and was truly an accomplished field zoologist, with many discoveries to his credit. Despite his credentials, Sanderson resented the academic zoologists with doctoral degrees who did not take his work seriously. Sanderson also apparently served as a spy for the Allies during World War II, using his cover as a field zoologist to gather information for British intelligence. He spent much of his career as a major media figure, doing interviews about wildlife and field zoology, first on the radio and then on the early television talk shows; writing articles for popular magazines; and publishing books about his exploits. But early in his career, he not only argued that cryptids were real, but wrote books and articles on a wide variety of other paranormal phenomena, such as the

Yeti, UFOs, Atlantis, and the Bermuda Triangle, and other "vile vortices."[28] Cryptozoologists are fond of pointing to Sanderson as a true field biologist who believed in cryptids, but he did almost all his fieldwork in the 1920s (as a teenager) and 1930s, when very little was known about many parts of the world and few major biological surveys of remote regions were undertaken.

Historian Brian Regal details the stories of a number of the other prominent figures in cryptozoological research, a few of whom were formally trained in fields of study relevant to their work in cryptozoology.[29] They include Grover Krantz (1931–2002), who was an accomplished anthropologist with expertise in human hand and foot mechanics. But he also spent most of his time obsessed with Bigfoot, which (as Regal documents in detail) caused tremendous problems at Washington State University, where he was a professor. Some of his troubles, however, were self-inflicted: he was notorious for his feuds not only with fellow academics, but also with other Bigfoot researchers; his writings were problematic because he failed to cite evidence to support his assertions; and his late work was rejected because his research was outdated, not taking into account the discoveries in genetics and fossil primates that made the existence of Bigfoot less and less probable. As Joshua Blu Buhs showed, Krantz also became more and more a gullible believer and less and less a critical skeptic of the bad evidence as his commitment to Bigfoot increased. He proclaimed that he could unerringly distinguish real from fake Bigfoot tracks—and then a source sent him a "track" that Krantz proclaimed as real, only to have its source show that it was another clever hoax that had taken very little skill to produce.[30] As Buhs and Regal describe it, Krantz retired from Washington State in 1998 an embittered man, as both the credibility of Bigfoot in the scientific community and his own standing as a scientist had plummeted.[31] He died still proclaiming that he had an infallible way to show that Bigfoot was real—rumored to involve fingerprint-like "dermal ridges" in Sasquatch tracks[32]—but refused to disclose the secret.

Almost all the "old guard," who founded and nurtured the "golden age of cryptozoology," as Regal called it,[33] are dead or inactive, and scientists who lent their names to the board of directors of the ISC are dead or very old. Even as early as 1993, cryptozoologists were being called an "endangered species."[34] The ISC is extinct—along with its journal, newsletter, and Web site—because it suffered from perpetual financial problems, the journal issues were often years late, and the field never expanded with more scholarship, as all professional societies must. With the death of the ISC, cryptozoology no longer has a formal professional organization with the pretense of serious scholarship. Loren Coleman still keeps one Web site active,[35] and many other sites are maintained by fans of Bigfoot and Nessie. A few cryptozoologists are in major academic positions, like Bigfoot advocate Jeff Meldrum at Idaho State University, but as both Buhs and Regal point out, much of the academic energy and scholarly force of the ISC and its supporters has vanished, and the field is once again the domain of amateurs.[36]

At the same time, the heart of the cryptozoological enterprise may be changing. The rise of skeptical cryptozoological scholarship by academics—including folklorist Michel Meurger, classical folklorist Adrienne Mayor, and paleozoologist Darren Naish—suggests a path forward for a kind of "post-cryptid cryptozoology": a folkloristic cryptozoology that investigates animal-themed mysteries without any particular expectation of discovering new species.[37] Like other areas of fringe science, the value and energy of amateur cryptozoological scholarship may depend on neglect by professional scholars. In that sense, the call for greater attention from mainstream academics may be self-destructive to cryptozoology, with the arrival of professionals displacing both the popular voices and the working assumptions that so far have unified the cryptozoology community.

Another side of the cryptozoology movement has a motive utterly opposed to that of the academically inclined researchers who wish to bring more science to cryptozoology: the creationists who wish to use cryptozoology to bring science down.[38] One cryptozoology Web site lists among the biographies of famous cryptozoologists those of such creationists as William Gibbons and Kent Hovind.[39] Creationists are the major players in the search for Mokele Mbembe and other cryptids said to resemble prehistoric reptiles, not because of their curiosity about nature, but because of their mistaken belief that the discovery of such cryptids will overthrow evolutionary biology. As Gibbons puts it,

> The very idea of dinosaurs—those "terrible lizards"—still living in some remote corner of the world today threatens to evoke sidesplitting laughter from the modern paleontologist. . . . Yet, there are a few dissenting voices within the scientific community, although often marginalized, that beg to differ—and for good reason.
>
> According to conventional evolutionary wisdom, the last of the dinosaurs became extinct over 65 million years ago during a period known as "the great dying." . . . So teach the colorful books and science publications that have shaped the thinking of the generations since Darwin. However, is this really the case?

> Throughout history, over 200 cultures around the world have left us with a rich and detailed record of monstrous creatures that they knew as "dragons." . . . From such evidence as the Behemoth of Job 41 to the historical records of 16th century England, and present-day eyewitness accounts, I propose that dinosaurs, or at least creatures that look remarkably like them, are still very much with us today in the dark and remote recesses of our modern, fast-paced world.[40]

This conviction is absurd and shows no understanding of science. To begin with, it is not difficult to find dinosaurs surviving into the present day. There may be some in your backyard right now. They are called birds. If additional groups of surviving dinosaurs (or of other prehistoric reptiles of interest to cryptozoologists, such as plesiosaurs or pterosaurs) were discovered in Loch Ness or the Congo Basin, they would just be additional, late-surviving descendents of groups known from fossils. There are many such "living fossils," including the coelacanth, three genera of lungfish, the horseshoe crab, monoplacophoran mollusks, trigoniid bivalves, and lingulid brachiopods,[41] not to mention sharks, crocodiles, and turtles. Further such discoveries would only cause paleontologists to reassess why those creatures did not leave a more complete fossil record, not overthrow the enormous body of evidence that evolution occurred (and is still occurring). The creationists are openly anti-scientific, trying to overthrow not only evolutionary biology, but paleontology, astronomy, anthropology, and any other field that conflicts with a literal reading of Genesis.[42]

The credentials of the creationist cryptozoologists are even less credible in field biology than are those of most amateur cryptid hunters. Gibbons has a degree in religious education from a seminary, and Hovind promoted untruths about the fossil record and evolution from his pulpit in Pensacola, Florida. He even had the gall to call himself Dr. Dino, even though he knows nothing about paleontology and his "doctorate" is from a purely religious school that does not even pretend to be able to grant secular academic degrees.[43] His dissertation is a short paper that contains no original research, but does include numerous errors in spelling, citation, and reasoning.[44] Dr. Dino's methods finally caught up with him: he was convicted of tax evasion in 2007 and was sentenced to ten years in federal prison.[45] The presence of such creationists in the cryptozoology community further damages its chances of being taken seriously as a legitimate field of scholarship.

WHY DO PEOPLE BELIEVE IN MONSTERS?

Roy Mackal, Bernard Heuvelmans, Richard Greenwell, Ivan Sanderson, Grover Krantz, and Karl Shuker are exceptions among cryptozoologists. They were among the few who had formal science education and training, even though most of them did not have training in a science that is specifically relevant to cryptozoology. What about the rest of the people who believe in cryptids? What can be said about them beyond the anecdotal descriptions of the Bigfoot subculture that begins this chapter?

Before looking further at the psychology of cryptozoology, let us briefly consider the wider topic of paranormal belief—"paranormal" being the umbrella under which cryptozoology is typically placed. (This is a widely adopted and, we think, useful convention, but it conflicts with standard definitions of the word "paranormal" as meaning something like "inexplicable by or incompatible with current science"—in effect, supernatural or outside physical law. By that standard, homeopathy should be considered paranormal, for example, while more classic "paranormal" subjects like Bigfoot, Atlantis, and UFOs should not. Sociologists tend to define the paranormal as beliefs and experiences that are "dually rejected—not accepted by science and not typically associated with mainstream religion"—or those ideas unified by the distinction that "traditional science regards their existence or validity as so improbable as to be all but impossible."[46] By these latter definitions—or by the even simpler "stuff you'd see on *The X-Files*" rule of thumb—cryptozoology is a paranormal topic.)

Rigorous, large-scale studies have been conducted to determine what the population of the United States as a whole, and what kinds of people in particular, think about paranormal subjects. Well-known examples include the series of polls of paranormal beliefs conducted by the Gallup organization since 1990 and the Baylor Religion Survey, which is a huge data set of answers to multiple-choice questions collected since 2005 that looks not only at religious and paranormal beliefs, but also at the demographic data behind them. These very broad surveys are complemented by in-depth studies of specific populations of paranormal believers, such as the psychologist Susan Clancy's experimental work with people who believe that they have been abducted by space aliens.[47] Across the board, the evidence indicates that the clichés about the fringe nature of paranormal belief are wrong: paranormal beliefs are not at all uncommon; they are not restricted to people who are socially marginalized; nor are they a sign of low intelligence or poor mental health.

It is important to understand that essentially everybody believes in things that scientists consider to be either unproven, implausible, or demonstrably false. When polled about even very short lists of ten or so ideas selected from the hundreds of paranormal and pseudoscientific notions critiqued in the skeptical literature, *large majorities of the population* readily affirm that they hold one or more of those paranormal beliefs. In 2005, Gallup's survey found that 73 percent of American adults affirm at least one paranormal belief from a list of ten; 57 percent believe in at least two of those ideas; and 43 percent believe in three or more.[48] These numbers are not unusual, but are consistent with other findings. In *Paranormal America*, a book based on the Baylor Religion Survey data, sociologists Christopher Bader, F. Carson Mencken, and Joseph Baker discuss their similar finding that most Americans agree with one or more items from a short list of paranormal beliefs:

> Outside the halls of the academy a broader stereotype is often applied to paranormal believers—people who believe in or have experienced the paranormal are "different." People who do not believe in the paranormal are perceived to be normal; those who believe in paranormal topics are considered weird, unconventional, strange, or deviant.
>
> There is a big problem with this simplistic assessment—believing in something paranormal has become the norm in our society. When asked if they believe in the reality of nine different paranormal subjects including telekinesis, fortune-telling, astrology, communication with the dead, haunted houses, ghosts, Atlantis, UFOs and monsters, over two-thirds of Americans (68%) believe in at least one. In a strictly numerical sense, people who do not believe in anything paranormal are now the "odd men out" in American society. Less than a third of Americans (32%) are dismissive of all nine subjects. What this means is that distinguishing between people who do and do not believe or experience the paranormal is increasingly less useful. Rather, people may be more readily distinguishing [*sic*] by *how much* of the paranormal they find credible.[49]

The demographics of paranormal belief in general and of any given belief in particular naturally vary with those factors that you would expect, such as age, sex, education, religion, political party affiliation, and income—but not necessarily *as* you might expect. For a counterintuitive example, respondents in the Baylor survey who identified themselves as adhering to "no religion" were considerably more likely than evangelical Protestants to affirm a belief in haunted houses.[50] Other predictors were perhaps more surprising, such as marital status and cohabitation: marriage predicts less involvement with the paranormal, while cohabitation predicts more. Married people claimed fewer paranormal beliefs than unmarried, and when asked whether they have had any of five paranormal experiences (such as consulting a psychic, using a Ouija board to contact spirits, or spotting a UFO), unmarried respondents claimed about two such experiences, while married respondents claimed less than one on average. But cohabitating people held almost twice as many paranormal beliefs as non-cohabitating people.[51]

Among gender differences, women were more likely than men in general to affirm belief in the paranormal ideas included in the Baylor survey—twice as likely to accept mediumistic communication with the dead, psychic powers, or astrology. Men held greater levels of belief in only one item: the claim that "some UFOs are probably spaceships from other worlds."[52] Why this should be is a matter of considerable discussion and speculation. "Women tend to want to improve themselves, to become better people," suggested Bader, who is also a director of the Association of Religion Data Archives. "Men tend to want to go out and capture something, to prove it's real."[53] However, women respondents in the Baylor survey were also more likely than men to accept the existence of cryptids—a finding that flies in the face of stereotypes about cryptozoology—although men followed close behind women on this item (as well as in belief in Atlantis). We recommend caution about jumping to conclusions about the role of gender in paranormal belief. The effects of gender vary with the year of the survey and the age of the respondent; and, as we explore further in the case of cryptozoology, the demographics within each paranormal subtopic are more complicated than they appear.

Pop culture stereotypes might lead us to mistakenly anticipate some correlations that turn out not to exist in the Baylor data. Rather than paranormalism predicting isolation from mainstream activities (or vice versa), the Baylor researchers found "no evidence that involvement in conventional activities deters belief in the paranormal." Involvement in one or more cultural organizations, fraternities, trade unions, sports, or civic or service groups had no discernable impact on the number of paranormal beliefs a respondent held or the number of paranormal experiences that person reported. Regardless of community involvement, people affirmed their belief in an average of about two of the paranormal items included on the survey.[54]

According to Bader, Mencken, and Baker, as age and income increase, belief in alien visitors, ghosts, psychic abilities, and other paranormal phenomena noticeably decreases (although paranormal beliefs are common at all ages and levels of income). Party affiliation is also related: independents report the most paranormal beliefs, followed by Democrats and Republicans.[55] Considered in combination, the researchers said, investments in a conventional lifestyle (such as college education, marriage, and mainstream religiosity) tend to reduce paranormal beliefs. These relationships could be plausibly described in more than one way. Among the possible explanations is the hypothesis favored by Bader, Mencken, and Baker that people are more likely to explore new ideas when they are less invested in conventional lifestyles: "As stakes in conformity increase, paranormal beliefs and experiences steadily and markedly decrease. A person with the highest stakes in conformity accepts on average only two paranormal beliefs. This is less than half the level of belief exhibited by those with the lowest stakes in conformity."[56] Nonetheless, even the most conventional people generally hold paranormal beliefs. As Mencken adds,

> All humans are seeking enlightenment and discovery. New information helps us to reduce risk in our lives, and to make better informed decisions. Many paranormal practices (psychics, mediums, communication with the dead, astrology, etc.) are about giving people an insight into their future. Those groups not bound to conventional religious systems are freer to explore these alternative systems in order to gain information that may help them improve their lives.

Both the undereducated poor and the hyper-educated cultural elites may feel that sort of freedom to explore—or, rather, may feel less bound to the status quo. "Since conventional religiosity is for and run by highly conventional people and provides many empirical rewards for this group," Mencken notes, "those from lower socioeconomic status groups will not gain many spiritual or conventional rewards from participating in conventional religion. Alternative beliefs systems can be empowering." On the other end of the spectrum, "those who are hyper-educated, or cultural elites, may condemn conventional norms of behavior as too bourgeosie."[57]

No matter how you slice it, the paranormal thrives in every category and level of society. A sizable literature of sociological research has attempted to make sense of why so many people accept so many paranormal claims. Some hypotheses include

- *Deviance*: People who strongly embrace paranormal ideas are thought to be outside the norms of society and so are defined by sociologists as "deviant." In sociology, deviance theory is focused largely on criminals and others who violate the rules of society, often in an extreme way (such as mass murder). Bader, Mencken, and Baker generally subscribe to the approach proposed by criminologist Travis Hirschi, in which deviant or unconventional behavior is less costly, and therefore a more readily selected option, to those who have invested less in conformity.[58] By this reasoning, people with lower levels of education, income, marriage attachments, church participation, or involvement with community activities should tend to subscribe to a larger number of paranormal beliefs. The Baylor data offer some support for this view, but with a number of caveats. Churchgoing correlates with paranormal belief, for example, but the relationship is not straightforward. (Those who attend church once a month claim more belief in paranormal ideas than those who never attend church, and also more than those who attend church more than once per month.)[59] Involvement in community activities has no relationship to paranormal belief. Furthermore, some groups of people who are very invested in the paranormal are also very conventional. Members of Bigfoot hunting groups, for example, tend to be so conventional in most ways that Bader, Mencken, and Baker "might even call them hyperconventional."[60] Moreover, if belief in the reality of some paranormal phenomena is the norm in our society, can those who hold those normal ideas truly be considered deviant?

- *Wish for control*: Our increasingly complex world is impossible for most of us to understand—from our enormous government to the instantaneously changing global economy to our technologically complicated electronic devices. As researchers who have studied religious cults and many other fringe movements have shown, humans have a deep longing for a simpler world that they can better comprehend—even if this hope is no longer realistic. Thus those who embrace an alternative worldview may gain the subjective benefit of feeling they have more meaning in and control over their lives.

- *"Small steps" hypothesis*: According to this model, people who already accept one kind of supernatural belief system (such as religion) are likely to accept others (such as paranormal ideas), since it is only a small step from embracing one to embracing many. The complication is that the largest subset of believers in the paranormal accept only a single paranormal belief. The next largest accepts only two, and so on. (In this sense, those Bigfoot researchers who are actively hostile to paranormal beliefs other than the existence of Bigfoot are very typical of paranormal believers.) In the Baylor Religion Survey of 2005, fully 19 percent of respondents held one paranormal belief, 16 percent held two paranormal beliefs, 10 percent held three—and the remainder declined steadily in the same pattern, with only 2 percent of respondents affirming nine paranormal beliefs. Gallup found a similar pattern that same year, with 16 percent of respondents claiming one paranormal belief and only 1 percent affirming ten such beliefs.[61] The "small steps" hypothesis is further complicated by the fact that paranormal ideas are not discrete, equally weighted units of belief, and the steps from one belief to another are not always equally difficult. Some paranormal beliefs imply or suggest others, or are at least easily compatible, while others conflict. (Premonitory dreams are conceptually more similar to fortune-telling than to alien abduction, for example.)

Whatever the range of factors at play, the psychology and sociology of paranormal belief are as kaleidoscopic and colorful as any other window into the human experience. A sharper focus on a specific subcategory or community of paranormal believers only complicates this perspective further—and further dispels simplistic stereotypes. Consider those who believe that they were abducted by aliens (surely Hollywood's favorite go-to cliché for wild-eyed, aluminum foil hat-wearing kooks). In 2003, Bader reported the results of his survey of alien abductees, which were consistent with previous research on new religious movements: abductees are not only much better educated than the general population, but also more likely to hold white-collar jobs.[62] Commenting on both this research and the Baylor Religion Survey data, Bader and his co-authors wrote in *Paranormal America*,

> Unless we choose to define someone as a fringe member of society simply because they claim to have been abducted by aliens or are chasing Bigfoot . . . abductees are *not* marginal people. Many of the people we met would be better described as elites.
>
> One way to understand why elites might be attracted to the paranormal is to think of paranormal beliefs and experiences as something that is "cutting edge." Whenever a new technology enters the market there are people who immediately embrace it, people who are excited by new things and ready to take risks. Marketers call such people "early adopters." Then there are the rest of us, who want to wait

> until a new idea or technology is fully proven before we jump aboard. Early adopters tend to be those . . . who have been continually exposed to new ideas throughout their lives via higher education and contacts with other educated people. They also have the resources to try new things.[63]

This sociological assessment of one sample from the hard-to-study alien abductee population is complemented by Clancy's psychological research with another group of abductees. "There's little evidence that this was a particularly psychopathological group," she found.[64] Abductees do tend to score highly in assessments of "fantasy proneness" (a normal personality type characterized by vivid imagination and high hypnotizability), and, intriguingly, Clancy's research revealed that they are more prone to creating false memories under laboratory conditions than is the general population—but these are variations on normal functioning.[65] "What can we say conclusively about this diverse group of abductees?" Clancy asked. "In the end, not much. Research on the topic is clearly in its infancy. What we *can* confidently say is that these people are not crazy."

Returning to cryptozoology, it is clear that this community is no easier to pigeonhole than is that of abductees. The Baylor Religion Survey offers a sense of the very wide interest that American adults have for cryptids. Asked in 2005, "Have you ever read a book, consulted a Web site, or researched the following topics: Mysterious animals, such as Bigfoot or the Loch Ness Monster?" 20 percent of Americans indicated that they had. Men exhibited a little more interest, with about 24 percent replying that they had researched cryptids in some fashion, but more than 17 percent of women had also done so. Interest was similarly high in every age group (highest in the 18 to 30 range, at 32 percent), political affiliation (highest among independents), and religious category (highest among those who claimed to adhere to "no religion"). Interest hovered in the 18 to 24 percent range in every income category, and between 19 and 23.5 percent in every level of education from high school onward.[66]

A lot of people dig monsters, whether they believe in them or not—but it happens that many *do* believe. When asked whether "Creatures such as Bigfoot and the Loch Ness Monster will one day be discovered by science," about 17 percent of the people in the sample agreed (14.2 percent) or strongly agreed (2.7 percent) with that statement. (Those tens of millions of believers in the United States have proved to be tempting targets for at least some preliminary research. For example, psychologists Matthew Sharps, Justin Matthews, and Janet Asten have attempted to link cryptozoological beliefs with the tendency to exhibit features of attention deficit hyperactivity disorder [ADHD]. They found some correlation with measures of the component hyperactivity or impulsivity in isolation, but "no significant relationship" between overall ADHD tendencies and cryptozoological beliefs.)[67]

At this level of involvement with cryptozoology—*belief* in the probability of cryptids, as opposed to motivation to research the topic—an interesting demographic reversal occurred in the Baylor Religion Survey data: more women than men agreed that cryptids will one day be discovered. It is a fairly slim margin (17.6 percent of women either agreed or strongly agreed, compared with 15.9 percent of men), but the difference is visible in the "strongly agree" range, so those respondents might plausibly be thought of as serious believers in cryptozoology. Further, women also dominated the undecideds, making them considerably less likely than men to assert that cryptids do *not* exist. Just under 53 percent of women disagreed or strongly disagreed that cryptids will one day be discovered, compared with almost 61 percent of men.[68]

The Baylor Religion Survey found that women continued to lead the way when asked a more direct question in 2007: "In your opinion, does each of the following exist? Bigfoot," with 17.5 percent of women agreeing that Bigfoot either "absolutely" (3.2 percent) or "probably" (14.3 percent) does exist. Fewer men expressed belief at either level of confidence, with 2.9 percent believing that Bigfoot absolutely does exist, and 11.7 percent thinking that it probably is real (with 14.6 percent of men accepting Bigfoot overall).[69]

A more recent poll by Angus Reid Public Opinion, conducted in 2012, found even higher levels of belief in Bigfoot in both Canada (21 percent) and the United States (29 percent) than had been found in 2007, with men in both countries only slightly more likely than women to give an affirmative answer to the question "Do you think Bigfoot (also known as Sasquatch) is real?" In Canada, 3 percent of women and 2 percent of men said that Bigfoot "definitely is real," while 15 percent of women and 21 percent of men said that it "probably is real." In the United States, 8 percent of women and 6 percent of men said that Bigfoot "definitely is real," while 19 percent of women and 25 percent of men said that it "probably is real." In both countries, women led not only among those who felt certain that Sasquatches exist, but also among respondents who answered "not sure." Among the 17 percent of Britons who gave a positive answer to a similar Angus Reid question, "Do you think the Loch Ness monster is real?" women were more likely than men to say that

Nessie "definitely is real" (3 percent of women compared with 2 percent of men) or "probably is real" (15 percent of women compared with 13 percent of men) or to respond that they were "not sure."[70]

If women are roughly as likely to believe in cryptids as are men (or possibly even more likely; the differences tend to be near the margins of error in relevant polls), why does cryptozoology have such a strong reputation as a "boy's topic"? The answer appears to hinge on the distinction between *belief in cryptids* and *participation in cryptozoology*. It is a distinction that also applies to other areas of belief, such as religion. Identifying with a religion or holding beliefs associated with that religion is not the same as participating in the rituals or practices of that religion, such as churchgoing. In addition to belief and practice, sociologists distinguish other dimensions of involvement with religion that can be usefully applied to paranormal belief communities, such as knowledge of the topic, personal experiences (consider eyewitness sightings of Bigfoot), and consequences suffered as a result of involvement (such as ridicule). In regard to cryptozoology, it seems that women are more likely to *believe*, while men are (for whatever reason) more likely to *practice*. This lopsidedness toward male involvement begins at the most basic level of participation in cryptozoology (reading a book or consulting a Web site) and increases at each more active level of engagement.

To explore the *subculture* of cryptozoology, sociologists Bader, Mencken, and Baker got to know the Bigfoot community of East Texas and even accompanied some dedicated Bigfoot hunters on a late-night search for the creature.[71] The sociologists soon found that they "were among very normal people talking about a very strange subject."[72] Bigfooters have formed their own subculture of people who believe strongly in the reality of Bigfoot and who spend a significant amount of time and resources researching and hunting Bigfoot and following the internal goings-on within the Bigfoot community itself. They hold their own meetings, speak their own jargon, share their own body of accepted knowledge, and have their own distinctive way of looking at the world. In all these characteristics, Bigfoot believers reflect the typical pattern of any subculture, from other dedicated cryptozoologists to ghost hunters to fans of any other subject—be it NASCAR or vampires, opera or scientific skepticism.

In 2009, Bader, Mencken, and Baker attended the annual Texas Bigfoot Research Conference in Tyler, Texas, where they were permitted to administer an anonymous questionnaire to the attendees.[73] They found the conference of nearly 400 Bigfoot enthusiasts to be much like any other convention of any other interest group. Celebrities of the cryptozoology world like Loren Coleman, Bob Gimlin, and Smoky Crabtree (known for the movie about his Bigfoot experience, *The Legend of Boggy Creek*) spent much of their time in the hallways and the auditorium posing for photographs and signing books. Exhibitors selling books, DVDs, T-shirts, and every other sort of Bigfoot merchandise filled the hallways. Most of the attendees knew the Bigfoot legends and evidence backward and forward, and they spoke in shorthand about the "Skookum cast," the "PG [Patterson-Gimlin] film," the "Ohio howl," and the "shoot/don't shoot" controversy (whether a Bigfoot hunter should actually shoot or not if he finds Bigfoot).[74] As sociologists have long pointed out, learning the argot, or distinctive lingo, of a subculture is part of the process of becoming a member of the subculture, and using the jargon is a way to distinguish insiders from outsiders and is a mark of acceptance when initiates master it.

Strikingly absent were the sort of colorful characters who might attend a typical comic-book or science-fiction gathering like Comic-Con or Dragon-Con. Although the researchers were by this time familiar with the Bigfoot subculture, they "were still surprised by the staid nature of the conference and the conventionality of most of those in attendance." The conferees were mostly conservatively dressed, middle-aged, middle-class white people who attended a day-long slate of presentations. The large majority—67 percent—were male.[75] Nor did attendees at the Texas Bigfoot Research Conference merely look conventional; by key measures, they were. They were better educated than average Americans, better paid, and more likely to be married.[76]

All this was disappointing to reporter Mike Leggett, who wrote about his experience at the conference that same year:

> Bigfoot is boring.
>
> Correction. Bigfoot conferences are boring. . . .
>
> I went to the 2009 Texas Bigfoot Conference expecting people in gorilla suits milling about among semi-crazed gangs of gonzo, tattooed, barrel-chested bean-dips. I found instead only a polite, older crowd of mildly sleepy true believers who only came alive at the mention of the TV show "MonsterQuest" or the movie "The Legend of Boggy Creek." I thought surely someone would be selling BLT—Bigfoot, lettuce and tomato—sandwiches and Abominable Snowman cones during the lunch break, but there were only Cokes and Subway sandwiches.[77]

Bader, Mencken, and Baker found the attendees even more conventional—*and 88 percent male*—when they restricted their focus to the subset who could be considered "Bigfoot hunters" by virtue of belonging to an organization devoted to Bigfoot field research. These Bigfoot hunters were notable not only for their high income and education level, but also for their conventionally high church attendance and their marital status (77 percent were married, compared with 57 percent of Americans in general in 2009).[78]

Psychologists and sociologists do not yet fully understand all the complex motivations that make people embrace unconventional beliefs like cryptozoology, although research is beginning to tease apart some of the questions.

IS CRYPTOZOOLOGY SCIENCE? CAN IT BE MADE SCIENTIFIC?

Cryptozoology has the reputation of being part of a general pseudoscientific fringe—just one more facet of paranormal belief. Given Bigfooters' humdrum appearance and attempt to portray themselves as serious scientific researchers trying to track down just another species of primate, it should not be surprising that they typically try to avoid any association with other paranormal beliefs, which they scorn as unscientific. Despite this sentiment, the dividing line between cryptozoology and the paranormal is unclear—and to some extent artificial.

While cryptozoology need not be considered "paranormal" by all definitions (cryptids are usually conceived as flesh-and-blood creatures, not as phenomena outside known physical laws), it is difficult to avoid the association when such visible figures in cryptozoology as Ivan Sanderson and Jon-Erik Beckjord have also advocated for a range of other paranormal phenomena. More important, it is fairly common for cryptozoological testimony to include paranormal or supernatural elements—much more common than most cryptozoologists prefer to admit. "Paranormal folkloric entities, whether ghosts, vampires, or lycanthropes, are not cryptozoological," insists cryptozoology micro-publisher Chad Arment,[79] but Loren Coleman discusses modern American sightings of creatures with a "classic werewolf look," which he includes in his book about Mothman—a UFO- and Men in Black–associated, flying, humanoid cryptid with glowing red eyes![80] Like Mothman, Bigfoot often has been described as having glowing red eyes or has been connected with other, even more mysterious lights, such as a luminous sphere carried in the hand or the glowing bronze light that rose from one Bigfoot's hiding place in the woods and shot away into the sky.[81]

The connection between Bigfoot and UFOs is especially strong. For example, Christopher Bader, F. Carson Mencken, and Joseph Baker discuss a well-known incident that occurred in the vicinity of Uniontown, Pennsylvania, in 1973, in which two huge ape-like creatures with glowing green eyes allegedly appeared alongside a radiant white dome 100 feet in diameter.[82] This is not an especially unusual report. In one 1974 case, two witnesses allegedly saw several Sasquatches in the vicinity of a bright red, hovering light that rotated like a police car's beacon.[83] In 1976, one man saw a Bigfoot in Montana and also spotted a hovering gray UFO half a mile away.[84] In 1983, logger Stan Johnson claimed that a certain Bigfoot Allone of the Rrowe family of Bigfeet had explained to him that the Bigfoot Peoples are from the fifth dimension, where they originally lived on the planet Centuris. When Centuris was about to be destroyed, the Bigfoot Peoples were rescued by beings from the nearby planet Arice. They then lived happily on Arice until the rise of an evil ruler caused some of them to come to Earth during the last Ice Age. Here they competed with dinosaurs and rampaging cave dwellers. (This story has all the confusion of prehistoric events found in pop-culture entertainment like *The Flintstones*, as well as creationist chronology. Perhaps not surprisingly, Johnson also claimed that the Bigfeet pray "to God, and to Jesus Christ.") There was a war on Earth between good and bad Bigfeet, both of which still live in remote regions of Earth.[85]

Some Bigfoot followers claim that Sasquatches can vanish into thin air[86] or that they are shape-shifters—or even that they are "paraphysical, inter-dimensional nature people that are profoundly psychic."[87] The late Jon-Erik Beckjord (a persona non grata among more conventional Bigfoot researchers) regularly appeared on radio talk shows, *Today*, and *Late Night with David Letterman*. During many public appearances, he asserted that Bigfoot is a shape-shifter that cannot be caught or shot and that can "manipulate the light spectrum they're in so that people can't see them." Bigfoot uses its telepathic powers to sense the presence of humans, and it shares a "space-time origin and connection with UFOs and come[s] from an alternate universe by a wormhole."[88]

(Supernatural elements are not restricted to the subtopic of Bigfoot, but are found intertwined with the lore of, testimony about, and modern search for many cryptids. For example, a video of what was purported to be the Loch Ness monster that made global media waves in 2007 was shot by a man who also claimed to have filmed fairies.[89] One of the top ten or so photographs of the Loch Ness monster was shot by a man who claimed to have successfully summoned Nessie using a team of psychics.[90] Second in Nessie fame to only the Surgeon's Photograph, Robert Rines's "flipper" pictures were apparently captured with the psychic assistance of a local dowser.[91] The vampiric chupacabra (goat sucker) is as much a part of UFOlogy as it is of cryptozoology. From fortune-telling mermen to Scandinavian lake serpents whose appearance foretells the rise and fall of royal dynasties, monsters and the supernatural have always gone hand in hand.)

Many cryptozoologists are unhappy that the boundaries between their area of interest and other paranormal subcultures are so thin and so frequently crossed—so unhappy that some prominent cryptozoology portals do not permit any mention of the paranormal, even for the purpose of relevant comparison.[92] The impulse to sanitize away the monster literature's supernatural elements is very old; even classical authors attempted to demythologize the lore of fabulous animals. Such selective filtering can lead, warned folklorist Michel Meurger, to "gross distortion" of the very testimony on which cryptozoology depends in the first place.[93] But it is easy to sympathize with cryptozoologists who refuse to entertain "any ludicrous paranormal, occultic or supernatural viewpoints when discussing the nature and origins of such animals" when they know all too well that "it is hard enough making a case for [cryptids] as flesh and blood biological entities without having to deal with quasi-scientific nonsense."[94] It is a difficult situation: skeptics justifiably complain when cryptozoologists sweep the supernatural aspects of cryptid testimony under the rug, and yet skeptics also justifiably complain that cryptozoology is not sufficiently scientific (as we are about to do here once again). How are cryptozoologists supposed to be taken seriously or to aspire to any sort of scientific rigor when they are lumped together with believers in ghosts and Atlantis and alien abduction? As Aaron Bauer and Anthony Russell put it: "Cryptozoology as a science suffers from an image problem. Many critics regard cryptozoology as the bailiwick of the fringe element, of credulous individuals with no real credentials as scientific researchers. Certainly, the field does attract a disproportionate number of adherents whose interests tend toward the bizarre, and even to the supernatural."[95]

Nor are cryptozoology's problems limited to poor public relations. The fact is that cryptozoology has deep and ongoing challenges of focus and quality control. These challenges are not lost on cryptozoologists themselves, as we see in a stinging editorial from John Kirk:

> No other scientific field sees the kind of half-cocked, pseudoscientific babble that cryptozoology seems to produce. . . . Once you jettison science, the whole field of cryptozoology collapses into a pit of speculation and conjecture. . . . We get to the finish line when we have completed a thorough and exhaustive examination of actual tangible pieces of evidence which are utterly irrefutable. This is the standard that applies to all other areas of science be they mathematics, biology, physics, chemistry or biology. Nothing else is acceptable. Now please . . . keep your "skepticals" on.[96]

Richard Greenwell was another among those who pushed cryptozoologists to raise their scientific standards and look more critically at the evidence. For example, he was skeptical about a photograph of the alleged monster of Lake Champlain and was not even sure that it shows a living thing.[97] His article on the black cat of Puerto Penasco debunks the myths that had grown around a story about a huge 5-foot-long black panther in Mexico.[98] Throughout the *ISC Newsletter*, he sprinkled quotes about the importance of science and the philosophy of science.

In 1985, Greenwell published an influential article in which he suggested a classification scheme for unknown animals—from the conventional creatures that are discovered and named every year to the extreme cryptids that most scientists reject. His classification scheme has seven categories: I–IV, which include animals that are known and described; and V–VII, which contain creatures that are unconventional. For instance, animals in category IV are known taxa that may survive, even though they are thought to be extinct in historic times. As examples, Greenwell gave the White-winged Petrel and the marsupial thylacine (which is known as the "Tasmanian wolf" or "Tasmanian tiger" and appears on the coat of arms of Tasmania).[99] The petrel was thought to be extinct but was rediscovered, while the last known thylacine died in a zoo in 1936 and wild Tasmanian tigers have not been seen since. These animals are unremarkable, since there is physical evidence that they existed—in the form of the petrels themselves and the last tigers in the zoo—and most scientists would consider it surprising but not shocking if the thylacine still survives.

But then Greenwell made the error of false equivalence of categories. Category V includes animals that are known from fossils, but may have survived into historic times. For a precedent, he gave the coelacanth, and then claimed that the plesiosaur fits that category, too. This reasoning may also include the alleged dinosaur Mokele Mbembe as well as Bigfoot or the Yeti as a surviving *Gigantopithecus*—but by any classification system, the existence of one "living fossil" does not imply the survival of another. Coelacanths live in very deep water in only a few areas along the perimeter of the Indian Ocean (from South Africa to Indonesia). Given how rare they are and how seldom they come near the surface, it is not surprising that they were not found until 1938. But plesiosaurs were surface-dwelling, air-breathing reptiles that lived in the tropical waters of the epicontinental seas and the Tethys Sea during the Cretaceous period (144–65 million years ago), and it is extremely unlikely that any survive. Not only is it geologically impossible that lakes like Loch Ness harbor them, but the excellent fossil record of marine vertebrates of all sizes, from seals to whales, includes not a single plesiosaur fossil in any rocks younger than 65 million years. Greenwell made a serious error in conflating the idea that just because coelacanths were first known from fossils, *any* extinct creature can conceivably have survived into the present. Some prehistoric survivor-type cryptids would be harder to overlook than others, in either the living world or the fossil record.

Despite Greenwell's attempts to raise the scientific standards of cryptozoology, it continues to be performed as it has always been performed, and its methods leave a lot to be desired. Almost any piece of evidence, no matter how weak or incredible, is taken seriously, and eyewitness accounts are given much more weight than modern psychological research shows they deserve.[100] As geologist and cryptozoology critic Sharon Hill explained,

> I've researched and published on why amateur investigation groups fail to reach the high bar of science. I see these groups doing what I call "sham inquiry." It sounds sciencey, it looks sciencey, and it can fool a lot of people into thinking it's scientific but there are clear reasons why it is not. "Sham inquiry" is about the process and why the results they get out of that process are inferior to scientific inquiry.
>
> The primary problem . . . is that cryptozoologists, by and large, assume that a mystery creature is out there for them to find. They begin with a bias. . . . They are not testing a hypothesis but instead seeking evidence to support their position. . . . They also begin with the wrong question. Instead of "what happened?" they ask "Is it a cryptid?". . .
>
> Some are worse than others, for sure. I admire many so-called cryptozoologists. . . . I don't admire when the basic ideals of science are ignored—good scholarship in research, quality data collection and documentation, proper publication, skepticism, and open criticism. Instead, the bulk of popular cryptozoology is a

> jumble of the same old poor-quality evidence, a ton of hype, rampant speculation and unfounded assumptions, even conspiracy theories, and, too often, paranormal explanations.
>
> I want to make two clarifications. First, amateurs and non-scientists CAN do science. And, second, cryptozoology CAN be a science. But right now, I don't see that occurring often. It takes a lot of effort to do this and resources that the average enthusiast does not have. Too much is missing to call cryptozoology a science at this moment in time.[101]

If its adherents want cryptozoology to be taken seriously as true science, rather than as pseudoscience or "sham inquiry," they have to begin to play by the rules of real science. At a minimum, they must

- *Rethink their fundamental assumptions*: As Hill pointed out, in response to an odd occurrence or sighting, the cryptozoologists' first thought is: Is it a cryptid? or, worse, Is it cryptid X? But they should start several steps back, and ask the more basic scientific question: What really happened? One of the most powerful rules of science is Occam's razor: *the simplest explanation is usually the best*, because it involves the fewest leaps or assumptions. Generally speaking, it is a good idea not to invoke additional "what if" forces or factors or creatures unless the evidence really demands it. Real scientists do not try to force the data to fit their expectations and biases, but consider all hypotheses that explain the circumstances and rule them out, one by one. This is neatly captured in another aphorism that has long been favored in medical circles: "When you hear hoofbeats in the night, think first of horses, not of zebras." Almost every cryptozoological "observation" or bit of "evidence" can be plausibly accounted for by simple explanations that involve animals and phenomena that we know to be real, such as bears, boat wakes, and practical jokes. Cryptozoologists tend to preferentially and prematurely discard these simpler explanations in favor of the more desirable, less parsimonious assumptions that support their views.

- *Test the null hypothesis*: Related to this point is a concept in science that comes from statistics: the null hypothesis. Most statistical tests begin with the assumption that the null hypothesis is true and indicate whether there is enough evidence to reasonably reject it. Applied to testable scientific claims (for example, plesiosaurs live in Loch Ness), the null hypothesis is that the hypothesis formed by the researcher is *not* true or that the effect or animal or relationship hypothesized does *not* exist. In an experiment, a researcher might treat one group of subjects differently from another group to determine if the treatment makes the groups different in another way. For example, he might leave two slices of bread in the sun for several hours, covering one with plastic wrap, and then compare the amount of mold that grows on each. The null hypothesis is that plastic wrap will not affect the amount of mold that grows. However, if the researcher has a good reason to think that it should and there is significantly more mold on the covered slice than on the other, he can reasonably reject the null hypothesis.

Applied more generally, the null hypothesis is that what we know about the world is accurate. We continue to accept that hypothesis until we have sufficient evidence to reject it in favor of an alternative hypothesis. In other words, the default scientific assumption is that any strange observation can be explained by known natural causes; the less likely explanation—cryptid activity or paranormal phenomena—should be accepted only when all natural causes can be ruled out.

- *Meet the burden of proof*: In the sciences, most conventional ideas require only modest evidence to show that they are reasonable. But the more extreme the idea, the greater the burden of proof required. As Marcello Truzzi said, "An extraordinary claim requires extraordinary proof," and Carl Sagan echoed, "Extraordinary claims require extraordinary evidence."[102] The evidence needed to support the claims of cryptozoologists does not have to be as "extraordinary" as that required to overturn the germ theory of disease, for example, but it must be sufficiently strong and solid to overcome the long history of failure of all organized attempts to find any cryptid or any part of any cryptid. It follows from the claim that eyewitnesses fairly often see cryptids that carcasses or bones of those cryptids should also fairly often appear—but they do not. The absence of physical remains is not just a lack of evidence for cryptids, but strong positive evidence that hypothesized cryptids do not exist. This evidence against cryptids could be overbalanced by robust evidence *for* cryptids, but that would take more than easily faked footprints, inconsistent "eyewitness accounts," or inconclusive photographs or videos.

- *Collect high-quality data*: There is still no credible physical evidence in the form of carcasses, bones, hair, or other biological samples to support the claims for the existence of cryptids. This is the barest minimum that scientists demand to establish the existence of any cryptid (or any new animal species). Even crisp, high-resolution photographs or videos are merely suggestive (when those are available, which is rare in a field that gives us the cliché of the distant, blurry "blobsquatch"). Cryptozoologists take eyewitness testimony, which is subject to error, far too seriously, while simultaneously failing to be serious about how they collect that testimony. They tend not to take sufficient care to question stories, even when those stories are clearly false; they fail to quarantine prize evidence that is associated with known hoaxers and to disclose when evidence is so contaminated; and they often exaggerate or misreport accounts when they compile them. It is far too rare for cryptozoologists to do the necessary hard work of consulting primary documents or of interviewing witnesses and double-checking their stories for consistency and cultural bias, which Benjamin Radford and Joe Nickell discovered in their investigations of a number of purported cryptids.[103] When they do interview witnesses, they may often lead the witnesses to desired testimony.

- *Publish work that meets scientific standards*: Scholars and scientists like Darren Naish, Leigh Van Valen, Christine Janis, Colin Groves, and Adrienne Mayor not only regularly publish mainstream academic papers, but also occasionally write articles on topics of interest to cryptozoologists (such as the possibility of prehistoric animals having survived into the historical past and influenced human mythology) or on the chance of presumedly extinct creatures, such as the marsupial thylacine, having held on in remote areas of the world. These scientists publish their papers in peer-reviewed scientific journals, and their work exhibits the appropriate degree of skepticism and caution. Most cryptozoologists, however, write only for their like-minded audience, publish in their own journals and on their own Web sites, and make no effort to write high-quality papers with proper documentation that could be accepted by a mainstream science journal.

- *Be open to criticism*: All scientists must acknowledge that peer review is a harsh process that bruises the ego and can be biased or unfair at times. But good scientists have a thick skin, a strong sense of the merits of their research, a desire to improve their methods or look for better approaches, and the will to persevere to get their best work published. Such peer criticism is essential because it guarantees that only the best scientific work eventually is published. When, in the face of criticism, cryptozoologists claim that scientists are not open-minded about their ideas and then retreat to their own comfort zones, they are doing themselves a disservice. If they accepted the criticism as a call to base their work on hard evidence and to meet the other standards of scientific research, they would find that the scientific community is remarkably receptive to publishing unconventional ideas.

- *Be skeptical of their own data*: Good scientists must be skeptical of their own work as well as that of others. They have to be willing to toss out data or results if they are not good enough to support their hypotheses. As Richard Feynman said, "The first principle is that you must not fool yourself—and you are the easiest person to fool." If cryptozoologists want to be taken seriously, they must undertake the hard task of weeding out the questionable sightings and inconclusive photographs and videos of cryptids; be more careful in their scholarship about what eyewitnesses actually said; and be honest when their preferred ideas are not supported by the facts.

WHY DOES CRYPTOZOOLOGY MATTER?

There is no solid evidence that any of the cryptids discussed in this book exist and much evidence that their existence is extremely unlikely. Nevertheless, a large number of people continue to believe that they are real, no matter how much evidence to the contrary is accumulated and how often the stories of sightings are debunked.

People are entitled to believe whatever they wish. As Michael Shermer has shown, such beliefs can serve a function in the human psyche, allowing people a sense of the mysterious and the magical and an escape from mundane reality.[104] In prescientific times, the myths about monsters bound the members of society by a set of shared beliefs and values. Similarly, Bigfoot hunters derive a sense of satisfaction, community, and purpose from their wanderings in the woods, even if every attempt to find their cryptid has failed. Like members of other subcultures (including the skeptical subculture in which the authors of this book participate), Bigfooters enjoy the camaraderie when they take part in Bigfoot conferences and buy Bigfoot merchandise.

But scientists do not have the option of accepting something on the basis of belief alone. Instead, they must follow the evidence and examine any source of data critically, from both the subjects' and their own biases. Daniel Cohen, who writes about both science and the paranormal, addressed the hard reality of scientific endeavor: "It is genuinely exciting to believe in ghosts or flying saucers or the Abominable Snowman or the Lost Continent of Atlantis. Real science is nowhere near so thrilling, no matter how well it is presented. A rigorous logical approach to evidence is hard and restrictive; it destroys the beloved romantic myths and is going to be resented. It is a terrible day for a child when he discovers that Santa Claus does not exist, and adults are not much different."[105]

You might ask: Why should we even care what cryptozoologists think? How are they doing society any harm? Whatever your personal feelings about cryptozoology, and whatever your assessment of the evidence for cryptids, the answer to this "harm" question may be less obvious than it appears. On the one hand, even if one or more cryptids exist, the field of cryptozoology could have social costs that cryptozoology enthusiasts discount (perhaps by eroding confidence in the methods of science or by tempting advocates into beliefs that damage their reputations). On the other hand, even if all beliefs in cryptids are uniformly mistaken, the field could have little social cost—or even a net benefit (perhaps by encouraging outdoor activity or inspiring interest in zoology). Before drawing hasty conclusions, it is worth considering these words of caution from psychologist Ray Hyman:

> Why should we care? This question, in various guises, inevitably comes up in press conferences and in talks by skeptics. I vividly recall a press conference in the early days of CSICOP [Committee for the Scientific Investigation of Claims of the Paranormal]. A reporter asked why the group was so concerned about the growing belief in the paranormal. What harm does it do if someone believes that Uri Geller can bend a spoon with his mind? Besides, aren't there more important problems to worry about, such as the population explosion, famine, the homeless, drugs, acid rain, nuclear proliferation, and so on?
>
> Some of my colleagues immediately jumped in to defend the importance of their mission. One referred to the Jim Jones tragedy in Guyana. Another pointed out that belief in the paranormal made Hitler and Nazism possible. One skeptic stood up and dramatically announced that he had in his briefcase several suicide letters by young students who believed they would come back in better incarnations after their deaths.

> To me such reflex reactions are no better than the pathetic arguments that believers make for the existence of poltergeists, prophetic dreams, ghosts, and the like. In no case could any of those skeptics' assertions be backed up with anything resembling scientific evidence.
>
> I believe we should be more careful about how we justify our concerns. We should not make rash claims or feel the necessity to sensationalize the possibilities.[106]

What are some plausible costs and consequences of cryptozoology? Or is that the wrong question? Hyman continued, "My own inclination is to admit that I do not know how to measure the amount of harm that comes from belief in the paranormal," but he offered some reasons for scholars to concern themselves with the study of paranormal topics: some paranormal claims could turn out to be true, in which case they would be very important; whether they are true or not, the study of paranormal beliefs can lead us to important insights about the human mind and belief in general; and paranormal views that are based on "intuition and other nonscientific methods . . . known to induce compelling, but illusory, beliefs" shine a spotlight on the general dangers posed by the weakness of critical thinking skills and the lack of scientific literacy. After all, as Hyman warned, "If members of our society—including generals, business executives, and political leaders—develop their beliefs about the paranormal on such an illusory foundation, what does that tell us about how they are making decisions that affect the state of the world?"[107]

The authors of this book are divided in our assessment of the net consequences of cryptozoological beliefs and enthusiasms. Daniel Loxton is quite sympathetic to cryptozoology; Donald Prothero is much more critical.

Coming out of cryptozoology himself, Loxton is inclined to regard monster hunting—even in the permanent absence of any genuine cryptids—in much the same terms as *The Hitchhiker's Guide to the Galaxy*'s description of Earth: "mostly harmless." Unlike many of the paranormal topics that skeptics critique, cryptozoology does not ask adherents to reject the laws of physics, partake in unproved or disproved medical treatments, live in terror of alien invaders, or sell their possessions in anticipation of the end of the world. Cryptozoology is not terribly expensive for most people—no more, certainly, than hunting or fishing—and it offers the tangible benefits of any literary hobby or fan community. It encourages the development of skills of reading and historical investigation, and it brings enthusiasts together. Is cryptozoology so different in this respect from the skeptical subculture in which we two participate? As Loxton has put it, whether communities form around the love of Bigfoot or skepticism or model trains, "finding commonality with other human beings is a good in itself—an *end* in itself. Indeed, in respect to this *particular* end, the 'skeptical' part of the skeptical community is largely beside the point."[108] If fans of monsters find a similar commonality in their shared love of old microfilm and camping trips, is that really so bad? Furthermore, Loxton argues that the love of cryptozoological mysteries may offer some of the same educational benefits that science advocates promote: love of the natural world and experience grappling with the nature of scientific evidence. Extrapolating from his own life experience, Loxton suspects that the cryptozoology literature may as easily and as often become a "gateway drug" for science literacy as it may be for attitudes of resistance to science:

> Scientific skepticism turns a critical eye on paranormal and fringe science claims, so it's usually framed by narratives of tension or conflict (as *X-Files* audiences will recall): either the stubborn, reductive skeptic versus the open-minded, intuitive believer; or, the responsible, science-based thinker versus the foolish kook.
>
> But I have been both, and my experience did not fit into that story of conflict.
>
> I was a believer in everything (a Fox Mulder if ever there was one!) who eventually became a "professional skeptic." And yet, my life as a skeptical investigator is a seamless continuation from my paranormal enthusiasm: from inquisitive nerd with a passion for weird mysteries, to inquisitive nerd with a passion for weird mysteries. Nothing changed, except the information I had to work with.[109]

Loxton suggests three reasons why science-minded skeptics may wish to support the research of cryptozoologists in principle (or at least the freedom to pursue such research), even as they strenuously critique it in particulars:

- Echoing Hyman, there's the off chance that cryptozoologists could be right. After all, the laws of physics do not have to be overturned in order for the Bigfoot hypothesis to be true—Bigfoot just has to be found. It is plausible on the face of it, although in our view exceedingly unlikely.

- Research into popular mysteries and paranormal beliefs provides a public good: satisfaction of the desire to have subjects of wide popular interest probed and examined. These fringe topics are typically neglected by mainstream scholarship; by studying and chronicling the paranormal, the most responsible writers perform a valuable public service. Indeed, when critical scholars *do* deign to investigate paranormal subjects, they inevitably build their investigations on top of preliminary work that was undertaken on the fringe. For that reason, as Gerald Durrell emphasized, cryptozoological footwork retains value even in the absence of any genuine cryptids:

> Up to now most zoologists have treated the whole subject of sea-serpents, abominable snowmen and similar creatures as something that is not quite nice. It's as though they feel there were some gigantic conspiracy afoot to undermine their ideas of what does and does not exist in the world. This attitude makes it extremely hard to get at the facts behind reports of these alleged creatures. Whether they exist or not, it was essential for someone to collect all the reports and sift them through for publication. If the animals *do* exist and *are* discovered, this book will prove a valuable piece of research. If, on the other hand, it is proved that they do *not* exist, the book loses none of its value, for it becomes an important contribution to zoological mythology.[110]

- Research into unconventional topics, such as the existence of Bigfoot, provides a barometer of the health of academic freedom. Indeed, such research is what academic freedom is designed to protect. By using that freedom to explore unorthodoxies, scholars who are sympathetic to cryptozoology or the paranormal provide a measure of insurance against academic dogma. As long as the quality and integrity of their work is strong, all of us are better served when scientists and other academics have the freedom to reexamine existing consensus views, dig deeper into unresolved mysteries, pursue long shots, or even waste their time flailing around on the fringes of science.

But is cryptozoology "mostly harmless," as Loxton believes? For his part, Prothero is not so sure. Whatever the romantic appeal of monster mysteries, cryptozoology as it exists today is unquestionably a pseudoscience. None of the cryptids it purports to study have been demonstrated to exist; the reality of most is exceptionally unlikely; and some, like the Loch Ness monster of popular legend, can be definitively rejected as untrue.

Granted, cryptozoology is less obviously dangerous than are some other pseudoscientific claims, such as the discredited but fiercely promoted speculation that routine vaccinations cause autism or the assertion that HIV does not cause AIDS (a belief that has been calculated to have caused 365,000 premature deaths in South Africa alone).[111] But aren't these and other fringe topics unified by a common pattern of pseudoscientific thinking? Rather than merely wasting time and resources, the widespread acceptance of the reality of cryptids may feed into the general culture of ignorance, pseudoscience, and anti-science. The more the paranormal is touted by the media as acceptable and scientifically credible—rather than subjected to the harsh scrutiny of the scientific method, the rigor of critical thinking, and the demand for real evidence—the more people are made vulnerable to the predations of con artists, gurus, and cult leaders. The more the creationist cryptozoologists manage to damage the understanding of science, the worse off we all are.[112]

This is a serious problem in a culture where critical thinking is in short supply and basic science literacy is rare. Study after study over the years has shown that the American public has an abysmally poor understanding of how the world really works. For example, a national survey commissioned by the California Academy of Sciences in 2008 found that

- Only 53% of adults know how long it takes for the Earth to revolve around the Sun.
- Only 59% of adults know that the earliest humans and dinosaurs did not live at the same time.
- Only 47% of adults can roughly approximate the percent of the Earth's surface that is covered by water.
- Only 21% of adults answered all three questions correctly.[113]

This level of understanding sounds almost too poor to believe, but it is consistent with the findings of many other surveys. In the United States, barely 50 percent of adults know that an electron is smaller than an atom.[114] Most adults cannot define concepts like cell, molecule, or DNA. Only about 33 percent of adults agree that humans share more than half their genes with mice; only 38 percent of adults understand that humans share more than half their genes with chimpanzees.[115] All but 16 percent agree that the center of Earth is very hot, but only 38 percent know that the Big Bang describes the early history of the universe.[116] Moreover, a surprisingly large number of American adults *still think that the sun revolves around Earth!* These people are not limited to just those few dedicated advocates who actively promote geocentrism for religious reasons,[117] but include a hefty 18 percent of the United States population at large. (The same misapprehension is held by similar numbers in Germany and the United Kingdom.)[118] No one knows how many American adults literally believe that the Earth is flat, but they include a subset of creationists who promote a flat Earth for biblical reasons.[119]

Another way to frame the question is to ask how the United States stacks up against other countries. Scientific illiteracy is a global challenge, but the United States faces some obstacles to public understanding that are almost uniquely American, such as resistance to the fact of biological evolution. This particular struggle is especially relevant in the context of the impact of cryptozoology; as we have seen, cryptozoology is frequently employed by creationists to erode confidence in evolution. In a 2005 ranking of thirty-four industrialized countries published in the journal *Science*, the United States scored almost dead last, with a mere 40 percent of citizens accepting evolution. By comparison, the citizens of Iceland, Denmark, Sweden, and France were fully *twice* as likely to affirm that evolution is true. Only Turkey scored worse than the United States; countries like Italy, Hungary, and the Czech Republic left the United States far behind.[120] In a closer comparison, Canadians are much more likely to accept evolution than are their American counterparts. Both the quality of evolution education and the level of public acceptance of evolution vary by province, but evolution is accepted by 58 percent of the population overall, with only 22 percent advocating creationism.[121] In addition, Canadians generally respect the separation of church and state better than do Americans, since the fundamentalists are not so numerous or powerful in their political system. In contrast, the Republican Party now follows a platform dictated largely by the fundamentalists in its political base, including planks that support creationism, ban abortion and stem-cell research, deny global warming, and ignore the limits of Earth's resources in the face of population growth.

It is a matter for particular concern for Americans that students in the United States lag significantly behind those in many other industrialized nations in science literacy. In 2009, for example, the Programme for International Student Assessment (PISA) of the Organisation for Economic Cooperation and Development found that among fifteen-year-olds, twenty-one countries scored higher in science literacy than the United States. At the top of the list were China and Finland, followed by (among others) Japan, New Zealand, Canada, Australia, the Netherlands, Germany, and the United Kingdom.[122] Other rankings in recent years give similar results. Although the exact order of the top-ranked countries might be shuffled a bit, science literacy in the United States chronically falls behind that in many other nations, with a dismal twenty-ninth-place finish in the 2006 PISA rankings.[123] Students in the United States even lag behind their counterparts in nations like Estonia and Slovenia that have a fraction of our wealth. That alone is a mark of disgrace for American society—that we spend so much money per child, and yet end up with such mediocre results. What does it say for the future economic well-being of the United States if it is outperformed by competitors such as China on such a crucial factor as understanding science? Innovation and scientific output are complex issues weaving many social and economic threads, but public understanding of science is a part of that tapestry.

A recent headline from CNN reads, "China Shoots Up Rankings as Science Power, Study Finds."[124] As the article summarizes, a study conducted by the Royal Society, the world's oldest and foremost scientific organization, found that although the United States is still the dominant scientific power in terms of scientific publications, China has experienced a "meteoric rise" in scientific publications and new research.[125] In 2003, fewer than 5 percent of scientific articles were published in China. By 2008, 10 percent were, putting China second only to the United States. Meanwhile, the American share of scientific publications dropped from 26 to 21 percent. According to Sir Chris Llewellyn Smith, chair of the advisory group for the study:

> The scientific world is changing and new players are fast appearing. Beyond the emergence of China, we see the rise of South-East Asian, Middle Eastern, North African and other nations. The increase in scientific research and collaboration, which can help us to find solutions to the global challenges we now face, is very welcome. However, no historically dominant nation can afford to rest on its laurels if it wants to retain the competitive economic advantage that being a scientific leader brings.[126]

China is shooting up in many rankings, making it the second-largest economic power as well. Seemingly unencumbered by global-warming deniers, stem-cell-research antagonists, or creationists who interfere with science policy, China is making huge investments in new technologies for a world that is facing the consequences of global warming and limited oil,[127] while the United States and the United Kingdom slip down the rankings of countries investing in green technology.[128] Germany and several Scandinavian countries have long led the world in their investments in green technology and their commitment to lowering energy use and reducing greenhouse-gas emissions—yet their economies are stronger than those of the United States and most of the nations in southern Europe.[129]

Americans still hold the lion's share of Nobel Prizes in the sciences and have since 1956, when the effect of Germany's experiment with Hitler, descent into anti-Semitism, and instigation of World War II caused a "brain drain" of scientists from Germany to the United States and other countries and ended German supremacy in science.[130] But how long can the American preeminence in science last when the pernicious influences of creationism, paranormalism, and dedicated anti-science lobbies continue to erode the scientific culture?

Some people deny that the United States, the dominant world power since 1945, could ever cede its place to another country. But dominance is tied to innovation, which is tied to science. As historians have pointed out, many other powerful societies with enormous economic reach and once-flourishing arts and sciences have declined. Only 150 years ago, the British Empire of Queen Victoria spanned the globe, but now Great Britain is a relatively minor player among international powers, as it lost most of its economic dominance and colonial empire during and after World Wars I and II. The once-mighty Soviet Union fell in just a few years in the 1990s, along with the regimes of the nations it controlled. For ten years, the United States has been wasting trillions of dollars on wars that have cost thousands of American lives, while running up huge budget deficits in a time of economic recession. Americans like to think of their country as exceptional, but that is not the lesson that history teaches.

Widespread paranormal and pseudoscientific thinking ought not to be dismissed as charming eccentricity; nor is the endless conveyor belt of monster media merely innocent entertainment. Paranormal hype makes money for its purveyors, but that easy profitability comes at the shared cost of encouraging anti-scientific and anti-rational thinking. Such thinking can be directly harmful, especially when it intersects with public health; moreover, the erosion of scientific understanding robs us of opportunities for success and understanding that we wish for our children. For these reasons, leading American scientists have asked us for decades to reach for higher standards, and asked us to care more deeply about truth. This stuff matters. As astrophysicist Neil deGrasse Tyson explains:

> If you are scientifically literate the world looks very different to you. It's not just a lot of mysterious things happening. There is a lot we understand out there. And that understanding empowers you to, first, not be taken advantage of by others who do understand it. And second there are issues that confront society that have science as their foundation. If you are scientifically illiterate, in a way, you are disenfranchising yourself from the democratic process, and you don't even know it.[131]

Carl Sagan was even more blunt: "We've arranged a global civilization in which the most critical elements profoundly depend on science and technology. We have also arranged things so that almost no one understands science and technology. This is a prescription for disaster."[132]

1. CRYPTOZOOLOGY

1 "Americans 'Find Body of Bigfoot,'" August 15, 2008, BBC News, http://news.bbc.co.uk/2/hi/americas/7564635.stm; "CNN—Has Bigfoot Been Found?" August 17, 2008, YouTube, http://www.youtube.com/watch?v=OMrKrphWAEk&feature=related; Ki Mae Heussner, "Legend of Bigfoot: Discovery? Try Hoax," August 15, 2008, ABC News, http://abcnews.go.com/Technology/story?id=5583488&page=1; "Bigfoot Body Found in Georgia!!!!" August 14, 2008, YouTube, http://www.youtube.com/watch?v=4NzB5xfTSDY (all accessed October 10, 2011).

2 "Georgia Bigfoot Tracker," August 14, 2008, YouTube, http://www.youtube.com/watch?v=e_zFoCo_Tnc (accessed October 10, 2011).

3 Writing about the collaboration between "fast-talking Tom Biscardi" and Bigfoot hoaxer Ivan Marx, reporter Richard Harris was already asking in 1973 whether the for-profit team was composed of "adventurers or con-men" ("The Bigfoot: Fact, Fiction, or Flim-Flam?" *Eureka [Calif.] Times-Standard*, June 8, 1973, 13).

4 Matthew Moneymaker, "The 2008 Dead Bigfoot Hoax from Georgia," Bigfoot Field Researchers Organization, http://www.bfro.net/hoax.asp#biscardi (accessed October 10, 2011).

5 "HOAX! Georgia Bigfoot Body Press Conference," August 15, 2008, YouTube, http://www.youtube.com/watch?v=pEinriwAPnQ (accessed October 10, 2011).

6 The image is available at "Georgia Gorilla: Bigfoot Body's First Photo," August 12, 2008, CryptoMundo, http://www.cryptomundo.com/cryptozoo-news/ga-gorilla-pic/ (accessed October 10, 2011).

7 "Alleged 'Bigfoot' Finders: Costume Was Filled with Roadkill," August 20, 2008, WSB-TV, http://www.wsbtv.com/news/news/alleged-bigfoot-finders-costume-was-filled-with-ro/nFBhF/ (accessed February 7, 2012).

8 Quoted in Paul Wagenseil, "Bigfoot Body Revealed to Be Halloween Costume," August 20, 2008, Fox News, http://www.foxnews.com/story/0,2933,406101,00.html (accessed October 10, 2011).

9 "Bigfoot Hoaxers Say It Was Just 'a Big Joke,'" August 21, 2008, CNN, http://edition.cnn.com/2008/US/08/21/bigfoot.hoax/ (accessed February 7, 2012).

10 Quoted in Steve Bass, "NGCSU Prof Shed Light on Bigfoot Hoax," November 6, 2008, *The Saint: The Online Student Newspaper for North Georgia College and State University*, http://www.ngcsuthesaint.com/2008/11/ngcsu-prof-shed-light-on-bigfoot-hoax/ (accessed October 10, 2011).

11 For an excellent essay on the way science is practiced and the way creationists subvert these processes, see Stephen Jay Gould, "An Essay on a Pig Roast," *Natural History*, January 1989, 14–25, reprinted in *Bully for Brontosaurus: Reflections in Natural History* (New York: Norton, 1991), 432–447.

12 Carl Sagan, *The Demon-Haunted World: Science as a Candle in the Darkness* (New York: Ballantine, 1996), 30.

13 Michael Shermer, *The Believing Brain: From Ghosts and Gods to Politics and Conspiracies: How We Construct Beliefs and Reinforce Them as Truths* (New York: Holt, 2011).

14 Michael Shermer, *Why People Believe Weird Things: Pseudoscience, Superstition, and Other Confusions of Our Time* (New York: Freeman, 1997), 48; Sagan, *Demon-Haunted World*, 10.

15 "Carl Sagan on Alien Abduction," *Nova*, February 27, 1996, http://www.pbs.org/wgbh/nova/space/sagan-alien-abduction.html (accessed December 22, 2011); Marcello Truzzi, "On the Extraordinary: An Attempt at Clarification," *Zetetic Scholar* 1 (1978): 11.

16 National Center for Science Education, "Project Steve," October 17, 2008, http://ncse.com/taking-action/project-steve (accessed October 10, 2011).

17 Cryptozoological Realms, "Cryptozoology Biographies," http://www.cryptozoology.net/english/biographies/biographies_index.html (accessed October 10, 2011).

18 See, for example, John Bindernagel, *North America's Great Ape: The Sasquatch* (Courtenay, B.C.: Beachcomber Books, 1998), x.

19 Loren Coleman and Jerome Clark, *Cryptozoology A to Z: The Encyclopedia of Loch Monsters, Sasquatch, Chupacabras, and Other Authentic Mysteries of Nature* (New York: Simon and Schuster / Fireside, 1999), 102.

20 See, for example, British Columbia Scientific Cryptozoology Club, "Mokele Mbembe," http://bcscc.ca/mokele.htm (accessed January 28, 2012).

21 Daniel Loxton asked the Southern Association of Colleges and Schools Commission on Colleges (the regional accreditation body for degree-granting higher-education institutions in Georgia and other southern states) whether Immanuel Baptist College has ever been accredited to grant academic degrees in Georgia. A representative responded, "Immanuel Baptist College has never been accredited by SACSCOC" (Terri Latimer, e-mail to Daniel Loxton, February 15, 2012).

22 *In the Matter of Warnborough College*, Docket Nos. 95-164-ST and 96-60-SF, United States Department of Education, August 9, 1996, http://oha.ed.gov/cases/1995-164st.html (accessed February 5, 2012).

23 Sarah Lyall, "Americans Say a College Near Oxford Duped Them," *New York Times*, October 2, 1995, http://query.nytimes.com/gst/fullpage.html?res=990CE1D91531F931A35753C1A963958260 (accessed February 5, 2012).

24 Ben S. Roesch and John L. Moore, "Cryptozoology," in *The Skeptic Encyclopedia of Pseudoscience*, ed. Michael Shermer (New York: ABC-Clio, 2002), 1:71–78.

25 Darren Naish, e-mail to Daniel Loxton, February 5, 2012.

26 Matt Cartmill, reviews of *Bigfoot Exposed: An Anthropologist Examines America's Enduring Legend*, by David J. Daegling, and *Sasquatch: Legend Meets Science*, by Jeff Meldrum, *American Journal of Physical Anthropology* 135, no. 1 (2008): 117–118; David J. Daegling, *Bigfoot Exposed: An Anthropologist Examines America's Enduring Legend* (Lanham, Md.: Altamira Press, 2004).

27 Thomas Henry Huxley, "Biogenesis and Abiogenesis" (1870), in *Critiques and Addresses* (London: Macmillan, 1873), 229.

28 Quoted in Michael Shermer, "Show Me the Body," *Scientific American*, May 2003, http://www.michaelshermer.com/2003/05/show-me-the-body/ (accessed October 10, 2011).

29 Elizabeth F. Loftus and Katherine Ketcham, *Witness for the Defense: The Accused, the Eyewitness, and the Expert Who Put Memory on Trial* (New York: St. Martin's Press, 1991); Austin Cline, "Eyewitness Testimony and Memory," About.com, http://atheism.about.com/od/parapsychology/a/eyewitness.htm (accessed October 10, 2011).

30 Daniel Simons, "Selective Attention Test," March 10, 2010, YouTube, http://www.youtube.com/watch?v=vJG698U2Mvo (accessed October 10, 2011).

31 A. Leo Levin and Harold Kramer, *Trial Advocacy Problems and Materials* (Mineola, N.Y.: Foundation Press, 1968), 269.

32 Elizabeth F. Loftus, *Memory: Surprising New Insights into How We Remember and Why We Forget* (New York: Addison-Wesley, 1980), 37.

33 Benjamin Radford and Joe Nickell, *Lake Monster Mysteries: Investigating the World's Most Elusive Creatures* (Lexington: University Press of Kentucky, 2006), 160–163.

34 Quoted in Elsbeth Bothe, "Facing the Beltway Snipers, Profilers Were Dead Wrong," *Baltimore Sun*, December 15, 2002, http://articles.baltimoresun.com/2002-12-15/entertainment/0212160297_1_van-zandt-serial-killers-snipers (accessed October 10, 2011).

35 David Rennie, "Sniper Police Deny "White Man in White Van" Theory Hampered Search," *Telegraph*, October 28, 2002, http://www.telegraph.co.uk/news/worldnews/northamerica/usa/1411536/Sniper-police-deny-white-man-in-white-van-theory-hampered-search.html (accessed November 21, 2012).

36 Shermer, *Why People Believe Weird Things*, 88–89.

37 Ibid., 89–91.

38 Susan A. Clancy, *Abducted: How People Come to Believe They Were Kidnapped by Aliens* (Cambridge, Mass.: Harvard University Press, 2005), 33.

39 The search for the cryptid Mokele Mbembe provides a stark example: the interrogation of Congolese villagers by Roy Mackal's armed party in 1981. When local informants de-

nied knowledge of the creature, Mackal "confronted" them with "a barrage of information" until they became "visibly disturbed, and some, in their confusion, admitted to a great deal more knowledge" (*A Living Dinosaur? In Search of Mokele-Mbembe* [Leiden: Brill, 1987], 160).

40 See, for example, Loren Coleman, "The Meaning of Cryptozoology: Who Invented the Term Cryptozoology?" 2003, The Cryptozoologist, www.lorencoleman.com/cryptozoology_faq.html (accessed February 12, 2011), and "Cryptozoology's Fathers," June 19, 2010, CryptoMundo, http://www.cryptomundo.com/cryptozoo-news/czfather/ (accessed May 5, 2012).

41 Bernard Heuvelmans, *Sur la piste des bêtes ignorées* (Paris: Librarie Plon, 1955); Lucien Blancou, *Géographie cynégétique du monde* (Paris: Presses Universitaires de France, 1959).

42 Bernard Heuvelmans, *In the Wake of the Sea Serpents*, trans. Richard Garnett (New York: Hill and Wang, 1968), 508.

43 The word "cryptozoological" appears at a line break and so is hyphenated: "crypto-zoological." It is unknown if the reviewer intended the word to have a hyphen, but he used it in much the same way that writers use it today. See Ralph Thompson, review of *The Lungfish and the Unicorn*, by Willy Ley, *New York Times*, April 22, 1941, 19; and Willy Ley, *The Lungfish and the Unicorn: An Excursion into Romantic Zoology* (New York: Modern Age Books, 1941).

44 A. C. Oudemans, *The Great Sea-serpent: An Historical and Critical Treatise. With the Reports of 187 Appearances . . . the Suppositions and Suggestions of Scientific and Non-scientific Persons, and the Author's Conclusions* (Leiden: Brill, 1892).

45 Coleman and Clark, *Cryptozoology A to Z*.

46 Darren Naish, "Monster Hunting? Well, No. No," October 10, 2007, Tetrapod Zoology, http://scienceblogs.com/tetrapodzoology/2007/10/monster_hunting_well_no_no.php (accessed October 10, 2011).

47 Bernard Heuvelmans, "How Many Animal Species Remain to Be Discovered?" *Cryptozoology* 2 (1983): 5.

48 Bernard Heuvelmans, "Annotated Checklist of Apparently Unknown Animals with Which Cryptozoology Is Concerned," *Cryptozoology* 5 (1986): 1–26.

49 Naish, "Monster Hunting?"

50 Darren Naish, "Multiple New Species of Large Living Mammal," June 1, 2007, Tetrapod Zoology, http://scienceblogs.com/tetrapodzoology/2007/06/multiple_new_species_of_large.php (accessed October 10, 2011).

51 Ralph M. Wetzel, Robert E. Dubos, Robert L. Martin, and Philip Myers, "*Catagonus*, an 'Extinct' Peccary, Alive in Paraguay," *Science* 189 (1975): 379–381.

52 N. F. Goldsmith and I. Yanai-Inbar, "Coelacanthid in Israel's Early Miocene? *Latimeria* Tests Schaeffer's Theory," *Journal of Vertebrate Paleontology* 17, supplement 3 (1997): 49A; T. Ørvig, "A Vertebrate Bone from the Swedish Paleocene," *Geologiska Föreningens i Stockholm Förhandlingar* 108 (1986): 139–141.

53 Michael A. Woodley, Darren Naish, and Hugh P. Shanahan, "How Many Extant Pinniped Species Remain to Be Described?" *Historical Biology* 20 (2008): 225–235.

54 Darren Naish, e-mail to Daniel Loxton, April 30, 2012.

55 Andrew R. Solow and Woollcott K. Smith, "On Estimating the Number of Species from the Discovery Record," *Proceedings of the Royal Society B* 272 (2005): 285–287.

56 So devastating has the beetle infestation become in the new global-warming climate regime that British Columbia's forest industry currently depends on the logging of dead trees that were killed five or ten years ago by beetles. With the looming exhaustion of these stands of dead trees as a result of urgent harvesting, fires, and rot, the province faces such a catastrophic shortage of trees that its government has considered opening up the remaining protected areas of forest for logging—not to save the industry, but simply to *delay* the predicted loss of 11,000 forestry jobs. See, for example, "Confidential Pine Beetle Report Warns of 'Economic and Social' Havoc," April 18, 2012, CBC News, http://www.cbc.ca/news/canada/british-columbia/story/2012/04/18/bc-timber-supply-mpb.html (accessed April 28, 2012).

57 James H. Brown and Brian A. Maurer, "Macroecology: The Division of Food and Space Among Species on Continents," *Science* 243 (1989): 1145–1150. Other important publications include James H. Brown, Pablo A. Marquet, and Mark L. Taper, "Evolution of Body Size: Consequences of an Energetic Definition of Fitness," *American Naturalist* 142 (1993): 573–584; James H. Brown and Brian A. Maurer, "Evolution of Species Assemblages: Effects of Energetic Constraints and Species Dynamics on the Diversification of the North American Avifauna," *American Naturalist* 130 (1987): 1–17; James H. Brown and Paul F. Nicoletto, "Spatial Scaling of Species Composition: Body Masses of North American Land Mammals," *American Naturalist* 138 (1991): 1478–1512; William A. Calder III, *Size, Function, and Life History* (Cambridge, Mass.: Harvard University Press, 1984); John Damuth, "Home Range, Home Range Overlap, and Species Energy Use Among Herbivorous Mammals," *Biological Journal of the Linnean Society* 15 (1981): 185–193; A. S. Harestad and F. L. Bunnell, "Home Range and Body Weight—A Reevaluation," *Ecology* 60 (1979): 389–402; Stan L. Linstedt, Brian J. Miller, and Steven W. Buskirk, "Home Range, Time, and Body Size in Mammals," *Ecology* 67 (1986): 413–418; Brian K. McNab, "Bioenergetics and the Determination of Home Range Size," *American Naturalist* 47 (1963): 133–140; Robert Henry Peters, *The Ecological Implications of Body Size* (New York: Cambridge University Press, 1983); Michael Reiss, "Scaling of Home Range Size: Body Size, Metabolic Needs and Ecology," *Trends in Ecology and Evolution* 3 (1988): 85–86; Marina Silva and John A. Downing, "The Allometric Scaling of Density and Body Mass: A Nonlinear Relationship for Terrestrial Mammals," *American Naturalist* 145 (1995): 704–727; and Robert K. Swihart, Norman A. Slade, and Bradley J. Bergstrom, "Relating Body Size to the Rate of Home Range Use in Mammals," *Ecology* 69 (1988): 393–399.

58 Johan T. du Toit, "Home Range–Body Mass Relations: A Field Study on African Browsing Ruminants," *Oecologia* 85 (1990): 301–303.

59 Robert H. MacArthur and Edward O. Wilson, *Theory of Island Biogeography* (Princeton, N.J.: Princeton University Press, 1967).

60 Donald R. Prothero and Robert H. Dott Jr., *Evolution of the Earth*, 7th ed. (Dubuque, Iowa: McGraw-Hill, 2004), chap. 1.

61 Lars Werdelin and William Joseph Sanders, eds., *Cenozoic Mammals of Africa* (Berkeley: University of California Press, 2010).

2. BIGFOOT

1 John Napier, *Bigfoot: The Yeti and Sasquatch in Myth and Reality* (New York: Dutton, 1973), 202–203.
2 Loren Coleman and Patrick Huyghe, *The Field Guide to Bigfoot and Other Mystery Primates* (San Antonio, Tex.: Anomalist Books, 2006), 88–89.
3 Loren Coleman, *Bigfoot! The True Story of Apes in America* (New York: Paraview, 2003), 27.
4 "Mythology of the Two Americas: Iroquois and Hurons," *Larousse Encyclopedia of Mythology* (New York: Prometheus, 1959), 437–438.
5 Anthony Wonderley, *Oneida Iroquois Folklore, Myth, and History: New York Oral Narrative from the Notes of H. E. Allen and Others* (Syracuse, N.Y.: Syracuse University Press, 2004), 98–99; Raymond Fogelson, "Stoneclad Among the Cherokees," in *Manlike Monsters on Trial: Early Records and Modern Evidence*, ed. Marjorie Halpin and Michael M. Ames (Vancouver: University of British Columbia Press, 1980), 134.
6 Grover Krantz, *Bigfoot Sasquatch Evidence* (Surrey, B.C.: Hancock House, 1999), 143.
7 Wayne Suttles, *Coast Salish Essays* (Vancouver: Talonbooks, 1987), 90.
8 Quoted in ibid., 91.
9 Royal British Columbia Museum, "Thunderbird Park: Present Park: Kwakwaka'wakw Heraldic Pole, 1953," Thunderbird Park: Place of Cultural Sharing, http://www.royalbcmuseum.bc.ca/exhibits/tbird-park/main.htm?lang=eng (accessed April 16, 2012).
10 The Dzunuk'wa figure from the Royal British Columbia Museum's Thunderbird Park appears, for example, on the cover of John Green, *Sasquatch: The Apes Among Us* (Surrey, B.C.: Hancock House, 2006), and in the interior of Christopher L. Murphy, *Meet the Sasquatch* (Surrey, B.C.: Hancock House, 2004), 17, and Jeff Meldrum, *Sasquatch: Legend Meets Science* (New York: Forge, 2006), 75.
11 Suttles, *Coast Salish Essays*, 78–79.
12 Coleman, *Bigfoot!*, 27.
13 Suttles, *Coast Salish Essays*, 74.
14 J. W. Burns, "Introducing B.C.'s Hairy Giants," *Maclean's*, April 1, 1929, reprinted in John Green, *The Sasquatch File* (Agassiz, B.C.: Cheam, 1973), 10–11.
15 John Green, *On the Track of the Sasquatch* (Agassiz, B.C.: Cheam, 1968), 3.
16 Quoted in Burns, "Introducing B.C.'s Hairy Giants," 10–11.
17 John Burns, as told to Charles Tench, "My Search for B.C.'s Giant Indians," *Liberty Magazine*, 1954, reprinted in Murphy, *Meet the Sasquatch*, 32.
18 Ibid.
19 Ibid., 31.
20 Green, *On the Track of the Sasquatch*, 3.
21 Ibid.
22 "Let's Not Forget the Sasquatchewan Trade," *Vancouver Sun*, May 7, 1957, 5.

23 "Wanted: Sasquatch—$5000 Reward," *Vancouver Sun*, May 4, 1957, 25.

24 "Sasquatch Hunt Delayed by Intrepid Lillooet Band," *Vancouver Sun*, May 10, 1957, 3.

25 Jack Brooks, "Sasquatch-Hunters Get Sasquatched," *Vancouver Sun*, May 24, 1957, 3, final edition.

26 Don Hunter, with René Dahinden, *Sasquatch: The Search for North America's Incredible Creature* (New York: Signet, 1975), 83.

27 "'Nothing Monstrous About Sasquatch,' Says Their Pal," *Vancouver Sun*, May 25, 1957, 14.

28 Although there are chimpanzee- or gorilla-like monster yarns that predate that of Roe, they are distinct in two ways: they are outside the direct lineage of Sasquatch tradition, and their descriptions differ in one or more important respects from the modern composite description of Bigfoot. For example, the creatures featured in the Ape Canyon story of 1924 are retroactively included in Bigfoot lore, but their description is very different: "Their ears are about four inches long and stick straight up. They have four toes" (*Oregonian* [Portland], July 3, 1924, reprinted in John Green, *The Best of Sasquatch Bigfoot* [Surrey, B.C.: Hancock House, 2004], 61).

29 Green, *On the Track of the Sasquatch*, 10.

30 Ibid., 3.

31 Quoted in ibid., 11–12.

32 Ibid.

33 Daniel Loxton, "Junior Skeptic 20: Bigfoot Part One: Dawn of the Sasquatch," *Skeptic* 11, no. 2 (2004): 96–105.

34 John Green to Daniel Loxton, August 10, 2004.

35 Ibid.

36 Green, *On the Track of the Sasquatch*, 10.

37 Ibid.

38 John Green to Daniel Loxton, August 20, 2004.

39 John Kirk to Daniel Loxton, July 21, 2004.

40 Quoted in Roger Patterson and Christopher Murphy, *The Bigfoot Film Controversy* (Surrey, B.C.: Hancock House, 2005), 98.

41 It is possible that one or more Bigfooters eventually did see photographs of Roe, after his death (perhaps supplied by his daughter), but I am not aware of any published photograph of him.

42 Coleman, *Bigfoot!*, 68–69.

43 John Green, "The Ape and the IM Index," July 11, 2009, Southeast Sasquatch Association Archive, http://southeastsasquatchassociationarchive.blogspot.com/2009/07/work-of-john-green.html (accessed February 28, 2010).

44 Ivan T. Sanderson, *Abominable Snowmen: Legend Come to Life* (New York: Pyramid Books, 1968), 153.

45 Quoted in "This Yeti Has Big Feet: Abominable Snowman on Klamath," *Independent Press-Telegram* (Long Beach, Calif.), October 5, 1958, 19.

46 "Promised Hoax Exposé of Mysterious Footprints Fails to Materialize," *Humboldt Standard* (Eureka, Calif.), October 14, 1958, 11.

47 Coleman, *Bigfoot!*, 80.
48 Loren Coleman, "The Ray Wallace Debate: Part I," February 12, 2006, CryptoMundo, http://www.cryptomundo.com/cryptozoo-news/ray-wallace-1/ (accessed April 1, 2010).
49 Joshua Blu Buhs, *Bigfoot: The Life and Times of a Legend* (Chicago: University of Chicago Press, 2009), 105.
50 Quoted in Green, *Best of Sasquatch Bigfoot*, 12.
51 "Eye-Witnesses See Bigfoot: Humboldt Sheriff's Office Has No Jurisdiction in Footprint Case; Scene of Activity in Del Norte," *Humboldt Standard, October* 15, 1958, 1.
52 Michael McLeod, *Anatomy of a Beast: Obsession and Myth on the Trail of Bigfoot* (Berkeley: University of California Press, 2009), 179.
53 David J. Daegling, *Bigfoot Exposed: An Anthropologist Examines America's Enduring Legend* (Walnut Creek, Calif.: Altamira Press, 2004), 212.
54 Ibid., 211.
55 Mark Chorvinsky, "New Bigfoot Photo Investigation," *Strange Magazine*, no. 13 (1994), http://www.bigfootencounters.com/articles/strange14.htm (accessed April 1, 2010).
56 Daegling, *Bigfoot Exposed*, 180.
57 Ibid.
58 Loren Coleman to Daniel Loxton, August 24, 2004.
59 Loren Coleman, "Bad Bigfoot Data Is Still Bad Bigfoot Data," March 30, 2012, CryptoMundo, http://www.cryptomundo.com/cryptozoo-news/bad-bf-data/ (accessed March 30, 2012).
60 Murphy, *Meet the Sasquatch*, 109.
61 Green, *Best of Sasquatch Bigfoot*, 15.
62 Loren Coleman, discussion comment following "Bad Bigfoot Data Is Still Bad Bigfoot Data," March 30, 2012, CryptoMundo, http://www.cryptomundo.com/cryptozoo-news/bad-bf-data/#comment-78426 (accessed March 30, 2012).
63 In early 2012, an online Bigfoot critic who posts under the name Kitakaze alleged that his investigations into the case of the Patterson–Gimlin film have given him "specific knowledge of the suit and it still exists today" ("Bob Heironimus on Pax TV's Lie Detector," Bigfoot Forums, http://bigfootforums.com/index.php?/topic/7624-bob-heironimus-on-pax-tvs-lie-detector/page_view_findpost_p_576952 [accessed April 21, 2012]). According to Kitakaze, the still-confidential whereabouts and circumstances of the suit will be revealed in his upcoming independent documentary, for release in "anywhere from a year to two or more" ("Kitakaze's Patty Suit Bombshell," Bigfoot Forums, http://bigfootforums.com/index.php?/topic/30016-kitakazes-patty-suit-bombshell/page_view_findpost_p_577280 [accessed April 21, 2012]).
64 Quoted in Greg Long, *The Making of Bigfoot: The Inside Story* (New York: Prometheus Books, 2004), 192.
65 Green to Loxton, August 10, 2004.
66 Napier, *Bigfoot*, 95.
67 Krantz, *Bigfoot Sasquatch Evidence*, 122.
68 Meldrum, *Sasquatch*, 135.

69 Napier, *Bigfoot*, 91–92.
70 Quoted in Daegling, *Bigfoot Exposed*, 126.
71 Sanderson, *Abominable Snowmen*, 150.
72 John Kirk, interview with Daniel Loxton, Kitsilano, B.C., March 27, 2004.
73 John Green, quoted in Long, *Making of Bigfoot*, 179.
74 Ibid., 123.
75 Ibid., 47, 430.
76 Quoted in ibid., 192.
77 Ibid., 430.
78 Benjamin Radford, "Bigfoot at 50: Evaluating a Half-Century of Bigfoot Evidence," *Skeptical Inquirer* 26, no. 2 (2002): 31.
79 Patterson and Murphy, *Bigfoot Film Controversy*, 30.
80 Daegling, *Bigfoot Exposed*, 116.
81 Krantz, *Bigfoot Sasquatch Evidence*, 32.
82 Long, *Making of Bigfoot*, 39, 109–110.
83 John Green, letter to the editor, *Skeptical Inquirer*, July 25, 2004 [forwarded to Daniel Loxton by Green, August 9, 2004].
84 Long, *Making of Bigfoot*, 47, 364.
85 Ibid., 385.
86 Roger Patterson, *Do Abominable Snowmen of America Really Exist?* (Yakima, Wash.: Franklin Press, 1966); Matt Crowley, "Roger Patterson's Plagiarism," January 11, 2012, Orgone Research, http://orgoneresearch.com/2012/01/11/roger-patterson's-plagiarism/comment-page-1/ (accessed March 8, 2012).
87 Quoted in Long, *Making of Bigfoot*, 70.
88 Quoted in ibid., 249.
89 Daegling, *Bigfoot Exposed*, 47.
90 Coleman, *Bigfoot!*, 127.
91 Most recent sources tend to name a local butcher, Joe Rhodes, as the discoverer of the tracks, perhaps because of Marx's reputation as a hoaxer. However, newspaper reports at the time identified Marx as the actual discoverer: "Rhodes suggested Marx might look for possible tracks of Big Foot when he was hunting, and during the early fall Marx spotted some tracks at a dump" ("Cold Freezes Hounds off Humanoid's Trail," *Montana Standard* [Butte], December 7, 1969, 20).
92 Hunter, with Dahinden, *Sasquatch*, 151.
93 Ibid., 153.
94 Krantz says, "The trail also began and ended on a steep slope coming out of and going back into Lake Roosevelt behind the Grand Coulie Dam" (*Bigfoot Sasquatch Evidence*, 43); John Green agrees that "the tracks came out of the water and went back to it" (*Year of the Sasquatch* [Agassiz, B.C.: Cheam, 1970], 50).
95 Hunter, with Dahinden, *Sasquatch*, 154–157.
96 Ibid., 156.
97 "Searchers Seek Big-footed Sasquatch," *Idaho State Journal* (Pocatello), February 1, 1970.

98 René Dahinden's account of the Metlow affair is detailed and fascinating, as recounted in Hunter, with Dahinden, *Sasquatch*, 159–165.

99 Ibid., 165–166.

100 "Legendary Big Foot Is Captured on Film," *Fairbanks Daily News-Miner*, November 14, 1970, 3.

101 Tom Page offered $25,000 for a copy of the film, as reported in Hunter, with Dahinden, *Sasquatch*, 168.

102 Peter Byrne, "The Hoaxed Ivan Marx Footage, as Told by Peter Byrne, Former Head of 'The Bigfoot Project,'" September 2003, Bigfoot Encounters, http://www.bigfootencounters.com/hoaxes/marx_footage.htm (accessed April 5, 2010).

103 "Footprints Add to Tale of Giant Creature Living in Washington," *Idaho State Journal*, February 17, 1971; "Tourtists Flock to See Sasquatch's Foot Prints," *Lebanon (Pa.) Daily News*, February 18, 1971.

104 Byrne, "Hoaxed Ivan Marx Footage."

105 Quoted in "Sasquatch Quashed Again?" *Walla Walla (Wash.) Union Bulletin*, April 9, 1971, 1.

106 Hunter, with Dahinden, *Sasquatch*, 169–170.

107 Quoted in "Sasquatch Tracks Were Made by Man," *Centralia (Wash.) Daily Chronicle*, April 1, 1971, 4.

108 Byrne, "Hoaxed Ivan Marx Footage."

109 Murphy, *Meet the Sasquatch*, 109.

110 Grover Krantz, interview with John Yager, KXLY-TV, 1992, quoted in Michael Dennett, "Bigfoot Evidence: Are These Tracks Real?" *Skeptical Inquirer* 18, no. 5 (1994): 499–500.

111 Daegling, *Bigfoot Exposed*, 84.

112 Napier, *Bigfoot*, 125.

113 Ibid., 124.

114 McLeod, *Anatomy of a Beast*, 126.

115 Green, *Year of the Sasquatch*, 66.

116 "Scoftic" is a derisive jargon term coined by Roger Knights in 2003. The word is now used by cryptozoological enthusiasts to describe those they view as unreasonable, dogmatic skeptics. According to Knights, "scoftic" refers to someone who "gives witness testimony no weight whatsoever, on ideological grounds, and who asserts numerous other bits of unreasonable dogma, such as that the quantity of reports is insignificant. Scofticism is thus fanaticism behind a pose of reasonableness" (quoted in Loren Coleman, "Is 'Scoftic' a Useful Term?" April 28, 2007, CryptoMundo, http://www.cryptomundo.com/cryptozoo-news/scoftic/ [accessed March 19, 2010]).

117 Green, *Year of the Sasquatch*, 66.

118 Green, *Best of Sasquatch Bigfoot*, 9.

119 Krantz, *Bigfoot Sasquatch Evidence*, 41.

120 Green, *Year of the Sasquatch*, 55.

121 Krantz, *Bigfoot Sasquatch Evidence*, 250.

122 See, for example, "Phantom Bigfeet and UFOs," in Janet Bord and Colin Bord, *Bigfoot Casebook Updated: Sightings and Encounters from 1818 to 2004* (Enumclaw, Wash.: Pine Winds Press, 2006), chap. 7.

123 Napier, *Bigfoot*, 198.

124 Ibid.

125 Krantz, *Bigfoot Sasquatch Evidence*, 5.

126 Jeffrey D. Lozier, P. Aniello and Michael J. Hickerson, "Predicting the Distribution of Sasquatch in Western North America: Anything Goes with Ecological Niche Modeling," *Journal of Biogeography* 36 (2009): 1623–1627.

127 Murphy, *Meet the Sasquatch*, 123.

128 John A. Bindernagel, *North America's Great Ape: The Sasquatch* (Courtenay, B.C.: Beachcomber Books, 1998), 28–29.

129 "Three-Legged Bear Walking Upright," February 11, 2010, CoolestOne.com., http://www.coolestone.com/media/1102/3_Legged_Bear_Walking_Upright/ (accessed March 4, 2010).

130 Green, *On the Trail of the Sasquatch*, 30.

131 Green, *Sasquatch File*, 48.

132 Ibid.

133 Radford, "Bigfoot at 50," 31.

134 Murphy, *Meet the Sasquatch*, 124–125.

135 Krantz, *Bigfoot Sasquatch Evidence*, 36.

136 Daegling, *Bigfoot Exposed*, 175.

137 Ibid.

138 Green, *On the Track of the Sasquatch*, 71.

139 Krantz, *Bigfoot Sasquatch Evidence*, 125–126.

140 Meldrum, *Sasquatch*, 261–262.

141 Daegling, *Bigfoot Exposed*, 207.

142 Benjamin Radford, "Science Looks for Bigfoot," September 3, 2005, Committee for Skeptical Inquiry, http://www.csicop.org/specialarticles/show/science_looks_for_bigfoot (accessed March 25, 2010).

143 Doug Haijcek, dir., *Sasquatch: Legend Meets Science* (Minneapolis: Whitewolf Entertainment, 2003).

144 Meldrum, *Sasquatch*, 270.

145 Lead author Melba Ketchum's claims to have DNA evidence to proved the existence of Bigfoot have been bouncing around the mainstream press since at least 2011 (and even earlier in the cryptozoological corners of the blogosphere). See, for example, Monisha Martins, "Sasquatch: Is It Out There?" August 16, 2011, Maple Ridge News, http://www.mapleridgenews.com/news/127905518.html (accessed February 14, 2013).

146 M. S. Ketchum, P. W. Wojtkiewicz, A. B. Watts, D. W. Spence, A. K. Holzenburg, D. G. Toler, T. M. Prychitko, F. Zhang, S. Bollinger, R. Shoulders, and R. Smith, "Novel North American Hominins: Next Generation Sequencing of Three Whole Genomes and Associated Studies," special issue, DeNovo Scientific Journal (2013): 1–15.

147 Quoted in Sharon Hill, "Ketchum Bigfoot DNA Paper Released: Problems with Questionable Publication," February 13, 2013, Doubtful News, http://doubtfulnews.com/2013/02/ketchum-bigfoot-dna-paper-released-problems-with-questionable-publication/ (accessed February 14, 2013).

148 Craig Woolheater, "Ketchum Sasquatch DNA Study Update: Questions Answered . . . ," February 13, 2013, CryptoMundo, http://www.cryptomundo.com/bigfoot-report/ketchum-sasquatch-dna-study-update/ (accessed February 14, 2013).

149 Benjamin Radford, "Bigfoot DNA Discovered? Not So Fast," February 14, 2013, LiveScience, http://www.livescience.com/27140-bigfoot-dna-study-questioned.html (accessed February 14, 2013).

150 Zen Faulkes, "Sasquatch DNA: New Journal or Vanity Press?" February 13, 2013, NeuroDojo, http://neurodojo.blogspot.ca/2013/02/sasquatch-dna-new-journal-or-vanity.html (accessed February 14, 2013).

151 Ketchum et al., "Novel North American Hominins," 1, 11.

152 "Dr. Melba Ketchum's Press Release About Bigfoot DNA," November 24, 2012, Before It's News, http://beforeitsnews.com/paranormal/2012/11/dr-melba-ketchums-press-release-about-bigfoot-dna-2445438.html (accessed February 14, 2013).

153 Steven Novella, "Bigfoot DNA," November 26, 2012, Skepticblog, http://www.skepticblog.org/2012/11/26/bigfoot-dna/ (accessed February 14, 2013).

154 John Timmer, "Bigfoot Genome Paper 'Conclusively Proves' That Sasquatch Is Real—And It Only Took Founding a New Journal to Get the Results Published," February 13, 2013, Ars Technica, http://arstechnica.com/science/2013/02/bigfoot-genome-paper-conclusively-proves-that-sasquatch-is-real/ (accessed February 14, 2013).

155 Green, *Year of the Sasquatch*, 51.

156 Hunter, with Dahinden, *Sasquatch*, 169–171.

157 Michael Dennett, "Evidence for Bigfoot? An Investigation of the Mill Creek 'Sasquatch Prints,'" *Skeptical Inquirer* 13, no. 3 (1989): 272.

158 Meldrum, *Sasquatch*, 110–112, 237–240.

159 Coleman, *Bigfoot!*, 40–42.

160 Ibid., 42.

161 Loren Coleman to Daniel Loxton, August 11, 2004.

162 Krantz, *Bigfoot Sasquatch Evidence*, 34.

163 Daegling, *Bigfoot Exposed*, 75.

164 Krantz, *Bigfoot Sasquatch Evidenc*e, 42.

165 Robert Carroll, *The Skeptic's Dictionary* (New York: Wiley, 2003), 65.

166 Tim Mendham, "The Carlos Hoax," in *The Second Coming: All the Best from the Skeptic, 1986–1990*, ed. Barry Williams and Richard Saunders (Sydney: Australian Skeptics, 2001), 26–28.

167 Adding complexity to this story, the Venezuelan-born performance artist who portrayed Carlos, Deyvi Pena, was arrested in 2011 after federal authorities discovered that he had been living in the United States under the assumed identity Jose Luis Alvarez. Pena pleaded guilty to passport fraud in early 2012. See Jon Burstein. "Artist Pleads Guilty to Passport Fraud," *Orlando (Fla.) Sun Sentinel*, March 14, 2012, http://articles.orlando

sentinel.com/2012-03-14/news/fl-jose-alvarez-amazing-randi-plea-20120314_1_passport-fraud-sentencing-identity (accessed April 17, 2012).

168 "Sasquatch Tracks Were Made by Man."

169 Daegling, *Bigfoot Exposed*, 257.

170 John Rael to Daniel Loxton, March 21, 2010. I was apprised of the hoax early and asked to contribute some discussion. I was uncomfortable with the derisive tone of the project, so I declined. More information on the hoax is available at "Bigfoot Gets Kicked in the Nuts," February 7, 2010, Skepticallypwnd, http://skepticallypwnd.com/?p=79 (accessed March 22, 2010).

171 Green, *Year of the Sasquatch*, 73.

172 "Sasquatch Wanted," *Washington Post*, March 2, 1970, A9.

173 Daegling, *Bigfoot Exposed*, 190.

174 Quoted in ibid., 193.

175 Quoted in Coleman, *Bigfoot!*, 236.

176 Krantz, *Bigfoot Sasquatch Evidence*, 10.

177 The James Randi Educational Foundation tested just such a claim in 2007. Rosemary Hunter "applied for the JREF's Million Dollar Challenge with the extraordinary claim that she can make people urinate with the power of her mind," but failed the preliminary test (Alison Smith, "Rosemary Hunter's Challenge Test," November 11, 2007, James Randi Educational Foundation, http://www.randi.org/site/index.php/jref-news/108-rosemary-hunters-challenge-test.html [accessed March 18, 2010]).

178 John Bindernagel to Daniel Loxton, November 29, 2004.

179 Daegling, *Bigfoot Exposed*, 194.

180 Jason Loxton to Daniel Loxton, March 19, 2010.

181 Quoted in Coleman, *Bigfoot!*, 236.

182 Bindernagel to Loxton, November 29, 2004.

183 Green, *Sasquatch*, 409–410.

184 Several such cases are described by Janet Bord and Colin Bord, including that of Richard Davis, alleged to have shot a Bigfoot in the chest with a revolver in 1975—at close range. According to the tale, Davis saw the bullet hit the creature, which then ran off. (This case also has a paranormal aspect, with Davis psychically prevented from emptying the remainder of the cylinder into the creature's chest.) See Bord and Bord, *Bigfoot Casebook Updated*, 140–141.

185 Daegling, *Bigfoot Exposed*, 194.

186 I have argued that all parties should be able to agree that it is possible for entire categories of paranormal claims to be completely bogus, with hundreds or thousands of supporting testimonials comprising nothing but mistakes and hoaxes. All that's needed to demonstrate this to the satisfaction of almost everybody is to go through a list of similar claims: Bigfoot, fairies, ghosts, alien abduction, mermaids, and so on. See Daniel Loxton, "An Argument That Should Never Be Made Again," February 1, 2010, Skepticblog, http://skepticblog.org/2010/02/02/an-argument-that-should-never-be-made-again/ (accessed February 28, 2010).

187 "Bizarre Encounters in Wahiawa, Hawaii," and "State by State Sightings List," Bigfoot Encounters, http://www.bigfootencounters.com/sbs/aikanaka.htm (accessed March 8, 2012).

188 Green, *Sasquatch*, 233.

189 Krantz, *Bigfoot Sasquatch Evidence*, 236.

3. THE YETI

1 Simon Welfare and John Fairley, *Arthur C. Clarke's Mysterious World* (New York: A&W Visual Library, 1980), 14.

2 Brian Regal, *Searching for Sasquatch: Crackpots, Eggheads, and Cryptozoology* (New York: Palgrave Macmillan, 2011), 31.

3 Reinhold Messner, *My Quest for the Yeti: Confronting the Himalayas' Deepest Mystery* (New York: Pan, 1998), viii.

4 Colonel [Charles K.], Howard-Bury, "The Attempt on Everest," *Times* (London), October 21, 1921, 11.

5 C. K. Howard-Bury, *Mount Everest: The Reconnaissance* (New York: Longmans, Green, 1922), 141.

6 Ibid.

7 Howard-Bury, "Attempt on Everest."

8 Howard-Bury, *Mount Everest*, 141.

9 Henry Newman, "On Everest: The 'Wild Men' Myth," *Leader* (Allahabad, India), November 6, 1921, 9.

10 Henry Newman, "The 'Abominable Snowmen'" [letter to the editor], *Times*, July 29, 1937, 15.

11 For useful introductions to the varied vocabulary of the Yeti, see, for example, Edmund Hillary and Desmond Doig, *High in the Thin Cold Air* (Garden City, N.Y.: Doubleday, 1962) 32; and Messner, *My Quest for the Yeti*, viii.

12 William L. Straus Jr., "Abominable Snowman," *Science* 123 (1956): 1024–1025. Pranavananda published his interpretation in many venues; see, for example, Swami Pranavananda, "Abominable Stories About the Snowman," *Times of India* (Mumbai [Bombay]), May 22, 1955, 5. Srimat Swami Pranavananda is identified as "Sreemat Swami Pranavananda, an Indian religious notable," in "The 'Abominable Snowman' Unmasked: The Red Bear Believed Guilty of a Himalayan Fraud," *Times*, July 3, 1956, 7.

13 Regal, *Searching for Sasquatch*, 32.

14 B. H. Hodgson, "On the Mammalia of Nepal," *Journal of the Asiatic Society of Bengal* 1 (1832): 339n.

15 John Napier, *Bigfoot: The Yeti and Sasquatch in Myth and Reality* (New York: Dutton, 1973), 36.

16 Ivan T. Sanderson, *Abominable Snowmen: Legend Comes to Life* (New York: Pyramid, 1968), 43–45. Sanderson first offered a page of introductory fluff, waxing nostalgic about nineteenth-century British officer-explorers and their "wealth of wisdom" and "extraordinarily keen interest in the world about them," and then gave a wildly inaccurate,

paraphrased version of Laurence Waddell's tale—both embellishing it with made-up details not found in the original ("tracks made by some creature walking on two legs and two bare feet . . . brought whoops of admiration from the Major's mountain-born porters") and omitting the key detail: *Waddell had assessed the tracks as coming from a bear.* The most charitable view would be that Sanderson was bluffing about his citation rather than deliberately misrepresenting the evidence, and he simply had not read the original. (He complained on the following page about the "time-consuming and frustrating" difficulty of trying to track down primary sources that are "either lost in some archive or truly lost forever.") Bernard Heuvelmans at least quoted Waddell, but similarly omitted the key fact that Waddell had considered these footprints to be those of a bear and blamed "an atmosphere of superstition" for the claims of hairy wild men (*On the Track of Unknown Animals*, trans. Richard Garnett [New York: Hill and Wang, 1959], 128).

17 L. A. Waddell, *Among the Himalayas* (1899; rept., Delhi: Pilgrims Book House, 1998), 223–224.

18 Napier, *Bigfoot*, 36–37.

19 William Woodville Rockhill, *The Land of the Lamas: Notes of a Journey Through China, Mongolia and Tibet* (New York: Century, 1891), 150–151.

20 Napier, *Bigfoot*, 35.

21 Ibid., 36.

22 Quoted in ibid., 39–40

23 Joe Nickell, *Tracking the Man-Beasts: Sasquatch, Vampires, Zombies, and More* (Amherst, N.Y.: Prometheus Books, 2011), 56.

24 Consider the case of Man Bahadur, a thirty-five-year-old Nepalese pilgrim who spent an extended period with Edmund Hillary's high-altitude medicine team (1960/1961). As described by team member and doctor Michael Ward, Bahadur

> stayed for 14 days at 15,300 ft and above, and throughout this period wore neither shoes nor gloves, and walked in the snow and on rocks in bare feet without any evidence of frostbite. He wore minimal clothing and had no sleeping bag or protective equipment other than a woollen coat. He was continuously monitored whilst spending four days without shelter between 16,500 ft and 17,500 ft, with night temperatures between -13°C [8.6°F] and -15°C [5°F], and day temperatures below freezing. Eventually he developed deep cracks in the skin of his toes, which became infected, and he returned to lower levels for this reason. Had any European members of the party followed this regime they would undoubtedly have become severely frostbitten and hypothermic. ("The Yeti Footprints: Myth and Reality," *Alpine Journal* [1999]: 86)

For photographs of Bahadur and his feet, see "The Highest Livers," *Life*, May 12, 1961, 92.

25 Napier, *Bigfoot*, 61.

26 Ernst Schäfer, *Dach der Erde: Durch das Wunderland Hochtibet Tibetexpedition 1934/1936* (*Roof of the World: Through the Wonderland of Upper Tibet, Tibet Expedition, 1934–1936*) (Berlin: Parey, 1938).

27 Christopher Hale, *Himmler's Crusade: The Nazi Expedition to Find the Origins of the Aryan Race* (Edison, N.J.: Castle Books, 2006), 53.

28 Ibid., 180.
29 Schäfer, *Dach der Erde*, 81–87 (translated by Hans-Dieter Sues).
30 Quoted in Messner, *My Quest for the Yeti*, 108.
31 Quoted in ibid.
32 "A Himalayan 'Snowman'? Alleged Signs. Strange Imprints at 16,000 Ft.," *Times of India*, December 9, 1936, 11.
33 F. S. Smythe, "Abominable Snowman. Pursuit in the Himalayas," *Times*, November 10, 1937, 15–16.
34 Balu [H. W. Tilman], "Are the 'Snowmen' Bears?" *Times*, November 13, 1937, 13. Tilman's climbing partner Eric Shipton explained that same year that in 1936 he had been "amused to find that these men [Mana porters] had nicknamed Tilman 'Balu Sahib' (Balu meaning a bear) owing to the speed with which he moves over steep, forested ground" ("Survey Work in the Nanda Devi Region," *Himalayan Journal* 9 [1937], http://www.himalayanclub.org/journal/survey-work-in-the-nanda-devi-region/ (accessed May 17, 2012). Frank Smythe pretty clearly knew that Tilman was Balu, as evidenced in "Mr Smythe's Reply," *Times*, November 16, 1937, 17. Tilman rather theatrically denied that he was Balu: "Mr Smythe's facility for putting two and two together and making five is seen . . . in his identification of Balu with Your obedient servant, H. W. Tilman" ("Abominable Snowman" [letter to the editor], *Times*, December 1, 1937, 12). Despite this denial, Tilman made Balu's argument the central plank in his defense of the Yeti in 1938. See H. W. Tilman, "Notes on the Abominable Snowman," in *Men and Mountaineering: An Anthology of Writings by Climbers*, ed. Showell Styles (New York: White, 1968), 105 (excerpted from *Mount Everest, 1938* [Cambridge: Cambridge University Press, 1948]).
35 Tilman, "Abominable Snowman," 12.
36 Ralph Izzard, *The Abominable Snowman Adventure* (London: Hodder and Stoughton, 1955), 14.
37 Tilman, "Notes on the Abominable Snowman," 105.
38 Izzard, *Abominable Snowman Adventure*, 44.
39 Hale, *Himmler's Crusade*, 58.
40 Quoted in ibid., 128.
41 Messner, *My Quest for the Yeti*, 107–122.
42 Napier, *Bigfoot*, 46–47.
43 Sanderson, *Abominable Snowmen*, 269.
44 Napier, *Bigfoot*, 47.
45 Edmund Hillary, "Abominable—and Improbable?" *New York Times Magazine*, January 24, 1960, 13; Ward, "Yeti Footprints," 81.
46 Napier, *Bigfoot*, 141.
47 Eric Shipton, "A Mystery of Everest: Footprints of the 'Abominable Snowman,'" *Times*, December 6, 1951, 5.
48 Quoted in "*Gigantopithecus*: The Jury-Rigged Giant Bigfoot," Pacific Northwest and Siberia Expedition, http://www.bermuda-triangle.org/html/gigantopithecus--_the_jury-rig.html (accessed October 22, 2011).

49 Peter Gillman, "The Most Abominable Hoaxer?" *Sunday Times Magazine* (London), December 10, 1989, 39–44, and "The Yeti Footprints," *Alpine Journal* (2001): 143–151.

50 The photograph appears as an unnumbered figure, whose caption reads, "The abominable snowman's trail, photographed by Eric Shipton in 1951 with (*inset*) a single footprint . . . ," in Heuvelmans, *On the Track of Unknown Animals*, facing 136.

51 Shipton passed away on March 28, 1977, as noted in "Obituary: Mr. Eric Shipton: Mountaineer, Explorer and Writer," *Times*, March 30, 1977, 19; Gillman reported that Sherpa San Tenzing was dead by 1989, in "Most Abominable Hoaxer?" 44; and Ward died in 2005, as recorded in Jim Perrin, "Obituary: Michael Ward," *Guardian*, October 27, 2005, http://www.guardian.co.uk/news/2005/oct/27/guardianobituaries.everest (accessed May 21, 2012.) In 1999, Ward wrote his rebuttal to Gillman: "Yeti Footprints," 81–83.

52 Ward, "Yeti Footprints," 81–83.

53 Quoted in Gillman, "Most Abominable Hoaxer?" 44.

54 See, for example, "The Abominable Snowmen," *New York Times*, December 27, 1951, 18; "Abominable Himalayan," *Life*, December 1951, 88; Wladimir Tschernezky, "A Reconstruction of the Foot of the 'Abominable Snowman,'" *Nature* 186, no. 4723 (1960): 496–497.

55 Ward, "Yeti Footprints," 83.

56 Gillman, "Yeti Footprints," 145.

57 Ibid., 146.

58 Ward, "Yeti Footprints," 85.

59 Gillman, "Yeti Footprints," 150.

60 Ibid., 149–150.

61 Gillman, "Most Abominable Hoaxer?" 42.

62 Gillman, "Yeti Footprints," 145.

63 Quoted in ibid., 150.

64 Edouard Wyss-Dunant, "The Himalayan Footprints: New Traces Seen by Swiss Expedition," *Times*, June 6, 1952, 5.

65 Edouard Wyss-Dunant, "The Yeti: Biped or Quadroped?" *The Mountain World, 1960/61*, ed. Malcolm Branes (Chicago: Rand-McNally, 1961), 252–259.

66 "Abominable Snowman: Hairy Beast Seized Him, Says Porter," *Globe and Mail* (Toronto), December 30, 1952, 11, second edition.

67 Joshua Blu Buhs, *Bigfoot: The Life and Times of a Legend* (Chicago: University of Chicago Press, 2009), 36.

68 "Everest Climbers Given Medal by Ike," *Joplin (Mo.) Globe*, February 12, 1954, 20, final edition.

69 "Now Seek Abominable Snowman Says Col. Hunt," *Daily Express* (London), June 16, 1953, 2.

70 Quoted in "Tenzing Saved His Life, Hillary Says in London," *Globe and Mail* (Toronto), July 4, 1953, 8.

71 P. McL., "A Man of the Mountains," *Winnipeg Free Press*, August 13, 1955, 28.

72 Richard Crichfield, "In the Land of the Abominable Snowman," *Sunday Herald Magazine* (Chicago), May 20, 1979, 10.

73 Izzard describes hiring "nearly 300 coolies" plus, as a result of a supply problem, "a second coolie team of 70 men to follow our main party" (*Abominable Snowman Adventure*, 95–96).
74 Quoted in ibid., 108–109.
75 Charles Stonor, *The Sherpa and the Snowman* (London: Hollis & Carter, 1955), 118.
76 Izzard, *Abominable Snowman Adventure*.
77 Ibid., 137–145.
78 Ibid., 198–201.
79 Quoted in ibid., 199.
80 Stonor, *Sherpa and the Snowman*, 38.
81 Quoted in Izzard, *Abominable Snowman Adventure*, 103.
82 Stonor, *Sherpa and the Snowman*, 78.
83 "'Abominable Snowman' in Tibetan Zoo," *Times of India*, November 15, 1953, 1.
84 Stonor, *Sherpa and the Snowman*, 30.
85 Ibid., 64–65.
86 Ibid., 204–205.
87 Loren Coleman, *Tom Slick: True Life Encounters in Cryptozoology* (Fresno, Calif.: Craven Street Books, 2002), 74, 36–37, 186, 45.
88 Regal, *Searching for Sasquatch*, 37–40.
89 Loren Coleman, *Tom Slick and the Search for the Yeti* (London: Faber and Faber, 1989), 178–203.
90 Regal, *Searching for Sasquatch*, 37–41.
91 Buhs, *Bigfoot*, 42. Tom Slick was reported to have left Kathmandu for the mountains around March 14, and to have returned to Kathmandu on April 19. See "Texan to Hunt Asian 'Snowman,'" *Lawton (Okla.) Constitution*, March 14, 1957, 13; "Slick About Convinced Giant 'Snowman' Exists," *Miami (Okla.) Daily News Herald*, April 19, 1957, 9; and "Texan Finds Footprints of 'Snowman,'" *Pacific Stars and Stripes* (Tokyo), April 20, 1957, 5. However, Byrne later wrote that they "spent three months in the mountains" during the 1957 expedition (*The Search for Bigfoot: Monster, Myth or Man?* [Washington, D.C.: Acropolis Books, 1975], 117).
92 "Slick About Convinced Giant 'Snowman' Exists," 9. Byrne, however, mentions only two sets of tracks: "I found one set of footprints, in the Chhoyang Khola at 10,000 feet, and Tom [Slick], working with a separate party in another area, found a second set" (*Search for Bigfoot*, 117).
93 Regal, *Searching for Sasquatch*, 42–43.
94 Tom Slick, "Abominable Snowman—No Longer a Legend," *Daily Boston Globe*, May 18, 1958, B22.
95 Buhs, *Bigfoot*, 43.
96 Peter Byrne, "Hope to Take Him Alive with Drug-Bullet Gun," *Daily Boston Globe*, May 19, 1958, 8.
97 Peter Byrne, "We Dress Up as Natives to Fool Wary Animals," *Daily Boston Globe*, May 20, 1958, 10.
98 Buhs, *Bigfoot*, 43–45. However, Byrne reports that the end date is less clear: "[O]ne by one, the various members had to leave and before four months had passed all, with the

exception of my brother and I, had departed. We stayed on for another five months, making nine months in all" (*Search for Bigfoot*, 118).

99 "Snowman Reported Seen Eating Himalayan Frogs," *Washington Post*, June 17, 1958, A3.

100 "Americans Find Cave of Abominable Snowman," *Daily Boston Globe*, April 30, 1958, 9.

101 Buhs, *Bigfoot*, 45.

102 Ibid., 49.

103 Regal, *Searching for Sasquatch*, 43.

104 "Tracing the Origins of a 'Yeti's Finger,'" December 27, 2011, BBC News, http://www.bbc.co.uk/news/science-environment-16264752 (accessed May 19, 2012).

105 Loren Coleman, "Jimmy Stewart and the Yeti," The Anomalist, http://www.anomalist.com/milestones/stewart.html (accessed October 22, 2011); Byrne, *Search for Bigfoot*, 120.

106 "Abominable Snowman" [season 3, episode 59], *MonsterQuest*, History Channel, October 25, 2009, "MonsterQuest: Abominable Snowman, Pt. 4," YouTube, http://www.youtube.com/watch?v=syLd4zx1BWU&feature=related (accessed October 22, 2011).

107 Regal, *Searching for Sasquatch*, 47.

108 Quoted in "Yeti Finger Mystery Solved by Edinburgh Scientists," December 27, 2011, BBC News, http://www.bbc.co.uk/news/uk-scotland-edinburgh-east-fife-16316397 (accessed May 19, 2012).

109 Tom Slick, "Expedition a Success, Proves Yeti Exists," *Daily Boston Globe*, July 26, 1958, 4.

110 Buhs, *Bigfoot*, 46.

111 Regal, *Searching for Sasquatch*, 46–47.

112 Napier, *Bigfoot*, 52.

113 Buhs, *Bigfoot*, 112.

114 For a discussion of the scientific team, see Michael Ward, "Himalayan Scientific Expedition, 1960–61," *Alpine Journal* (1961): 343. The team's 150 porters are noted in Buhs, *Bigfoot*, 112.

115 Edmund Hillary and Desmond Doig, *High in the Thin Cold Air* (Garden City, N.Y.: Doubleday, 1962) 130–131.

116 Ibid., 82–83.

117 Ibid., 103.

118 Ibid., 131.

119 Ibid., 131–132.

120 Ibid., 88.

121 The loan agreement for the scalp specified that "Sir Edmund Hillary and the members of the expedition put into effect an appeal to the sponsors of the expedition, mountaineering societies, and the general public to establish and support a school for the benefit of the people of Khunde and Khumjung villages." With help from the World Book Encyclopedia, the Indian Aluminum Company, and the International Red Cross, and labor from Hillary and his friends from the expedition and local village, the school was soon made a reality. See Hillary and Doig, *High in the Thin Cold Air*, 88. That school was only the beginning. "At the request of Sherpa residents," Hillary reflected in 2003, "we helped establish 27 schools, two hospitals, and a dozen medical clinics—plus quite a few

bridges over wild rivers. We constructed several airfields and rebuilt Buddhist monasteries and cultural centers. We planted a million seedlings in Sagarmatha National Park to replace the vast number of trees destroyed for firewood and used to build the small hotels that came with the growth of tourism" ("My Story," *National Geographic*, May 2003, 40).

122 Napier, *Bigfoot*, 59.

123 Ibid., 61–62.

124 Ibid., pl. 3.

125 Quoted in "British Climber Says He Saw Abominable Snowman," *Stars and Stripes* (Darmstadt), June 13, 1970, 4.

126 Quoted in "Names and Faces," *Boston Globe*, June 8, 1970, 2.

127 Zhou Guoxing, "The Status of the Wildman Research in China," Bigfoot Encounters, http://www.bigfootencounters.com/biology/zhou.htm (accessed May 19, 2012).

128 "Yeti Is Just a Plain, Old Brown Bear, Says Chinese Expert," World Tibet Network News, January 13, 1998, http://www.tibet.ca/en/newsroom/wtn/archive/old?y=1998&m=1&p=13_4 (accessed May 19, 2012).

129 Quoted in "Dogs from U.S. Will Hunt the Yeti in Nepal: Search for Snowman Opens This Month," *Chicago Daily Tribune*, January 4, 1958, A8.

130 Regal, *Searching for Sasquatch*, 144–147.

131 Ibid., 151–156.

132 Anthony B. Wooldridge, "First Photos of the Yeti: An Encounter in North India," *Cryptozoology* 5 (1986): 63–76, and "An Encounter in Northern India," Bigfoot Encounters, http://www.bigfootencounters.com/articles/wooldridge.htm (accessed October 22, 2011).

133 Quoted in Michael Dennett, "Abominable Snowman Photo Comes to Rocky End," *Skeptical Inquirer* 13, no. 2 (1989): 118–119; Jerome Clark, *Unexplained! Strange Sightings, Incredible Occurrences, and Puzzling Physical Phenomena* (Farmington Hills, Mich.: Visible Ink Press, 1999), 599–600.

134 "The Snow Walker Film Footage," Bigfoot Encounters, http://www.bigfootencounters.com/films/snowwalker.htm (accessed October 22, 2011).

135 Messner, *My Quest for the Yeti*, 7–8.

136 Ibid., 98–100.

137 Ibid., 129–130, 142.

138 Ibid., 146.

139 Ibid., 154, 156.

140 Ted Chamberlain, "Reinhold Messner: Climbing Legend, Yeti Hunter," National Geographic Adventure, http://www.nationalgeographic.com/adventure/0005/q_n_a.html (accessed October 22, 2011).

141 Messner, *My Quest for the Yeti*, 156.

142 Charles Haviland, "'Yeti Prints' Found Near Everest," December 1, 2007, BBC News, http://news.bbc.co.uk/2/hi/south_asia/7122705.stm (accessed October 22, 2011).

143 Alastair Lawson, "'Yeti Hairs' Belong to a Goat," October 13, 2008, BBC News, http://news.bbc.co.uk/2/hi/south_asia/7666900.stm (accessed October 22, 2011).

144 "Abominable Snowman" [season 3, episode 59], *MonsterQuest*, History Channel, October 25, 2009, "MonsterQuest: Abominable Snowman, Pt. 1," YouTube http://www.youtube.com/watch?v=TYwO_k7vDmg (accessed October 22, 2011).

145 Messner, *My Quest for the Yeti*, 157.

4. NESSIE

1 As Ronald Binns explains, "In fact there had been a road along the north shore of Loch Ness since the end of the eighteenth-century. Tourists and motor vehicles had been passing along it well before 1933. . . . All that happened in 1933 was that sections of the north shore road were improved. The road was resurfaced along its entire length and there were a number of repairs carried out" (*The Loch Ness Mystery Solved* [London: Open Books, 1983]), 63.

2 Tony Shiels, *Monstrum! A Wizard's Tale* (London: Fortean Times, 1990), 69.

3 "The Quaternary (the last two million years)," Scottish Geology, http://www.scottishgeology.com/geology/geological_time_scale/timeline/quaternary.htm (accessed June 29, 2010).

4 More specifically, both the River Ness and the Caledonian Canal open into the innermost section of Moray Firth, called Beauly Firth. For its part, the length of the River Ness varies, depending on what is counted. Loch Ness empties first into a man-made body of water called Loch Dochfour (created when the water level of Loch Ness was raised during the construction of the Caledonian Canal). In total, the river route from Loch Ness proper to the mouth of the River Ness at Beauly Firth is a little over 8 miles. Sources that estimate the length of the River Ness at 6 or 7 miles are excluding the length of Loch Dochfour.

5 Binns, *Loch Ness Mystery Solved*, 72.

6 Quoted in ibid., 63–65.

7 "The Sea Serpent in the Highlands," *Times* (London), March 6, 1856, 12, reprinted from the *Inverness Courier*.

8 Lewis Spence, "Mythical Beasts in Scottish Folklore," *Scotsman* (Edinburgh), March 4, 1933, 15.

9 Carol Rose, *Giants, Monsters, and Dragons: An Encyclopedia of Folklore, Legend, and Myth* (New York: Norton, 2000), 109, 205, 388.

10 Thomas Hannan, "Each Usage: The Water Horse," *Scotsman*, October 25, 1933, 10.

11 Spence, "Mythical Beasts in Scottish Folklore," 15.

12 Hannan, "Each Usage," 10.

13 John Graham Dalyell, *The Darker Superstitions of Scotland, Illustrated from History and Practice* (Edinburgh: Waugh and Innes, 1834), 543–544, 682.

14 W. M. Parker and S. M. Young, letters to the editor, *Scotsman*, October 24, 1933, 17.

15 John Noble Wilford, "Legends of the Lochs: Quests by Saints and Science," *New York Times*, June 5, 1976, 8.

16 Michel Meurger, with Claude Gagnon, *Lake Monster Traditions: A Cross-Cultural Analysis* (London: Fortean Times, 1988), 122–123, 126.

17 Ibid., 122.

18 Quoted in ibid., 212.

19 "Sea Serpent Hoax of 1904 Is Bared," *New York Times*, April 25, 1934, 17.

20 *Inverness Courier*, October 8, 1868. Incidentally, this is an early use of the word "monster" in relation to Loch Ness, contradicting the common claim that Alex Campbell coined the term for Nessie in 1933. Campbell's use of the word was, however, instrumental in the creation of the modern legend of a Loch Ness "monster."

21 Adrian Shine, *Loch Ness* (Drumnadrochit: Loch Ness Project, 2006), 7.

22 Reuters, "Travel Bargain Hoax—and So Is 'Monster,'" *Los Angeles Times*, April 2, 1972, 2; "Operation Cleansweep 2001," Loch Ness and Morar Project, http://www.lochnessproject.org/loch_ness_sundberg.htm (accessed July 5, 2010).

23 "A Scene at Lochend," *Inverness Courier*, July 1, 1852.

24 The newspaper quotes him as having said, "Dia mu'n cuairt duian, 'a iad na h-eich-uisg' tk'enn!" Catriona Parsons was kind enough to tackle a translation for this book: "I'm pretty confident this inscription is a garbled version of the Gaelic phrase 'Dia mu'n cuairt dhuinn, 's iad na h-eich uisg' a' tighinn!'" she concluded. Literally, "(Let) God be around us, and the water-horses [kelpies] coming!" My thanks to the Gaelic Council of Nova Scotia.

25 The three men were Ian Milne, R. C. M. Macdougall, and G. D. Gallon, all of whom were Inverness locals. See Rupert T. Gould, *The Loch Ness Monster* (1934; repr., Secaucus, N.J.: Citadel Press, 1976), 36.

26 Quoted in "What Was It? A Strange Experience on Loch Ness," *Northern Chronicle* (Inverness), August 27, 1930.

27 The keeper's story, itself a contender for "first recorded Nessie sighting," is an important case about which little seems to be known. Is this anonymous "keeper" the same person as William Miller, who is discussed in Gould, *Loch Ness Monster*, 35? Miller is identified as a "keeper," and his alleged sighting in 1923 matches the description and timescale of that of the keeper mentioned seven years later in "What Was It?"

28 Piscator, letter to the editor, *Inverness Courier*, August 29, 1930.

29 "Lake Mystery in Scotland," *Kokomo (Ind.) Tribune*, October 8, 1933, 12.

30 Most books identify her as "Mrs. Mackay" or even "Mrs. John Mackay," although she was the primary witness. I take her first name from Tony Harmsworth, who interviewed her in 1986, *Loch Ness, Nessie and Me: The Truth Revealed* (Drumnadrochit: Harmsworth, 2010), 91–93.

31 Today, this building is home to the Loch Ness Exhibition Center and its associated research effort (Loch Ness and Morar Project).

32 Alex Campbell, "Strange Spectacle on Loch Ness: What Was It?" *Inverness Courier*, May 2, 1933.

33 Gould, *Loch Ness Monster*, 39–40.

34 "Loch Ness 'Monster': Ship Captain's Views on Occurrence," *Inverness Courier*, May 12, 1933.

35 Dick Raynor, "A Brief Overview of 'Nessie' History," Loch Ness Investigation, http://www.lochnessinvestigation.org/history.html (accessed June 25, 2010).

36 Binns, *Loch Ness Mystery Solved*, 208.

37 There is some uncertainty about this date. Most sources date the Mackays' sighting to April 14, 1933 (which was the date provided by Gould in 1934). However, Campbell's original news item dates the sighting to "Friday of last week," which would be April 28, 1933. Compare Gould, *Loch Ness Monster*, 39, with Campbell "Strange Spectacle on Loch Ness."

38 Dennis Dunn, "*King Kong* Caps the Lot," *Daily Express* (London), April 19, 1933, 3.

39 "*King Kong* and the Schoolboy," *Scotsman*, May 9, 1933, 6.

40 George Spicer, letter to the editor, *Inverness Courier*, August 4, 1933.

41 Tim Dinsdale, *The Loch Ness Monster* (London: Routledge and Kegan Paul, 1961), 41.

42 Shine, *Loch Ness*, 9.

43 Quoted in F. W. Holiday, *The Great Orm of Loch Ness: A Practical Inquiry into the Nature and Habits of Water-Monsters* (New York: Norton, 1969), 30–31.

44 Quoted in Gould, *Loch Ness Monster*, 44.

45 Quoted in Holiday, *Great Orm of Loch Ness*, 31. Note, however, that Spicer's initial letter to the *Inverness Courier* estimated the creature's length as "six to eight feet long." He swiftly revised this measure upward to what became the canonical description of 25 or 30 feet long. In 1934, Spicer told Gould, "After having ascertained the with of the road, and giving the matter mature thought in every way, I afterwards came to the conclusion that the creature I saw must have been at least 25 feet in length" (quoted in Gould, *Loch Ness Monster*, 46).

46 Quoted in Holiday, *Great Orm of Loch Ness*, 30–31.

47 Spicer, letter to the editor.

48 Gould, *Loch Ness Monster*, 46.

49 Shine, *Loch Ness*, 7.

50 Adrian Shine, e-mail to James Loxton (*Junior Skeptic* research assistant), June 22, 2006.

51 Gould, *Loch Ness Monster*, 26.

52 Denis Lyell, letter to the editor, *Scotsman*, April 30, 1938, 17.

53 "Loch Ness Monster: The Plesiosaurus Theory," *Scotsman*, October 17, 1933, 9.

54 "Loch Ness 'Monster': Ship Captain's Views on Occurrence."

55 Duke of Portland, letter to the editor, *Scotsman*, October 20, 1933, 11.

56 Gould, *Loch Ness Monster*, 30.

57 Amy Willis, "Loch Ness Monster: The Strangest Theories and Sightings," August 26, 2009, *Telegraph* (London), http://www.telegraph.co.uk/news/newstopics/howabout that/6095673/Loch-Ness-Monster-the-strangest-theories-and-sightings.html (accessed June 28, 2010).

58 See, for example, Henry Bauer, *The Enigma of Loch Ness: Making Sense of a Mystery* (Urbana: University of Illinois Press, 1986), 51; and Binns, *Loch Ness Mystery Solved*, 50–51. In addition, in early 2010 I made my own failed attempt to locate the story using the searchable "ProQuest Historical Newspapers" archive and other newspaper archives.

59 Bauer, *Enigma of Loch Ness*, 159.

60 Charles Thomas, "The 'Monster' Episode in Adomnan's *Life* of St. Columba," *Cryptozoology* 7 (1988): 40–41. The Latin in the original article has been omitted.

61 Ibid., 40.
62 Ibid., 43.
63 Ibid., 44.
64 Robert Bakewell, *Introduction to Geology* (New Haven, Conn.: Howe, 1833), 213.
65 Philip Henry Gosse, *The Romance of Natural History* (London: Nisbet, 1861), 357–360. Gosse, it should be noted, is best known for his book *Omphalos: An Attempt to Untie the Geological Knot* (London: Van Voorst, 1857), which underlies much of creationist thinking to this day.
66 *Elasmosaurus* was a plesiosaur with an exceptionally long neck that lived in the late Cretaceous period (98–65 million years ago). After lopping off the top of the elasmosaur's head with a machete, the hero bizarrely—and successfully—transplants a young companion's brain into the beast. Eventually, its human intelligence fading, the creature (I can't help but think of it as "Elasmosaurenstein") turns on the protagonist and devours him. See Wardon Allan Curtis, "The Monster of Lake LaMetrie" (1899), Gutenberg Consortia Center, http://ebooks.gutenberg.us/WorldeBookLibrary.com/lametrie.htm (accessed September 12, 2011).
67 Arthur Conan Doyle, *The Lost World* (1912), Gutenberg Consortia Center, http://www.gutenberg.org/files/139/139-h/139-h.htm (accessed September 12, 2011).
68 Rupert T. Gould, *The Case for the Sea-Serpent* (London: Allan, 1930; New York: Putnam, 1934).
69 Quoted in Binns, *Loch Ness Mystery Solved*, 25–26.
70 Ibid., 24.
71 Philip Stalker, "Loch Ness Monster: A Puzzled Highland Community," *Scotsman*, October 16, 1933, 11.
72 Philip Stalker, "Loch Ness Monster: The Plesiosaurus Theory," *Scotsman*, October 17, 1933, 9.
73 Alex Campbell to Ness Fishery Board, October 28, 1933, quoted in Gould, *Loch Ness Monster*, 110–112.
74 Quoted in Dinsdale, *Loch Ness Monster*, 126.
75 Binns, *Loch Ness Mystery Solved*, 79–80.
76 Stalker, "Loch Ness Monster: The Plesiosaurus Theory," 9.
77 F. W. Memory, "The Monster Is a Seal: Conclusions of the 'Daily Mail' Mission," *Daily Mail* (London), ca. December 18, 1933–January 19, 1934, undated clipping, Loch Ness and Morar Project Archive, http://www.lochnessproject.org/adrian_shine_archiveroom/paperspdfs/LOCH_NESS_1933damail.pdf (accessed October 10, 2011).
78 Rupert Gould, "The Loch Ness 'Monster': A Survey of the Evidence," *Times*, December 9, 1933, 13.
79 "The Monster of Loch Ness: New Accounts from Eye-witnesses," *Times*, December 18, 1933, 9.
80 Douglas Russell, letter to the editor, *Scotsman*, October 20, 1933, 15.
81 Nicholas Witchell, *The Loch Ness Story* (Harmondsworth: Penguin, 1975), 72.
82 *Daily Express* (Glasgow), December 16, 1933, 8.
83 *Daily Express*, May 10, 1934, 6; "Trips to See the 'Monster,'" *Scotsman*, October 31, 1933.
84 "Buses to the 'Monster,'" *Scotsman*, March 10, 1934, 11.

85 "I'm the Monster of Loch Ness" (sung by Leslie Holmes in a British Pathé film released on January 25, 1934), British Pathé, http://www.britishpathe.com/record.php?id=28215 (accessed June 1, 2010).
86 "Loch Ness in the Films," *Scotsman*, December 11, 1934, 8.
87 *Daily Express*, November 13, 1933, 10; September 3, 1934, 20; January 25, 1934, 7.
88 "Sandy, the Loch Ness Monster," *Times*, January 30, 1934, 12.
89 *Daily Express*, January 9, 1934, 11.
90 *Scotsman*, February 28, 1934, 14.
91 Dennis Dunn, "Monster Bobs Up Again . . . Hotels Doing Fine," *Daily Express*, April 24, 1934, 3.
92 *Scotsman*, December 16, 1950, 6.
93 Steuart Campbell, *The Loch Ness Monster: The Evidence* (Amherst, N.Y.: Prometheus, 1997), 37–38.
94 Gould, *Loch Ness Monster*, 23.
95 Elwood Baumann, *The Loch Ness Monster* (New York: Franklin Watts, 1972), 12.
96 Roy P. Mackal, *The Monsters of Loch Ness* (Chicago: Swallow Press, 1976), 95–96.
97 Campbell, *Loch Ness Monster*, 38.
98 Mackal, *Monsters of Loch Ness*, 95.
99 Holiday, *Great Orm of Loch Ness*, 26.
100 Mike Dash, "Frank Searle's Lost Second Book," December 27, 2009, *Dry as Dust: A Fortean in the Archives*, Charles Fort Institute, Blogs.forteana.org/node/95 (accessed July 4, 2010).
101 Quoted in Holiday, *Great Orm of Loch Ness*, 26–27.
102 Mackal, *Monsters of Loch Ness*, 95.
103 Quoted in "Mr. Wetherell and a Broadcast," *Daily Express*, December 23, 1933.
104 Binns, *Loch Ness Mystery Solved*, 28.
105 Quoted in "Hunter's Story of Finding of Spoor," *Scotsman*, December 26, 1933, 10.
106 Strix [Peter Fleming], *Spectator*, April 5, 1957, quoted in Binns, *Loch Ness Mystery Solved*, 28–29.
107 David Martin and Alastair Boyd, *Nessie: The Surgeon's Photograph Exposed* (London: Thorne, 1999), 34.
108 "Monster Mystery Deepens," *Daily Mail*, January 4, 1934.
109 Martin and Boyd, *Nessie*, 27.
110 The "Surgeon's Photograph" is really two photographs: the famous image and a less impressive image said to show the creature diving. Cryptozoologists criticize skeptics for concentrating on the famous image and ignoring the second, but I will also follow that pattern. In my opinion, the second photo offers little to no additional information. It is consistent with the same model sinking or tipping over or with another object—but the image is too poor to answer for itself.
111 Mackal, *Monsters of Loch Ness*, 98.
112 Roy Chapman Andrews, *This Business of Exploring* (New York: Putnam, 1935), 59–60.
113 "The Making of a Monster," *Sunday Telegraph* (London), December 7, 1975, 6, quoted in Martin and Boyd, *Nessie*, 14.
114 Martin and Boyd, *Nessie*, 17.

115 Quoted in ibid., 20.

116 Robert Kenneth Wilson to Constance Whyte, 1955, quoted in ibid., 62.

117 Robert Kenneth Wilson to Maurice Burton, April 27, 1962, quoted in ibid., 57.

118 Major Egginton to Nicholas Witchell, November 3, 1970, quoted in ibid., 68–69.

119 Ibid., 50.

120 Nicholas Witchell, *The Loch Ness Story: Revised and Updated Edition* (London: Corgi, 1991), 47.

121 Martin and Boyd, *Nessie*, 77.

122 Major Egginton to Nicholas Witchell, November 9, 1970, quoted in ibid., 72.

123 Quoted in ibid., 70.

124 Denise Wilson to Tim Dinsdale, 1979, quoted in ibid., 58.

125 Quoted in ibid., 59.

126 Quoted in ibid., 77–78.

127 Ibid., 83; Paul H. LeBlond and M. J. Collins, "The Wilson Nessie Photograph: A Size Determination Based on Physical Principles," *Scottish Naturalist* 100 (1988): 95–108.

128 Quoted in Benjamin Radford and Joe Nickell, *Lake Monster Mysteries: Investigating the World's Most Elusive Creatures* (Lexington: University Press of Kentucky, 2006), 17.

129 Witchell, *Loch Ness Story*, 75–77.

130 Quoted in ibid.

131 Witchell, *Loch Ness Story: Revised*, 82–83. The man who Stuart took to see the props is identified by Witchell as "another Loch Ness resident, who is known to me and with whom I confirmed the following details." His name is revealed as Richard Frere in Campbell, *Loch Ness Monster*, 44.

132 Mackal, *Monsters of Loch Ness*, 102.

133 Quoted in Radford and Nickell, *Lake Monster Mysteries*, 19.

134 Mackal, *Monsters of Loch Ness*, 13.

135 Dinsdale, *Loch Ness Monster*, 5–6.

136 Ibid., 81.

137 Ibid., 82.

138 Ibid., 95–96.

139 Ibid., 100.

140 Ibid., 104.

141 Ibid., 110.

142 Ibid.

143 Mavis Cole, "Summer's Here and Britain Sees Loch Ness Monster," *Chicago Daily Tribune*, June 14, 1960, B6.

144 Donald White, "The Great Monster Hunt," *Boston Globe*, December 6, 1970, G57.

145 Tim Dinsdale, *Project Water Horse* (London: Routledge and Kegan Paul, 1975), 159–160.

146 White, "Great Monster Hunt."

147 Dinsdale, *Project Water Horse*, 11.

148 A detailed discussion of the analysis of the film is in Binns, *Loch Ness Mystery Solved*, 107–125. See also "Photographic Interpretation Report: Loch Ness," in "Did Tim Dinsdale Film Nessie?" Cryptozoo-oscity, http://cryptozoo-oscity.blogspot.com/2010/05/did-tim-dinsdale-film-nessie.html (accessed September 14, 2010).

149 "The Dinsdale Film," http://henryhbauer.homestead.com/DinsdaleFilm.html (accessed September 14, 2010).
150 Shine, *Loch Ness*, 12.
151 Binns, *Loch Ness Mystery Solved*, 121.
152 Ibid., 108.
153 Quoted in Henry Allen, "On the Monster Trail," *Washington Post*, April 16, 1972, E9.
154 Dinsdale, *Project Water Horse*, 154–155.
155 Ibid., 6–7.
156 Quoted in Allen, "On the Monster Trail."
157 Tony Harmsworth, "Tim Dinsdale's Loch Ness Monster Film," Loch Ness Information, http://www.loch-ness.org/filmandvideo.html (accessed September 16, 2010).
158 Loren Coleman and Patrick Huyghe, *The Field Guide to Lake Monsters, Sea Serpents, and Other Mystery Denizens of the Deep* (New York: Penguin / Tarcher, 2003), 21.
159 Dunn, "Monster Bobs Up Again."
160 Dinsdale, *Loch Ness Monster*, 13.
161 Loren Coleman, "Why Do Plesiosaurs Have Long Necks?" September 2, 2009, Cryptomundo, http://www.cryptomundo.com/cryptozoo-news/plesiosaurs-necks/ (accessed June 23, 2010).
162 Loren Coleman and Jerome Clark, *Cryptozoology A to Z: The Encyclopedia of Loch Monsters, Sasquatch, Chupacabras, and Other Authentic Mysteries of Nature* (New York: Simon and Schuster / Fireside, 1999), 140.
163 Michael Thompson-Noel, "The Day I Saw Five Loch Ness Monsters," *Financial Times* (London), September 13, 1997, 17.
164 Bernard Heuvelmans, "Review of the Monsters of Loch Ness," *Skeptical Inquirer* 2, no. 1 (1977): 110.
165 Karl Shuker, *The Beasts That Hide from Man: Seeking the World's Last Undiscovered Animals* (New York: Paraview Press, 2003), 187.
166 Dinsdale, *Loch Ness Monster*, 232–233.
167 Shine, *Loch Ness*, 26.
168 Adrian J. Shine, "Postscript: Surgeon or Sturgeon?" *Scottish Naturalist* 105 (1993): 271–282.
169 Bauer, *Enigma of Loch Ness*, 56.
170 Mackal, *Monsters of Loch Ness*, 200–201.
171 Dinsdale, *Loch Ness Monster*, 81–82.
172 John Kirk, *In the Domain of the Lake Monsters* (Toronto: Key Porter Books, 1998), 115, 119.
173 Mackal, *Monsters of Loch Ness*, 29.
174 Bentley Murray, letter to the editor, *Scotsman*, October 30, 1933, 13.
175 A. T. C., letter to the editor, *Scotsman*, November 4, 1933, 15.
176 "Loch Ness Monster Theory," *Inverness Courier*, September 5, 1933.
177 "The Loch Ness 'Monster': Is It a Giant Eel?" *Inverness Courier*, September 15, 1933.
178 W. P., letter to the editor, *Scotsman*, November 2, 1933, 11.
179 Alan MacLean, letter to the editor, *Scotsman*, November 16, 1933, 11.
180 John Moir, letter to the editor, *Scotsman*, November 6, 1933, 13.
181 Holiday, *Great Orm of Loch Ness*.

182 Joe Nickell, "Lake Monster Lookalikes," June 2007, Committee for Skeptical Inquiry, http://www.csicop.org/sb/show/lake_monster_lookalikes/ (accessed September 10, 2010).

183 This effect has often been disputed by cryptozoology enthusiasts, but it certainly happens. See, for example, Loren Coleman, "Real Otter Sense," June 16, 2007, CryptoMundo, http://www.cryptomundo.com/cryptozoo-news/real-otter-sense/ (accessed September 10, 2010).

184 Otters have, for example, been filmed at the loch by longtime Nessie researcher Dick Raynor, as he reported in "Otters in Loch Ness," Loch Ness Investigation, http://www.lochnessinvestigation.com/otters.htm (accessed April 10, 2012).

185 Mackal, *Monsters of Loch Ness*, 136.

186 Gould, *Loch Ness Monster*, 74.

187 Binns quotes an article from the *Inverness Courier*, but seems not to provide a citation for it, in *Loch Ness Mystery Solved*, 20.

188 Campbell, "Strange Spectacle on Loch Ness."

189 "Loch Ness 'Monster' Again," *Inverness Courier*, June 9, 1933.

190 Gould, *Loch Ness Monster*, 142.

191 Because the word "porpoise" has both formal and informal meanings, it is not clear what type of animals are described in this story. They may have been true porpoises, or they could just as well have been dolphins. See "Captured by Nature," *Daily Mail*, September 16, 1914, 3.

192 Gordon Williamson, "Seals in Loch Ness, " *Scientific Reports of the Whales Research Institute*, no. 39 (1988): 151–157.

193 Dick Raynor, "Seal in Urquhart Bay, 1999," Loch Ness Investigation, http://www.lochnessinvestigation.com/Seals.html (accessed June 25, 2010).

194 Dick Raynor, "What do you think of Dr. Rines and his ideas of tracking and filming a moving object in the loch?" Loch Ness Investigation, http://www.lochnessinvestigation.com/cyberspace.html (accessed June 25, 2010).

195 "The 'Monster'—Scientists' Views After Seeing Film," *Scotsman*, October 5, 1934, 9.

196 "Film That 'Removes Doubt' to Be Shown in Scotland," *Scotsman*, November 3, 1936, 10. This appears to contradict a quote from Irvine, not clearly sourced, that the creature had "a huge body . . . over 30 feet long" (quoted in Witchell, *Loch Ness Story*, 52–53).

197 "The Monster of Loch Ness: New Accounts from Eye-witnesses," *Times*, December 18, 1933, 9.

198 Quoted in F. W. Memory, "The 'Monster' Again: New Support for Seal Theory," *Daily Mail*, ca. December 18, 1933–January 19, 1934, undated clipping, Loch Ness and Morar Project Archive, http://www.lochnessproject.org/adrian_shine_archiveroom/paperspdfs/LOCH_NESS_1933damail.pdf (accessed January 13, 2012).

199 Quoted in F. W. Memory, "There *Is* a Seal in Loch Ness," *Daily Mail*, ca. December 18, 1933–January 19, 1934, undated clipping, Loch Ness and Morar Project Archive, http://www.lochnessproject.org/adrian_shine_archiveroom/paperspdfs/LOCH_NESS_1933damail.pdf (accessed January 13, 2012).

200 Compare "What Was It? A Strange Experience on Loch Ness," *Northern Chronicle*, August 27, 1930, with Gould, *Loch Ness Monster*, 37–38.

201 Quoted in Witchell, *Loch Ness Story*, 29–30.

202 It is interesting to note, however, that those who claim to have been abducted by aliens are more prone than control subjects to form false memories in a laboratory setting. See Susan A. Clancy, Richard J. McNally, Daniel L. Schacter, Mark F. Lenzenweger, and Roger K. Pitman, "Memory Distortion in People Reporting Abduction by Aliens," *Journal of Abnormal Psychology* 111 (2002): 455–461.

I know of no similar research involving people to claim to have seen cryptids, but it is intriguing to speculate that this population might also exhibit increased vulnerability to false memories. In either event, a significant rate of false memory is common to us all.

203 "Sea Serpent in the Highlands."

204 Mackal, *Monsters of Loch Ness*, 99. The claim of twenty-one photographs conflicts with news reports at the time, such as "Five Photographs Secured by Searchers—Seen 21 Times," *Scotsman*, August 9, 1934, 9. However, Mackal claims to have examined all twenty-one photographs from Mountain's expedition.

205 Binns, *Loch Ness Mystery Solved*, 37–38.

206 The Loch Ness Phenomena Investigation Bureau (LNPIB; now Loch Ness Investigation Bureau [LNIB]) was formed in 1962, and Adrian Shine says that its camera surveillance began the same year. See "Loch Ness Timeline," Loch Ness and Morar Project, http://www.lochnessproject.org/adrian_shine_archiveroom/loch_ness_archive_timeline.htm (accessed July 5, 2010). Holiday says that the camera rigs were built for the 1964 season, and mentions earlier work by the LNPIB, in *Great Orm of Loch Ness*, 64–65.

207 Holiday, *Great Orm of Loch Ness*, 65–67.

208 Mackal, *Monsters of Loch Ness*, 200.

209 Ibid., 45–48.

210 Dick Raynor, "Submarine Investigations," Loch Ness Investigation, www.lochnessinvestigation.org/Subs.html (accessed July 3, 2010).

211 Rikki Razdan and Alan Kielar, "Sonar and Photographic Searches for the Loch Ness Monster: A Reassessment," *Skeptical Inquirer* 9, no. 2 (1984–1985): 153–154. Based on the shape of the fin, Sir Peter Scott, a noted naturalist and a co-founder of the Loch Ness Phenomena Investigation Bureau, gave the creature the scientific name *Nessiteras rhombopteryx* (monster of Ness with the diamond-shaped fin) so it could be registered as an endangered species. See Peter Scott and Robert Rines, "Naming the Loch Ness Monster," *Nature* 258, no. 5535 (1975): 466–468. But it was soon pointed out that the name can be anagrammed into "monster hoax by Sir Peter S," as reported in "Loch Ness Monster Shown a Hoax by Another Name," *New York Times*, December 19, 1975, 78. To counter this insinuation with another anagram, Rines "came up with the antidote—'Yes, both pix are monsters—R'" (Dinsdale, *Loch Ness Monster*, 171).

212 *Mysteries of the Unknown: Mysterious Creatures* (Alexandria, Va.: Time-Life Books, 1988), 73.

213 Campbell, *Loch Ness Monster*, 67.
214 Harmsworth, *Loch Ness, Nessie and Me*, 181.
215 Witchell, *Loch Ness Story: Revised*, 165.
216 Campbell, *Loch Ness Monster*, 72–74.
217 Witchell, *Loch Ness Story: Revised*, 168.
218 Gould, *Loch Ness Monster*, 120.
219 Binns, *Loch Ness Mystery Solved*, 35.
220 David S. Martin, Adrian J. Shine, and Annie Duncan, "The Profundal Fauna of Loch Ness and Loch Morar," *Scottish Naturalist* 105 (1993): 119.
221 "Nessie Dead or Alive," Loch Ness and Morar Project, http://www.lochnessproject.org/loch_ness_reflections_news_links/loch_ness_traps.htm (accessed July 5, 2010).
222 Quoted in Holiday, *Great Orm of Loch Ness*, 201.
223 Binns, *Loch Ness Mystery Solved*, 148–149.
224 Shine, "Loch Ness Timeline."
225 Quoted in "BBC 'Proves' Nessie Does Not Exist," July 27, 2003, BBC News, http://news.bbc.co.uk/2/hi/science/nature/3096839.stm (accessed July 5, 2010).
226 Karen DeYoung, "Sonar Search for 'Nessie' Reveals 3 Wobbly Scratches," *Washington Post*, October 12, 1987, A1.
227 For a thorough discussion, see Adrian J. Shine and David S. Martin, "Loch Ness Habitats Observed by Sonar and Underwater Television," *Scottish Naturalist* 100 (1988): 111–199.
228 Ibid.
229 Dick Raynor, "The River Ness—From Loch Ness to the Sea," Loch Ness Investigation, http://www.lochnessinvestigation.org/RiverNessJourney.html (accessed September 19, 2010).
230 Quoted in "Commander Gould to Investigate," *Scotsman*, November 10, 1933, 11.
231 Meurger and Gagnon, *Lake Monster Traditions*, 135–139.
232 Mackal, *Monsters of Loch Ness*, 25.
233 Shine and Martin, "Loch Ness Habitats Observed by Sonar and Underwater Television."

5. THE EVOLUTION OF THE SEA SERPENT

1 Benjamin Radford, *Tracking the Chupacabra: The Vampire Beast in Fact, Fiction, and Folklore* (Albuquerque: University of New Mexico Press, 2011).
2 Sherrie Lynne Lyons, *Species, Serpents, Spirits, and Skulls: Science in the Margins in the Victorian Age* (Albany: State University of New York Press, 2009).
3 Ray Bradbury, "The Fog Horn," in *The Golden Apples of the Sun* (New York: Morrow, 1997), 1–9, members.fortunecity.com/ymir1/beastfro9.html (accessed February 2, 2012).
4 Dennis Loxton to Daniel Loxton, April 8, 2005.
5 Depending on the movie, Godzilla's height ranges from a measly 167 to 334 feet tall, according to Robert Biondi, "So Just How Big Is Godzilla? A Model Builder's Guide to Godzilla's Size Changes," *Kaiju Review*, no. 4 (1993).

6 Emily Vermeule, *Aspects of Death in Early Greek Art and Poetry* (Berkeley: University of California Press, 1979), 183.
7 The Holy Bible, Revised Standard Version (Glasgow: Collins, 1952).
8 For an overview, see J. V. Kinnier Wilson, "A Return to the Problems of Behemoth and Leviathan," *Vetus Testamentum* 25, fasc. 1 (1975): 1–14.
9 John K. Papadopoulos and Deborah Ruscillo, "A *Ketos* in Early Athens: An Archaeology of Whales and Sea Monsters in the Greek World," *American Journal of Archaeology* 106, no. 2 (2002): 213.
10 Ibid., 216.
11 Ibid., 187–227.
12 Katharine Shepard, *The Fish-Tailed Monster in Greek and Etruscan Art* (New York: Privately printed, 1940; repr., Landisville, Pa.: Coachwhip, 2011), 78.
13 Papadopoulos and Ruscillo, "*Ketos* in Early Athens," 216–219.
14 John Boardman, "'Very Like a Whale': Classical Sea Monsters," in *Monsters and Demons in the Ancient and Medieval Worlds: Papers Presented in Honor of Edith Porada*, ed. Anne E. Farkas, Prudence O. Harper, and Evelyn B. Harrison (Mainz: Zabern, 1987), 73–84.
15 Diego Cuoghi, "The Art of Imagining UFOs," ed. Daniel Loxton, trans. Daniela Cisi and Leonardo Serna, *Skeptic* 11, no. 1. (2004): 43–51.
16 Bernard Heuvelmans, *In the Wake of the Sea-Serpents* (New York: Hill and Wang, 1968), 508. For further discussion, see Loren Coleman, "The Meaning of Cryptozoology: Who Invented the Term Cryptozoology?" 2003, www.lorencoleman.com/cryptozoology_faq.html (accessed February 12, 2011).
17 Heuvelmans, *In the Wake of the Sea-Serpents*, 96–99.
18 Boardman, "Very Like a Whale,"78.
19 A. C. Oudemans, *The Great Sea-Serpent* (1892; repr., Landisville, Pa.: Coachwhip, 2007), 89.
20 Adrienne Mayor, "Paleocryptozoology: A Call for Collaboration Between Classicists and Cryptozoologists," *Cryptozoology* 8 (1989): 12–26.
21 *Pausanias Description of Greece*, trans. W. H. S. Jones (London: Heinemann, 1918), Paus. 1.44.8.
22 Pliny the Elder, *The Natural History*, trans. H. Rackham, Loeb Classical Library 353 (Cambridge, Mass.: Harvard University Press, 1940), 3:bk. 9.5, 173.
23 Richard B. Stothers, "Ancient Scientific Basis of the 'Great Serpent' from Historical Evidence, " *Isis* 95, no. 2 (2004): 223–226.
24 Aristotle, *History of Animals*, trans. D'Arcy Wentworth Thompson (Sioux Falls, S.Dak.: NuVision, 2004), bk. 4, chap. 8.
25 Ibid., bk. 8, chap. 28.
26 Heuvelmans, *In the Wake of the Sea-Serpents*, 82.
27 Apollonius of Rhodes, *Jason and the Golden Fleece*, trans. Richard Hunter (New York: Oxford University Press, 1995), 102.
28 W. R. Branch and W. D. Hacke, "A Fatal Attack on a Young Boy by an African Rock Python *Python sebae*," *Journal of Herpetology* 14, no. 3 (1980): 305–307.

29 Richard Stothers, "Ancient Scientific Basis of the 'Great Serpent' from Historical Evidence," *Isis* 95, no. 2 (2004): 228–229.

30 Pliny used the word *Dracones*; John Bostock's translation of *Natural History* (1855) renders it as "dragons," while H. Rackham's translation (1940) substitutes "serpents."

31 Pliny the Elder, *Natural History*, 3:bk. 8.11, 25–27.

32 Edward Topsell, *The Historie of Serpents; or, the Second Booke of Living Creatures* (1608; LaVergue: ECCO Print Editions, 2011), 168.

33 Heuvelmans, *In the Wake of the Sea-Serpents*, 82.

34 Indeed, during my research I realized that this connection has occasionally been made. For example, the question was raised in this 2006 cryptozoology forum thread, in which poster Fel Faralei wondered, "It's a long shot but from what I've learn't about Caddy it seems to fit the description of a hippocampus quite well. . . . What if they were real only less beautiful and horse could it be caddy?" (http://www.cryptozoology.com/forum/topic_view_thread.php?tid=5&pid=372181 [accessed May 28, 2011]).

35 I based my drawing on eyewitness sketches and a composite drawing in Paul H. LeBlond and Edward L. Bousfield, *Cadborosaurus: Survivor from the Deep* (Victoria, B.C.: Horsdal & Schubart, 1995). In addition, I referred to artworks of Cadborosaurus by cryptozoological enthusiasts, including David John and Darren Naish.

36 Helen Scales, *Poseidon's Steed: The Story of Seahorses, from Myth to Reality* (New York: Gotham Books, 2009), 20–22.

37 Shepard, *Fish-Tailed Monster in Greek and Etruscan Art*, 26.

38 Scales, *Poseidon's Steed*, 31–33.

39 Publius Vergilius Maro, *Fourth Georgic of Virgil*, trans. R. M. Millington (London: Longmans, Green, Reader, and Dyer, 1870), 59.

40 Michael J. Curley, trans., *Physiologus: A Medieval Book of Nature Lore* (Chicago: University of Chicago Press, 2009), xvi–xix.

41 Ibid., xxi–xxvi.

42 James Carlill, trans., *Physiologus*, in *The Epic of the Beast*, ed. William Rose (London: Routledge, 1900), 189, 199–200, 231, reprinted in *The Book of Fabulous Beasts: A Treasury of Writings from Ancient Times to the Present*, ed. Joseph Nigg (New York: Oxford University Press, 1999), 116.

43 Curley, trans., *Physiologus*, xxvi–xxxiii.

44 *The Etymologies of Isidore of Seville*, trans. Stephen A. Barney (Cambridge: Cambridge University Press, 2006), 260.

45 Richard Barber, trans., *Bestiary: Being an English Version of the Bodleian Library, Oxford M.S. Bodley 764 with All the Original Miniatures Reproduced in Facsimile* (London: Folio Society, 1992), 10.

46 Ian Whitaker, "North Atlantic Sea-Creatures in the *King's Mirror* (*Konungs Skuggsjá*)," *Polar Record* 23 (1986): 3–13.

47 Peter Gilliver, "J. R. R. Tolkien and the *OED*," Oxford English Dictionary, http://www.oed.com/public/tolkien/ (accessed May 25, 2011).

48 Paul Henri Mallet, *Northern antiquities: or, An historical account of the manners, customs, religion and laws, maritime expeditions and discoveries, language and literature of the an-*

cient Scandinavians, trans. Thomas Percy, rev. I. A. Blackwell (London: Henry G. Bohn, 1847), 514.

49 George Ripley and Charles A. Dana, eds., *The New American Cyclopedia: A Popular Dictionary of General Knowledge* (New York: Appleton, 1869), 14:470–471.

50 According to Samuel Hibbert, "The faith in the Edda of the great serpent that Thor fished for, did not, as Dr Percy conceives, give rise to the notion of the sea-snake, but a real sea-snake was the foundation of the fable" (*A Description of the Shetland Islands: Comprising an Account of Their Geology, Scenery, Antiquities, and Superstitions* [Edinburgh: Constable, 1822], 565).

51 Oudemans, *Great Sea-Serpent*, 298.

52 Snorri Sturluson, prologue to *The Prose Edda*, trans. Arthur Gilchrist Brodeeur (New York: American-Scandinavian Foundation, 1916).

53 Charles Mackay, *Memoirs of Extraordinary Popular Delusions and the Madness of Crowds* (1841; New York: Barnes and Noble, 2002), 105.

54 Albertus was also associated with astrology and alchemy, which were at that time respectable areas of intellectual inquiry. Still, his alchemical interests were probably exaggerated. As Mackay noted, many magical tales were associated with Albertus and Aquinas, including the rumor that Albertus could control the weather: "Such stories as these shew the spirit of the age. Every great man who attempted to study the secrets of nature was thought a magician" (ibid., chap. 4).

55 Albert the Great, *Man and the Beasts: De animalibus, Books 22–26*, trans. James J. Scanlan (Binghamton, N.Y.: Medieval and Renaissance Texts and Studies, 1987), 350.

56 Olaus Magnus, *A Description of the Northern Peoples: Rome 1555*, ed. Peter Foote, trans. Peter Fisher and Humphrey Higgens (London: Hakluyt Society, 1998), 1124.

57 Ray Gibson "Nemertean Genera and Species of the World: An Annotated Checklist of Original Names and Description Citations, Synonyms, Current Taxonomic Status, Habitats and Recorded Zoogeographic Distribution," *Journal of Natural History* 29, no. 2 (1995): 271–561; Adriaan Gittenberger and Cor Schipper, "Long Live Linnaeus, *Lineus longissimus* (Gunnerus, 1770) (Vermes: Nemertea: Anopla: Heteronemertea: Lineidae), the Longest Animal Worldwide and Its Relatives Occurring in the Netherlands," *Zoologische Mededelingen* 82, no. 7 (2008): 59–63.

58 Olaus, *Description of the Northern Peoples*, 1128.

59 Oudemans, *Great Sea-Serpent*, 91.

60 "Recreations in Natural History . . . Ancient Flying Dragons, Pterosaurians, &c.," *New Monthly Magazine and Humorist*, pt. 3 (1843): 38–39.

61 For example, cryptozoologists regularly discard Bigfoot cases with which paranormal or inexplicable elements are associated (such as glowing eyes, telepathy, or a UFO). Some cryptozoology Web sites even have policies that forbid discussions of the paranormal, which can badly distort the eyewitness record—and destroys cryptozoology's foundation. For more on this issue, see Daniel Loxton, "An Argument That Should Never Be Made Again," February 2, 2010, Skepticblog, http://www.skepticblog.org/2010/02/02/an-argument-that-should-never-be-made-again/ (accessed October 4, 2011).

62 Erich Pontoppidan, *The Natural History of Norway: Part II* (London: Printed for A. Linde, 1755), 207–208.

63 Michel Meurger, with Claude Gagnon, *Lake Monster Traditions: A Cross-Cultural Analysis* (London: Fortean Times, 1988), 24.

64 Ibid., 28.

65 Pierre Belon, *De aquatilibus* (Paris: Charles Estienne, 1553), bk. 1:26–27.

66 Conrad Gesner, *Historiae animalium*, bk. 4, *Qui est de piscium & aquatilium animantium natura* (Zurich, 1558), 433 (translated by Donald Prothero and Doug Henning).

67 Conrad Gesner, *Nomenclator aquafilium animantium* (1560), quoted in Oudemans, *Great Sea-Serpent*, 92.

68 Heuvelmans, *In the Wake of the Sea-Serpents*, 99–100.

69 Ambroise Paré, *On Monsters and Marvels*, trans. Janis L. Pallister (Chicago: University of Chicago Press, 1982), 112.

70 Historians warn of the distorting effects of labeling figures of the past as belonging to modern movements. For example, consider the first warning expressed by Rebekah Higgitt (curator and historian of science at the Royal Observatory Greenwich & National Maritime Museum): "Do not ever call anyone a scientist who would not have recognised the term. The word . . . was not actually *used* until the 1870s. If we use the term to describe anyone before this date we risk loading their views, status, career, ambitions and work with associations that just do not exist before this date" ("Dos and Don'ts in History of Science," April 17, 2011, Teleskopos, http://teleskopos.wordpress.com/2011/04/17/dos-and-donts-in-history-of-science/ [accessed September 7, 2011]). The modern tradition of scientific skepticism dates to the twentieth century. However, centuries of earlier thinkers tried their hands at similar debunking projects. They might not have felt a kinship with my work, but I feel a connection to theirs!

71 Sir Thomas Browne, *Pseudodoxia Epidemica: Or, Enquiries into Commonly Presumed Truths* (Oxford: Benediction Classics, 2009), 242.

72 There is some controversy about the attribution of the *Codex canadensis*. Here, I have deferred to François-Marc Gagnon of the Gail and Stephen A. Jarislowsky Institute for Studies in Canadian Art, who concludes, "Until there is proof to the contrary, he can be considered the author of the *Codex canadensis*" ("About Louis Nicolas: Biographical Notes on Louis Nicolas, Presumed Author of the *Codex canadensis*," Library and Archives Canada, www.collectionscanada.gc.ca/codex/026014-1200-e.html [accessed May 2, 2011]).

73 Quoted in Meurger, with Gagnon, *Lake Monster Traditions*, 212.

74 Heuvelmans, *In the Wake of the Sea-Serpents*, 95.

75 Hans H. Lilienskiold, The Northern Lights Route, http://www.ub.uit.no/northernlights/eng/lillienskiold.htm (accessed May 25, 2011).

76 Hans Lilienskiold, *Speculum boreale eller den finmarchiske beschrifwelsis* (1698), quoted in Ole Lindquist, "Whales, Dolphins and Porpoises in the Economy and Culture of Peasant Fishermen in Norway, Orkney, Shetland, Faeroe Islands and Iceland, ca. 900–1900 A.D., and Norse Greenland, ca. 1000–1500 A.D." (Ph.D. diss., University of St Andrews, 1994), 2:sec. A.18, 1003–1006, www.fishernet.is/is/hvalveidar/19/35 (accessed May 25, 2011).

77 Heuvelmans, *In the Wake of the Sea-Serpents*, 114.
78 Pontoppidan, *Natural History of Norway*, iv.
79 Ibid., 196.
80 Ibid.
81 Ibid.
82 Ibid.
83 Ibid., 196–197.
84 Henry Lee, *Sea Monsters Unmasked* (London: Clowes, 1883), 63.
85 Meurger, with Gagnon, *Lake Monster Traditions*, 17.
86 Heuvelmans, *In the Wake of the Sea-Serpents*, 112.
87 Lee, *Sea Monsters Unmasked*, 2–3.
88 Daniel Loxton, "Bishop Pontoppidan Versus the Tree Geese," February 8, 2011, Skepticblog, http://www.skepticblog.org/2011/02/08/bishop-pontoppidan-versus-the-tree-geese/ (accessed February 12, 2011).
89 Pontoppidan, *Natural History of Norway*, 208.
90 Heuvelmans, *In the Wake of the Sea-Serpents*, 34.
91 Pliny the Elder, *Natural History*, 3:bk. 9.4, 171.
92 *A True and Perfect Account of the Miraculous Sea-Monster or Wonderful Fish Lately Taken in Ireland* (London: Printed for P. Brooksby and W. Whitwood, 1674).
93 "Wernerian Natural History Society," *Philosophical Magazine* 33 (1809): 90–91.
94 Rupert T. Gould, *The Case for the Sea-Serpent* (London: Allan, 1930; New York: Putnam, 1934), 239.
95 James Ritchie, "More About Monsters: The Sea-Serpent of Stronsay," *Times* (London), December 16, 1933, 15.
96 "Here's First Genuine Sea Monster Captured: Odd Fish Stumps Scientist," *Los Angeles Times*, March 3, 1934, 3; "France Has Sea Monster and the Body to Prove It," *New York Times*, March 1, 1934, 1.
97 "Sea 'Monster' Is Declared a Basking Shark: French Scientists Say It Is Not an Odd Fish," *New York Times*, March 16, 1934, 18.
98 Associated Press, "'Sea Serpent' Body Is Found in Pacific," *New York Times*, November 23, 1934, 21.
99 Associated Press, "Canada's 'Sea Serpent' Found to Be Only Shark," *New York Times*, November 27, 1934, 23.
100 John Saar, "Fishermen Made a Monstrous Mistake," *Washington Post*, July 21, 1977, 2.
101 John Saar, "'Plesiosaurus' Find: Monster Mystery Surfaces in Japan," *Los Angeles Times*, July 21, 1977, B6.
102 Ellis, *Monsters of the Sea* (New York: Lyons Press, 2001), 69.
103 John Goertzen, "New *Zuiyo Maru* Cryptid Observations: Strong Indications It Was a Marine Tetrapod," *CRS Quarterly* 38, no. 1 (2001): 19–29, http://www.creationresearch.org/crsq/articles/38/38_1/Cryptid.htm (accessed October 6, 2011).
104 Christopher McGowan, *The Dragon Seekers: How an Extraordinary Circle of Fossilists Discovered the Dinosaurs and Paved the Way for Darwin* (New York: Basic Books, 2001), ix.
105 For example, Thomas Jefferson suspected that mammoths or mastodons might survive in his time in the western interior of the North American continent. Avidly interested in

fossils, but unconvinced about extinction, Jefferson hoped on the basis of Native American fossil lore that the Lewis and Clark expedition might encounter living specimens. (Today, we might call Jefferson's speculation an example of cryptozoology.) For a discussion of Jefferson's thoughts, see Adrienne Mayor, *Fossil Legends of the First Americans* (Princeton, N.J.: Princeton University Press, 2005), 55–61.

106 William D. Conybeare, "On the Discovery of an Almost Complete Skeleton of the Plesiosaurus," *Philosophical Transactions of the Geological Society of London*, 2nd ser., 1 (1824): 381–389; McGowan, *Dragon Seekers*, 78–84.

107 William Hooker, "Additional Testimony Respecting the Sea-Serpent of the American Seas," *Edinburgh Journal of Science* 6 (1827): 126–133.

108 Robert Bakewell, *Introduction to Geology* (New Haven, Conn.: Howe, 1833), 213.

109 Benjamin Silliman, footnote in ibid., 214.

110 John Ruggles Cotting, *A Synopsis of Lectures on Geology, Comprising the Principles of the Science* (Taunton, Mass.: Published for the author, 1835), 58–59.

111 Heuvelmans, *In the Wake of the Sea-Serpents*, 276.

112 Lyons, *Species, Serpents, Spirits, and Skulls*, 19.

113 Gould, *Case for the Sea-Serpent*, 96, 99.

114 This request for a report was recorded in the Admiralty records, according to ibid., 96.

115 Peter M'Quhae, "The Great Sea Serpent" [letter to the editor], *Times*, October 14, 1848, 3.

116 Gould, *Case for the Sea-Serpent*, 97–101.

117 Ibid., 107.

118 Ibid., 105–106.

119 F. G. S., "To the Editor of the Times," *Times*, November 2, 1848, 3.

120 Philip Henry Gosse, *The Romance of Natural History* (London: Nisbet, 1863), 347.

121 [Richard Owen], review of *On the Origin of Species*, by Charles Darwin, *Edinburgh Review* 111 (1860): 487–532.

122 Charles Darwin to Joseph Dalton Hooker, August 4, 1872, Darwin Correspondence Project, http://www.darwinproject.ac.uk/entry-8449 (accessed February 19, 2012).

123 Quoted in Oudemans, *Great Sea-Serpent*, 219.

124 Richard Owen, "The Sea Serpent," *Times*, November 14, 1848, 8.

125 Louis Agassiz, "Extract from the Thirteenth Lecture of Professor Agassiz, Delivered in Philadelphia, Tuesday Evening, March 20, 1849," in Eugene Batchelder, *A Romance of the Fashionable World* (Boston: French, 1857), 153–157.

126 Louis Agassiz, "Eugene Batchelder, Esq.—," in ibid., 152–153.

127 Edward Newman, in *Zoologist* 7 (1849): 2356.

128 Willy Ley, *Willy Ley's Exotic Zoology* (1959; New York: Bonanza Books, 1987), 224.

129 Newman, in *Zoologist*, x–xi.

130 Gould, *Case for the Sea Serpent*, 84–85. I was not surprised to see this. Gould was among the more thorough and responsible of the early-twentieth-century proponents of both the sea serpent and the Loch Ness monster.

131 Gosse, *Romance of Natural History*, 368.

132 An instantly created river implies previous rain; sand implies previous erosion; tree rings imply seasons past; an animal's body implies gestation, growth, respiration, and so on. Consider fur: if God created, say, bunny rabbits, their silky pelts imply a prior period of hair growth. As Philip Henry Gosse put it,

> We have passed in review before us the whole organic world: and the result is uniform; that no example can be selected from the vast vegetable kingdom, none from the vast animal kingdom, which did not at the instant of its creation present indubitable evidences of a previous history. This is not put forth as a *hypothesis*, but as a *necessity*; I do not say that it was *probably* so, but that it was *certainly* so; not that it *may have been thus*, but that it *could not have been otherwise*. (*Omphalos: An Attempt to Untie the Geological Knot* [London: Van Voorst, 1857], 335)

133 Stephen Jay Gould, "Adam's Navel," in *The Flamingo's Smile: Reflections in Natural History* (New York: Norton, 1985), 100.

134 Philip Henry Gosse, *The Aquarium: An Unveiling of the Wonders of the Deep Sea* (London: Van Voorst, 1856).

135 Gosse, *Romance of Natural History*, 298.

136 Gosse and Darwin were socially acquainted and exchanged friendly correspondence on orchid cultivation and variation, as well as other topics. Darwin was especially impressed by Gosse's experiments with marine aquariums: "I saw Mr Gosse the other night & he told me that he had now the same several sea-animals & algæ living & breeding for 13 months in the same artificially made sea water! Does not this tempt you? it almost tempts me to set up a marine vivarium" (Charles Darwin to J. S. Henslow, March 26, 1855, Darwin Correspondence Project, http://www.darwinproject.ac.uk/entry-1655 [accessed September 10, 2011]).

137 Gosse, *Romance of Natural History*, 364.

138 Ibid., 358–359.

139 Jules Verne, *Journey to the Center of the Earth*, trans. Willis T. Bradley (New York: Ace Books, 1956), 187–188.

140 Edgar Rice Burroughs, *The Land That Time Forgot*, chaps. 4 and 5, http://www.literature.org/authors/burroughs-edgar-rice/the-land-that-time-forgot/chapter-04.html (accessed September 12, 2011).

141 Gould, *Case for the Sea Serpent*, 277.

142 "The Hydrargos sillimanii, or Great Sea Serpent," *New York Daily Tribune*, August 29, 1845, 1.

143 "Orange County Milk," *New York Observer and Chronicle*, January 25, 1845, 14.

144 "Hydrargos sillimanii, or Great Sea Serpent."

145 Douglas E. Jones, "Doctor Koch and His 'Immense Antediluvian Monsters,'" *Alabama Heritage* 12 (1989): 2–19.

146 Ibid.

147 Jeffries Wyman, "Hydrarchos sillimani," *Proceedings of the Boston Society of Natural History* 2 (1848): 65–68.

148 Jones, "Doctor Koch and his 'Immense Antediluvian Monsters.'"

149 James D. Dana, "On Dr. Koch's Evidence with Regard to the Contemporaneity of Man and Mastodon in Missouri," *American Journal of Science and Arts* 9 (1875): 335–346.

150 Ibid.

151 Quoted in George E. Gifford Jr., "Twelve Letters from Jeffries Wyman, M.D.: Hampden-Sydney Medical College, Richmond, Virginia, 1843–1848," *Journal of the History of Medicine and Allied Sciences* 20, no. 4 (1965): 320–322.

152 Dr. Lister, in *Proceedings of the Boston Society of Natural History* 2 (1848): 94–96.

153 Albert Koch, *Journey Through a Part of the United States in the Years 1844 to 1846*, trans. Ernst A. Stadler (Carbondale: Southern Illinois University Press, 1972), 104.

154 Ibid., 32.

155 Oddly enough, thanks to the rules of taxonomic priority, it is *Basolisaurus* that remains enshrined in the nomenclature. For Owen's analysis, see Richard Owen, "On the Teeth of the Zeuglodon (Basilosaurus of Dr. Harlan)," *Proceedings of the Geological Society of London* 3 (1842): 23–28.

156 Jones, "Doctor Koch and his 'Immense Antediluvian Monsters.'"

157 Brian Switek, *Written in Stone: Evolution, the Fossil Record, and Our Place in Nature* (New York: Bellevue Literary Press, 2010), 152, 277.

158 Robert Silverberg, *Scientists and Scoundrels: A Book of Hoaxes* (New York: Crowell, 1965), 67.

159 Google Books: Ngram Viewer, http://goo.gl/uxsjN (accessed September 22, 2011).

160 Owen, "Sea Serpent," 8.

161 George R. Price, "Science and the Supernatural," *Science* 122 (1955): 363.

162 "The Great Sea Serpent," *Eclectic Magazine of Foreign Literature* 17, no. 2 (1849): 234.

163 Michael Shermer, "Miracle on Probability Street," August 2004, http://www.michaelshermer.com/2004/08/miracle-on-probability-street/ (accessed August 3, 2011).

164 Captain Elkanah Finney, description of a sighting in 1815, recorded in 1817, quoted in Oudemans, *Great Sea-Serpent*, 128.

165 Edward Newman, "Notes on the Zoology of Spitsbergen," *Zoologist* 24 (1864): 202–203.

166 *Province* [Vancouver], March 9, 1943, 1, quoted in LeBlond and Bousfield, *Cadborosaurus*, 70.

167 Ley, *Willy Ley's Exotic Zoology*, 213.

168 Oudemans, *Great Sea-Serpent*, 53.

169 Loren Coleman, "Otter Nonsense," June 5, 2007, CryptoMundo, http://www.cryptomundo.com/cryptozoo-news/otter-nonsense/ (accessed September 24, 2011).

170 Photograph by Tony Markle, reprinted in Loren Coleman, "Real Otter Sense," June 16, 2011, CryptoMundo, http://www.cryptomundo.com/cryptozoo-news/real-otter-sense/ (accessed September 24, 2011).

171 Alex Campbell to Ness Fishery Board, October 28, 1933, quoted in Rupert Gould, *The Loch Ness Monster* (1934; Secaucus, N.J.: Citadel Press, 1976), 110–112.

172 McTaggart-Cowan went on to serve as head of the Department of Zoology at the University of British Columbia, and then the dean of graduate studies. See "Ian McTaggart-Cowan, Part 2," Heritage Conservation Trust Foundation, http://www.hctf.ca/AboutUs/mctaggart2.html (accessed September 26, 2011).

173 Quoted in Ray Gardner, "Caddy, King of the Coast," *Maclean's*, June 15, 1950, 43.
174 Oudemans, *Great Sea-Serpent*, 345.
175 Gosse, *Romance of Natural History*, 338–340.
176 Indeed, these cases are so similar that a few have suggested the possibility of plagiarism. Gould wrote, "The accounts given by Smith and Herriman are so singularly alike that one might be pardoned for thinking that one was copied from the other" (*Case for the Sea-Serpent*, 138–139). However, he felt that this "very remarkable string of coincidences" was unlikely to be anything more sinister than coincidence. Heuvelmans was more suspicious, saying that the story Captain Smith told of his adventure aboard the *Pekin* "seems most unlikely to have happened to him at all" (*In the Wake of the Sea-Serpents*, 201–202).
177 Quoted in Oudemans, *Great Sea-Serpent*, 342.
178 Quoted in "Notes," *Nature*, October 13, 1881, 565, http://books.google.ca/books?id=TMMKAAAAYAAJ (accessed September 24, 2011).
179 Pontoppidan, *Natural History of Norway*, 201.
180 Oudemans, *Great Sea-Serpent*, 396.
181 Archie H. Wills, "All in a Lifetime" (ca. 1985), 61–62 (hand-bound, typed manuscript), Central Branch, Greater Victoria Public Library, Victoria, B.C.
182 Ibid., 62.
183 The *Victoria Daily Times* managed not to resort to layoffs, but only because the staff accepted two waves of across the board wage cuts. See ibid., 61.
184 "A Cure for the Sea Serpent," *Critic*, July–December 1885, 153.
185 Quoted in Gardner, "Caddy, King of the Coast," 24.
186 See, for example, Bruce S. Ingram, "The Loch Ness Monster Paralleled in Canada, 'Cadborosaurus,'" *Illustrated London News*, January 6, 1934, 8; "Animals: Cup & Saucer," *Time*, October 16, 1933; and Bert Stoll, "Sea Serpent Appears Off Vancouver Island," *New York Times*, February 11, 1934.
187 Archie Wills, in *Victoria Daily Times*, June 8, 1959, 12, clipping, Archie H. Wills Fonds, University of Victoria Archives, Victoria, B.C.
188 Gardner, "Caddy, King of the Coast," 43.
189 "Yachtsmen Tell of Huge Serpent Seen off Victoria," *Victoria Daily Times*, October 5, 1933, clipping, Archie H. Wills Fonds.
190 "Not Humpback Says Langley," *Victoria Daily Times*, October 18, 1933.
191 Quoted in Ingram, "Loch Ness Monster Paralleled in Canada."
192 Ibid.
193 R. G. Rhodes, "The Diplodocus" [letter to the editor], *Victoria Daily Times*, October 16, 1933.
194 Quoted in "Says Serpent Not Conger Eel," *Victoria Daily Times*, October 17, 1933.
195 Quoted in "Caddy," *Victoria Daily Times*, October 20, 1933.
196 "The 8th Wonder of the World! Starts today for three days. Adventure that Leaps Far Beyond the Bounds of Imagination! Unique! Thrilling! Startling!" proclaimed the advertisement in the *Victoria Daily Times*, May 20, 1933, 9.
197 Quoted in Gardner, "Caddy, King of the Coast," 42.

198 Wills, "All in a Lifetime," 62–63.

199 G. Clifford Carl, *Guide to Marine Life of British Columbia* (Victoria: Royal British Columbia Museum, 1963), 134.

200 Wills, "All in a Lifetime," 62.

201 Quoted in Gardner, "Caddy, King of the Coast," 24.

202 Jim McKeachie, "$200 Offered for Caddy Photo," *Victoria Daily Times*, March 31, 1951, and "Doctor Carl Enters First 'Photograph' of Caddy in Times' $200 Competition," *Victoria Daily Times*, April 16, 1951, clippings, Archie H. Wills Fonds.

203 Michael A. Woodley, Darren Naish, and Hugh P. Shanahan, "How Many Extant Pinniped Species Remain to Be Described?" *Historical Biology* 20, no. 4 (2008): 225–235.

204 John Kirk, "BCSCC Member Sights Cadborosaurus-like Animal in Fraser River," *BCSCC Quarterly*, September 2010, 6. Kirk has also reported several sustained daylight sightings of 60-foot, serpentine Ogopogo—"I have had 11 sightings" (John Kirk, e-mail to Daniel Loxton, April 21, 2012)—and to have shot several sequences of video in association with those sightings. In one dramatic report, Kirk claimed to have seen two Ogopogos traveling together "like a series of arches. I could see the other shore through the gap between the underside of the humps and the water's surface. By being able to see under the arched humps I knew I was definitely not seeing a boat wake or an unusual wave effect, but rather the enormous forms of two huge aquatic animals which were members of the Ogopogo brood" (*In the Domain of the Lake Monsters* [Toronto: Key Porter Books, 1998], 18–20; see also 44–26, 52–53, 69–74, 80).

205 As of this writing, other skeptical investigators and I have seen only those segments of the video that appeared in previews for *Alaskan Monster Hunt*. At first glance, the "creatures" in the film appear more like a boat wake than a family of Caddys, but I'll reserve judgment until I've had a chance to review the entire video in detail.

206 Owen, "Sea Serpent," 8.

207 Edward L. Bousfield and Paul H. LeBlond, "Preliminary Studies on the Biology of a Large Marine Cryptid in Coastal Waters of British Columbia," *American Zoologist* 32 (1992): 2A, and "An Account of *Cadborosaurus willsi*, New Genus, New Species, a Large Aquatic Reptile from the Pacific Coast of North America," *Amphipacifica*, suppl. 1 (1995): 3–25. Bousfield was a retired research associate at the Royal Ontario Museum, and LeBlond was a professor of oceanography at the University of British Columbia.

208 Darren Naish, "*Cadborosaurus* and the Naden Harbour Carcass: Extant Mesozoic Marine Reptiles, or Just Bad Bad Science?" September 2, 2006, Tetrapod Zoology, http://darrennaish.blogspot.com/2006/09/cadborosaurus-and-naden-harbour_02.html (accessed September 26, 2011).

209 LeBlond and Bousfield, *Cadborosaurus*, 79.

210 "I don't agree with their conclusions," wrote Darren Naish, "but I do respect the guts and determination involved in publishing these ideas" ("A Baby Sea-Serpent No More: Reinterpreting Hagelund's Juvenile *Cadborosaurus*," September 26, 2011, Tetrapod Zoology, http://blogs.scientificamerican.com/tetrapod-zoology/2011/09/26/baby-sea-serpent-no-more/ [accessed September 28, 2011]).

211 LeBlond and Bousfield, *Cadborosaurus*, 55.

212 "Baby Sea Serpent Found in Stomach of Whale," *Los Angeles Times*, July 11, 1937, 3.

213 Jim Cosgrove, e-mail to Daniel Loxton, July 11, 2002. At the time, Cosgrove, a marine biologist, was the acting manager of natural history at the museum.

214 Or is there? I have long regarded the case of the Naden Harbour carcass as permanently unsolvable, so I was very surprised when Darren Naish contacted me during the last stages of preparing this manuscript to let me know that he and his colleagues believe that they have solved the mystery (Darren Naish, e-mail to Daniel Loxton, September 28, 2011). I look forward to reading their argument in print!

215 This bit of fakelore tells of "a beautiful Indian maiden, Cadboro," whose lover was transformed into a monster. See, for example, T. W. Paterson, "Sea Serpents Might Exist," *Daily Colonist*, May 23, 1965, 10. However, Cadboro Bay was actually named for a ship that belonged to the Hudson's Bay Company around 1842: the brigantine *Cadboro*, said to have been the first European vessel to anchor in the bay. The British Columbia Geographical Names Office adds, "Note that her log book and journal (copies of which are in the Provincial Archives) confirm the spelling 'Cadboro'" (BCGNIS Query Results, http://archive.ilmb.gov.bc.ca/bcgn-bin/bcg10?name=38867 [accessed September 28, 2011]).

216 Associated Press, "Amy, *Cadborosaurus*," *San Francisco Examiner*, November 3, 1933, clipping, Archie H. Wills Fonds.

217 Bob Davis, "Bob Davis Reveals: Authentic Sea Serpent Due to Arrive Any Moment off Canada," *New York Sun*, October 31, 1940, clipping, Archie H. Wills Fonds.

218 Archie Wills to Paul LeBlond and John R. Sibert, August 16, 1970. My thanks to Paul LeBlond for sharing his correspondence with Archie Wills—essential original sources that are not otherwise part of the public record.

219 William Carpenter, "On the Influence of Suggestion in Modifying and Directing Muscular Movement, Independently of Volition," *Notices of the Meetings of the Royal Institution*, March 12, 1852, 147–153.

220 Loren Coleman and Patrick Huyghe, *The Field Guide to Lake Monsters, Sea Serpents, and Other Mystery Denizens of the Deep* (New York: Penguin / Tarcher, 2003), 26.

221 Jack Nord, letter to the editor, *Victoria Daily Times*, October 11, 1933, Archie H. Wills Fonds.

222 Robin Baird, "Elephant Seals Around Southern Vancouver Island," *Victoria Naturalist* 47, no. 2 (1990): 6–7.

223 Ray Wormald, "Caddy Has Three Heads," [*Victoria Daily Times*], April 6, 1951, clipping, Archie H. Wills Fonds.

224 Stephen Cosgrove, e-mail to Daniel Loxton, April 17, 2012.

225 *Myths and Monsters*, broadcast on CBC, October 29, 1950, script or transcript, Archie H. Wills Fonds.

226 Cameron McCormick, "A Baby Cadborosaur No More. Part 4: What Is 'Cadborosaurus'?" September 21, 2011, The Lord Geekington, http://www.thelordgeekington.com/2011/09/baby-cadborosaur-no-more-part-4-what-is.html (accessed September 28, 2011).

227 T. H. Huxley, "The Sea Serpent," *Times*, January 11, 1893, 12.

228 Loren Coleman and Patrick Huyghe, *The Field Guide to Bigfoot and Other Mystery Primates* (San Antonio, Tex.: Anomalist Books, 2006).

229 Many authors confidently credit this sighting to Hans Egede; others either are vague about whether he personally witnessed this monster or assert that it was instead his son Poul who was the witness. Oudemans, for example, credits this as "the Sea-Serpent, as seen by Hans Egede, drawn by Bing" (*Great Sea-Serpent*, fig. 19), presumably after Pontoppidan, who specifies, "Mr. Bing, one of the missionaries . . . took a drawing of it (*Natural History of Norway*). Ellis concurs: "Since Egede was known as a sober, reliable observer, this picture thus became one of the earliest illustrations of a sea monster based upon a reliable eyewitness account" (*Monsters of the Sea*, 44). Others, however, clarify the record: "Hans Egede (1686–1758) was not a witness and his account is second-hand based presumably on the recollection of his son Poul (1708–1789) which was published separately later (P. Egede, 1741)" (Charles G. M. Paxton, Erik Knatterud, and Sharon L. Hedley, "Cetaceans, Sex and Sea Serpents: An Analysis of the Egede Accounts of a "Most Dreadful Monster" Seen off the Coast of Greenland in 1734," *Archives of Natural History* 32, no. 1 [2005]: 1–9). It is in Poul Egede's book *Continuation af den Grønlandski Mission: Forfattet i form af en Journal fra Anno 1734 til 1740* (1741) that a well-known early sketch of a sea serpent (drawn, according to Paxton and his colleagues, "by Egede senior," on the basis that the map is signed "H. E.") is found. The authorship of the map remains uncertain, though, according to Charles Paxton, e-mail to Daniel Loxton, April 22, 2012.

230 Pontoppidan, *Natural History of Norway*, 199.

231 A young gentleman of the Customs, in *Morning Chronicle and London Advertiser*, August 24, 1786, 3.

232 *Morning Post and Daily Advertiser*, September 9, 1786, 3.

233 Charles Gould, *Mythical Monsters* (London: Allen, 1886), 263.

234 Heuvelmans, *In the Wake of the Sea-Serpents*, 572–573.

235 Coleman and Huyghe, *Field Guide to Lake Monsters*, 73.

236 Heuvelmans, *In the Wake of the Sea-Serpents*, 277.

237 Ibid., 284.

238 Charles Paxton, "The Monster Manual," Fortean Times, no. 256, July 2010, www.forteantimes.com/features/commentary/4003/the_monster_manual.html (accessed February 28, 2011).

239 A meme is a small unit of "idea"—a piece of culture, like a catchy tune or the notion of teacup—that can be passed along from one host (a mind, a book, or any other place where an idea can survive and replicate) to another. Just as genes can be connected into successful organisms, so small ideas (memes) can be connected into successful collaborations. A classic example is the collaboration of the idea of "god" with the idea "tell people." These two ideas mutually reinforce, spreading farther and more quickly together than apart. It is an idea that is most usefully considered as a metaphor. Richard Dawkins introduced the concept of the meme in his book *The Selfish Gene* (New York: Oxford University Press, 1976), although others had proposed similar ideas in the past.

240 Owen, "Sea Serpent," 8.

241 Archie Wills, "Cynics," *Victoria Daily Times*, October 11, 1933.
242 Edward Cadogan, letter to the editor, *Times*, December 20, 1933, 8.
243 Dennis Loxton to Daniel Loxton, April 8, 2005.

6. MOKELE MBEMBE

1 "The Last Dinosaur" [season 3, episode 52], *MonsterQuest*, History Channel, June 24, 2009, "MonsterQuest: The Last Dinosaur, Pt. 1," YouTube, http://www.youtube.com/watch?v=7CeHwgMtvXQ (accessed November 2, 2011).
2 Martin G. Lockley, *A Guide to Dinosaur Tracksites of the Colorado Plateau and the American Southwest* (Denver: University of Colorado, Department of Geology, 1986), and *Tracking Dinosaurs: A New Look at an Ancient World* (Cambridge: Cambridge University Press, 1991); Martin G. Lockley and David G. Gillette, *Dinosaur Tracks and Traces* (Cambridge: Cambridge University Press, 1989); Martin G. Lockley and Adrian P. Hunt, *Dinosaur Tracks and Other Fossil Footprints of the Western United States* (New York: Columbia University Press, 1995); Martin G. Lockley, Barbara J. Fillmore, and Lori Marquardt, *Dinosaur Lake: The Story of the Purgatoire Valley Dinosaur Tracksite Area* (Denver: Colorado Geological Survey, 1997); Martin G. Lockley and Christian Meyer, *Dinosaur Tracks and Other Fossil Footprints of Europe* (New York: Columbia University Press, 2000).
3 William J. Gibbons, *Mokele-Mbembe: Mystery Beast of the Congo Basin* (Landisville, Pa.: Coachwhip, 2010), 133.
4 "Last Dinosaur, Pt. 1."
5 Gibbons, *Mokele-Mbembe*, 162.
6 For example, Rory Nugent describes this conversation with the Boha village witch doctor (near Lake Tele, in the heart of alleged Mokele Mbembe habitat): "My talk of a living dinosaur doesn't interest him. He believes Mokele-Mbembe is a powerful deity that constantly changes appearance, varying by divine whim and human perception. People have come to him with wildly differing descriptions of Mokele-Mbembe, and he believes them all, sure that no one would risk their own well-being by lying about the gods." Nugent further quotes the man as saying, "Sometimes people say Mokele-Mbembe is small, like a goat, and sometimes they say he is bigger than the tallest tree" (*Drums Along the Congo: On the Trail of Mokele-Mbembe, the Last Living Dinosaur* [Boston: Houghton Mifflin, 1993], 163–164).
7 This idea seems to have originated from James Powell's visit in 1976 to Cameroon, where local informants told him that *hippos* (not dinosaurs) used riverside caves "for dens and hibernated in them (but did not wall themselves up) during the great rains (not the dry season, as in current Mokele Mbembe mythology)." Hearing this, Powell arbitrarily decreed, "While nsok-nyen normally refers to the hippopotamus, it would probably also be applied to any large aquatic animal reminiscent of an elephant"—an inexcusable bit of cultural license, given that his "informants' response to the picture of a long-necked dinosaur was that they were unfamiliar with the animal, had never seen anything like it, and were quite sure it did not occur in the area" ("On the Trail of the Mokele-Mbembe: A Zoological Mystery," *Explorer's Journal*, June 1981, 86–87).

8 His Web site discusses how his planned "discoveries" of cryptids will undermine evolution. See William J. Gibbons, "Welcome to Creation Generation.Com . . . We Are Still Under Construction," Creation Generation: Genesis 1:1," http://www.creationgeneration.com (accessed November 2, 2011).

9 Robert Mullin's author biography at Crimson Moon Press (the home of his Christian allegorical science-fiction novel) describes him as a "seven-year editor with Creation Research Society" (Crimson Moon Press: Where Apologetic and Fiction Meet, http://crimsonmoonpress.com/Authors.html [accessed May 5, 2012]).

10 Gibbons, *Mokele-Mbembe*.

11 "Inside Story: Mullin on Mokele-Mbembe," June 28, 2009, CryptoMundo, http://www.cryptomundo.com/cryptozoo-news/mullin-mm/ (accessed May 5, 2012).

12 Gibbons, *Mokele-Mbembe*, 205.

13 Abbé Lievain Bonaventure Proyart, *History of Loango, Kakongo, and Other Kingdoms in Africa* (1776), in *A General Collection of the Best and Most Interesting Voyages and Travels in All Parts of the World*, trans. and ed. John Pinkerton (London: Longman, Hurst, Rees, Orme, and Brown, 1814), 16:557.

14 For example, Roy Mackal cited "a strange animal" painted on a rock in Kirumi Isumbirira, Iramba, Tanzania, that previous scholars had identified as mythical or fabulous. Mackal offered this as "cave rock art that could represent an unknown, long-necked, long-tailed, four-legged animal. This evidence should not be taken as more than suggestive because positive identification or conclusions certainly cannot be drawn. But the evidence cannot be dismissed" (*A Living Dinosaur? In Search of Mokele-Mbembe* [Leiden: Brill, 1987], 9–10). Perhaps—but this Tanzanian rock art also cannot be linked in any way to the Mokele Mbembe of Lake Tele (1,200 miles away), except by wishful thinking. African rock-art specialist Fidelis Taliwawa Masao warns that the meaning and interpretation of Central Tanzanian rock painting—including that cited by Mackal, which Masao labeled "Semi naturalistic paintings of animals"—is extremely conjectural. While Mackal claimed that the figure "definitely is not a giraffe," he had neither the training nor the grounds to assert this. Certainly, it *looks* more like a giraffe than a sauropod, and as Masao pointed out, "the giraffe is the most common single [animal] species represented" in rock art in this region ("Possible Meaning of the Rock Art of Central Tanzania," *Paideuma* 36, Afrika-Studien II [1990]: 189–199).

15 See, for example, Willy Ley, *Willy Ley's Exotic Zoology* (1959; New York: Bonanza Books, 1987), 66–74.

16 Surviving ichthyosaurs were advanced as an explanation for sightings of sea serpents in Robert Bakewell, *Introduction to Geology* (New Haven, Conn.: Howe, 1833), 213. This scenario was brought to life in Jules Verne, *Journey to the Center of the Earth*, trans. Willis T. Bradley (1864; New York: Ace Books, 1956), 187–188. For discussion of the influence of nineteenth-century fossil plesiosaur discoveries on cryptozoology, see chapters 4 and 5.

17 "Big Thunder Saurian Viewed and Approved," *New York Times*, February 17, 1905, 9.

18 "Old and Young Call to See the Dinosaur," *New York Times*, February 20, 1905, 12.

19 R. W. W., "Mr. Carnegie's Imitation Dinosaur 'Makes a Hit' in England," *New York Times*, June 4, 1905, X7.

20 "Diplodocus Dinner in Berlin," *New York Times*, May 14, 1908, 4.

21 "Hunting the Dinosaurus," *Chicago Daily Tribune*, July 17, 1910, G7.

22 R. Jay Gangewere, "Dippy's Anniversary," *Carnegie Magazine*, http://www.carnegiemuseums.org/cmag/bk_issue/1999/julaug/feat1.html (accessed April 25, 2012).

23 "The Dinosaurs of East Africa," *Bulletin of the American Geographical Society* 45, no. 3 (1913): 193–196. The taxonomic history of Eberhard Frass's sauropod fossils is quite complicated, but note that the name he selected, *Gigantosaurus*, is no longer recognized.

24 Carl Hagenbeck was the premier supplier of exotic game animals for zoological gardens, traveling menageries, and circuses. During his career, he imported more than 1,000 lions, over 1,000 bears, 300 elephants, 150 giraffes—and countless other animals besides. Hagenbeck developed lower-stress animal-training techniques and in 1876 patented a bar-less approach to constructing zoo enclosures. This became the design theme for the enormously popular private zoo that Hagenbeck opened on May 7, 1907—and the same principles influenced the construction of zoos throughout the world. See Herman Reichenbach, "Carl Hagenbeck's Tierpark and Modern Zoological Gardens," *Journal of the Society for the Bibliography of Natural History* 9, no. 4 (1980): 574, 577–582.

25 Carl Hagenbeck, *Beasts and Men: Being Carl Hagenbeck's Experiences for Half a Century Among Wild Animals*, trans. Hugh S. R. Elliot and A. G. Thacker (New York: Longmans, Green, 1912), 95–97.

26 Hagenbeck was also the first to import the African black rhinoceros (*Diceros bicornis*), the Sumatran rhinoceros (*Dicerorhinus sumatrensis*), the southern elephant seal (*Mirounga leonine*), and the leopard seal (*Hydrurga leptonyx*), among others. See Reichenbach, "Carl Hagenbeck's Tierpark and Modern Zoological Gardens," 580–581.

27 "A Memorable Book on the Traits of Wild Beasts," *New York Tribune*, January 29, 1910, 8; "'Sea Serpent' of the Forest," *Times of India* (Mumbai [Bombay]), January 6, 1910, 8.

28 "Brontosaurus Still Lives," *Washington Post*, January 23, 2010, M1.

29 "Prehistoric Cement Zoo," *Washington Post*, December 26, 1910, 6.

30 Both terms actually appeared in the early Western press coverage of the alleged African *Brontosaurus*. For example, Hagenbeck himself asserted that the region where the dinosaur dwelled was "infested with bloodthirsty savages," while one headline asked, "Are Scions of the Prehistoric Monsters That Once Roamed the Earth Still to Be Found Within Little-Known Parts of the Dark Continent?" ("Searching for Saurians in African Jungle," *Washington Post*, February 19, 1928, SM9). Of course, it is unfair to judge people of the past by modern standards. Hagenbeck was speaking of literally violent encounters, and even today the suspicious-sounding phrase "dark continent" is sometimes used without any intention to refer to race. For a recent discussion of the controversial term, see Ombudsman, "Should NPR Have Apologized for 'Dark Continent'?" February 27, 2008, NPR Omdudsman, http://www.npr.org/blogs/ombudsman/2008/02/should_npr_have_apologized_for.html (accessed April 27, 2012).

31 "British East Africa," *Uganda Herald* (Kampala), February 14, 1913, 12.

32 "Startling Rumour," *Bulawayo [Buluwayo, Zimbabwe] Chronicle*, December 3, 1909, 3.

33 "A Fearsome Beast," *Rhodesia [Zimbabwe] Herald* (Harare), December 17, 1909, 9.

34 "The 'Brontosaurus': More Hearsay," *Bulawayo Chronicle*, January 14, 1910, 7.

35 "Across Africa by Motor Boat," *Nyasaland Times* (Blantyre, Malawi), May 11, 1911, 5.

36 Untitled article, *East African Standard* (Nairobi, Kenya), July 15, 1911, 13.

37 Quoted in Bernard Heuvelmans. *On the Track of Unknown Animals*, trans. Richard Garnett (New York: Hill and Wang, 1959), 450.

38 Ibid., 441.

39 "Hunting the Dinosaurus," *Chicago Daily Tribune*, July 17, 1910, G7. (The spelling of his surname varies by source, spelled either "Brookes" or "Brooks.")

40 Quoted in "Sure the Diplodocus Is Not Yet Extinct," *New York Times*, October 2, 1910.

41 May Bosman, "Does Prehistoric Monster Still Live?" *Atlanta Constitution*, May 22, 1921, G8. This seems to have been a projection of the then-current African *Brontosaurus* idea on preexisting South African regional lore about the Groot Slang, a giant snake that was itself an expression of an effectively global folkloric "big snake" monster template projected since classical antiquity on regions from Libya to India.

42 Quoted and discussed in Ley, *Willy Ley's Exotic Zoology*, 71–72.

43 Heuvelmans, *On the Track of Unknown Animals*, 461.

44 Quoted in "The Congo Monster. Story Ridiculed," *Beira [Mozambique] News*, December 16, 1919.

45 "Startling Rumour."

46 "A Tale from Africa: *Semper Aliquid Novi*," *Times* (London), November 17, 1919, 11.

47 "Much Interest Aroused," *Rhodesia Herald*, November 25, 1919, 5.

48 "Dragon of the Prime," *Times*, December 12, 1919, 13.

49 "No Brontosaurus There: 'Wholly Fabulous Report,'" *Beira News*, December 23, 1919, 3.

50 Quoted in "Missionary's Leg Pulled," *Rhodesia Herald*, December 23, 1919, 5.

51 "The Brontosaurus Myth," *Dar-es-Salaam [Tanzania] Times*, March 19, 1921, 6.

52 Wentworth D. Gray, "The Brontosaurus" [letter to the editor], *Times*, February 23, 1920, 10.

53 "Game Hunter's Encounter," *Rhodesia Herald*, December 23, 1919, 23.

54 I mention Lester Stevens and his dog partly because Heuvelmans chose to feature his hunt for the "giant reptile" as the chapter opener on the topic in his book, and partly because the expedition was both widely reported and silly. Heuvelmans gave his name as "Captain Leicester Stevens," not "Lester Stevens" (*On the Track of Unknown Animals*, 434). See also "Experts Who Believe in It," *Bulawayo Chronicle*, December 26, 1919, 5.

55 Ley, *Willy Ley's Exotic Zoology*, 62–74; Ivan T. Sanderson, "There Could Be Dinosaurs," *Saturday Evening Post*, January 3, 1948, 17, 53, 56; Heuvelmans, *On the Track of Unknown Animals*, 434–484.

56 Powell, "On the Trail of the Mokele-Mbembe."

57 Mackal, *Living Dinosaur*.

58 Hans Schomburgk, quoted in Ley, *Willy Ley's Exotic Zoology*, 69. Schomburgk was in a position to make an informed complaint about the difficulty of ethnozoological investigation, having been the first Westerner to confirm the existence and describe for science the pygmy hippopotamus.

59 Powell does pose a question: "Was Michel sincere, or had his African creativity been fired by the publicity he had received in Heuvelmans' book? I shall probably never know" ("On the Trail of the Mokele-Mbembe, 87–88). However, he seems to have been

very impressed with his informant's sincerity, and gives no account of close questioning about the complete reversal of the man's story.

60 "The Last Dinosaur" [season 3, episode 52], *MonsterQuest*, History Channel, June 24, 2009, "MonsterQuest: The Last Dinosaur, Pt. 2," YouTube, http://www.youtube.com/watch?v–jfYqA4noqow (accessed May 4, 2012).

61 Gibbons, *Mokele-Mbembe*, 73, 83.

62 Tokuhara Takabayashi, "The First Japanese-Congolese Mokele-Mbembe Expeditions," *Cryptozoology: Interdisciplinary Journal of the International Society of Cryptozoology* 7 (1988): 67.

63 Gibbons, *Mokele-Mbembe*, 94.

64 For example, Congolese biologist Marcellin Agnagna led several tours to Boha village in the Congo, and Pierre Sima led several tours to Langoue village in Cameroon. Whatever the significance, both Agnagna and Sima claimed personal Mokele Mbembe sightings. See Gibbons, *Mokele-Mbembe*, 57–58, 170–171.

65 Mackal, *Living Dinosaur?* 59.

66 Mackal gives the name as "Antoine Meombe" in *Living Dinosaur?* 61; Powell gives it as "Miobe Antoine" in "On the Trail of the Mokele-Mbembe," 84–91.

67 Mackal, *Living Dinosaur?* 73.

68 Ibid., 74.

69 Powell, "On the Trail of the Mokele-Mbembe."

70 Ibid.

71 Ibid.

72 Mackal, *Living Dinosaur?* 160.

73 Ibid., 103.

74 Gibbons, *Mokele-Mbembe*, 74.

75 Redmond O'Hanlon, *No Mercy: A Journey into the Heart of the Congo* (New York: Vintage Books, 1998), 375.

76 John H. Acorn, "Good-Humored Adventure in the Congo," *Skeptical Inquirer* 19, no. 3 (1995): 46–48.

77 William Gibbons, "Was a Mokele-Mbembe Killed at Lake Tele?" The Anomalist Archive: High Strangeness Reports, http://www.anomalist.com/reports/mokele.html (accessed May 2, 2012).

78 Gibbons, *Mokele-Mbembe*, 109.

79 Mackal, *Living Dinosaur?* 93–95.

80 Quoted in Dembart Lee, "Proof of Dinosaur Sighting May Hinge on What Develops," *Los Angeles Times*, December 23, 1981, A16.

81 "Developed Film Fails to Show Any Dinosaur," *Los Angeles Times*, December 23, 1981, A3.

82 Gibbons, *Mokele-Mbembe*, 81.

83 Gibbons writes that Agnagna threatened "to have us arrested as spies," and implies that Agnagna may have stolen the expedition's undeveloped film (ibid., 81, 83).

84 Takabayashi, "First Japanese-Congolese Mokele-Mbembe Expeditions," 66–69. (This article also describes a previous visit by Takabayashi and two colleagues to speak with

missionary Eugene Thomas in 1986, and another Japanese expedition in 1987 that included Agnagna.)

85 O'Hanlon, *No Mercy*.

86 Nigel Burton, "Real Life Dinosaur Hunt," *Northern Echo* (Darlington, England), June 5, 1999, 7.

87 John Kirk, *In the Domain of the Lake Monsters* (Toronto: Key Porter Books, 1998), 258.

88 Gibbons, *Mokele-Mbembe*, 86.

89 Ibid., 85–100.

90 Ibid., 96.

91 "Megatransect Across Africa," *National Geographic*, http://ngm.nationalgeographic.com/ngm/data/2001/08/01/sights_n_sounds/media.5.2.html (accessed May 5, 2012).

92 Gibbons, *Mokele-Mbembe*, 112–125.

93 The search benefited from $15,000 from the British Broadcasting Corporation, and $50,000 from Canadian wealth manager and Young Earth creationist Paul Rockel. See ibid., 127–128. For a statement by Rockel on evolution, see "Where Are the Facts?" [letter to the editor], *Waterloo Region Record* (Ontario, Canada), January 10, 2011, http://www.therecord.com/opinion/letters/article/318832--where-are-the-facts (accessed May 2, 2012).

94 Gibbons, *Mokele-Mbembe*, 139.

95 Ibid., 146.

96 On an episode of the podcast from Divine Intervention Radio, Milt Marcy is introduced as a creationist with an active involvement in the notorious Paluxy River "man tracks" (a mixed set of misidentified dinosaur footprints, inorganic indentations, and outright fakes that are found alongside dinosaur trackways, interpreted by a subset of "creation scientists" as proof that human beings lived at the same time as non-avian dinosaurs). See "Living Dinosaurs? Cryptozoologists and Dinosaur Hunters, Dr. William Gibbons and Milt Marcy," Divine Intervention, podcast, March 28, 2008, http://divineintervention.typepad.com/divine_intervention/2008/03/episode-6-livin.html. (accessed May 5, 2012). The story of Marcy's offer to fund Gibbons is in Gibbons, *Mokele-Mbembe*, 151–152.

97 Gibbons, *Mokele-Mbembe*, 158–159.

98 Ibid., 165.

99 Ibid., 168–171.

100 Peter Beach is described as "a Biblical creationist" who promotes the idea that not only sauropods but also pterosaurs have survived to this day. See Nathaniel Coleman, "A Brave Biologist: Living Dinosaurs and Pterosaurs," http://www.livepterosaur.com/brave-biologist/ (accessed May 4, 2012). As a creationist speaker, Brian Sass bills himself as "B.SC. Paleobiology. Creation Generation & Genesis Park Research Teams." See "Dinosaur Hunter to Speak in Pangman [Saskatchewan] This Sunday," http://www.pangman.ca/wp-content/uploads/2010/03/Brian-Sass-News-Re-lease.pdf (accessed May 4, 2012).

101 Gibbons, *Mokele-Mbembe*, 180–183.

102 "Inside Story: Mullin on Mokele-Mbembe"; Gibbons, *Mokele-Mbembe*, 187–189.

103 Neil Mandt and Michael Mandt, "Wild Man & Swamp Dinosaur" [season 2, episode 3], *Destination Truth*, Syfy, March 19, 2008 (Ping Pong Productions).
104 "Last Dinosaur, Pt. 2."
105 Gibbons, *Mokele-Mbembe*, 197.
106 "Inside Story: Mullin on Mokele-Mbembe."
107 Gibbons, *Mokele-Mbembe*, 187–188.
108 Peter Dodson, Anna K. Behrensmeyer, Robert T. Bakker, and John S. McIntosh, "Taphonomy and Paleoecology of the Dinosaur Beds of the Jurassic Morrison Formation," *Paleobiology* 6 (1980): 208–232.
109 Bruce H. Tiffney, "Land Plants as Food and Habitat in the Age of Dinosaurs," in *The Complete Dinosaur*, 2nd ed., ed. M. K. Brett-Surman, Thomas R. Holtz Jr., and James O. Farlow (Bloomington: Indiana University Press, 2012), 569–588.
110 Paul C. Sereno, Allison L. Beck, Didier B. Dutheil, Hans C. E. Larsson, Gabrielle H. Lyon, Bourahima Moussa, Rudyard W. Sadlier, Christian A. Sidor, David J. Varricchio, Gregory P. Wilson, and Jeffrey A. Wilson, "Cretaceous Sauropods from the Sahara and the Uneven Rate of Skeletal Evolution Among Dinosaurs," *Science* 286 (1999): 1342–1347.
111 Donald R. Prothero, *After the Dinosaurs: The Age of Mammals* (Bloomington: Indiana University Press, 2006); Lars Werdelin and William Joseph Sanders, eds., *Cenozoic Mammals of Africa* (Berkeley: University of California Press, 2010).
112 Mackal, *Living Dinosaur?* 152.
113 "Living Dinosaurs? Cryptozoologists and Dinosaur Hunters."
114 Gibbons, *Mokele-Mbembe*, 63–66.
115 Gibbons, "Welcome to Creation Generation.Com"; "Mokele-mbembe: The Living Dinosaur!" Cryptozoological Realms, http://www.mokelembembe.com; Cryptozoology Research Team: Tracking Down the World's Last Natural Mysteries, http://www.livingdinos.com; William J. Gibbons, "In Search of the Congo Dinosaur," Institute for Creation Research, http://www.icr.org/article/306/ (all accessed November 3, 2011).
116 William Gibbons and Kent Hovind, *Claws, Jaws and Dinosaurs* (Pensacola, Fla.: CSE, 1999), 3.
117 Ibid., 34.
118 Donald R. Prothero, *Evolution: What the Fossils Say and Why It Matters* (New York: Columbia University Press, 2007).
119 Gibbons, "Welcome to Creation Generation.Com."
120 N. F. Goldsmith and I. Yanai-Inbar, "Coelacanthid in Israel's Early Miocene? *Latimeria* Tests Schaeffer's Theory," *Journal of Vertebrate Paleontology* 17, supplement 3 (1997): 49A; T. Ørvig, "A Vertebrate Bone from the Swedish Paleocene," *Geologiska Föreningens i Stockholm Förhandlingar* 108 (1986): 139–141.
121 Prothero, *Evolution*, 50–85.
122 Gibbons, "Welcome to Creation Generation.Com."
123 Prothero, *Evolution*, xvii.
124 Ibid., 24–49.

7. WHY *DO* PEOPLE BELIEVE IN MONSTERS?

1 Elizabeth Landau, "Why Do We Need to Look for Bigfoot?" CNN Health, June, 21, 2010, http://articles.cnn.com/2010-06-21/health/bigfoot.psychology.monsters_1_bigfoot-goat-sucker-monsters?_s=PM:HEALTH (accessed November 25, 2011).

2 Baylor Religion Survey, 2005, Q76F by Q74K [custom table, showing comparison], http://www.thearda.com/includes/crosstabs.asp?file=BRS2005&v=390&v2=377&p=2&s=on (accessed February 16, 2012).

3 Christopher D. Bader, F. Carson Mencken, and Joseph O. Baker, *Paranormal America: Ghost Encounters, UFO Sightings, Bigfoot Hunts, and Other Curiosities in Religion and Culture* (New York: New York University Press, 2011), 106–111.

4 Joshua Blu Buhs, *Bigfoot: The Life and Times of a Legend* (Chicago: University of Chicago Press, 2009), 174–175.

5 Karen Stollznow, e-mail to Daniel Loxton, February 16, 2012; Blake Smith, e-mail to Daniel Loxton, February 16, 2012.

6 Buhs, *Bigfoot*, 172.

7 Quoted in ibid.

8 Quoted in ibid., 177.

9 Ibid., 177–178.

10 Brian Regal, *Searching for Sasquatch: Crackpots, Eggheads, and Cryptozoology* (New York: Palgrave Macmillan, 2011).

11 Gerald Durrell, introduction to Bernard Heuvelmans, *On the Track of Unknown Animals*, trans. Richard Garnett (New York: Hill and Wang, 1959), 20.

12 Regal, *Searching for Sasquatch*.

13 Samuel Kneeland, in *Proceedings of the Boston Society of Natural History*, 1874, 338; Willy Ley, *Willy Ley's Exotic Zoology* (1959; New York: Bonanza Books, 1987).

14 Regal, *Searching for Sasquatch*, 71–74.

15 See, for example, John Napier, *Bigfoot: The Yeti and Sasquatch in Myth and Reality* (New York: Dutton, 1973), 105–107.

16 Ibid., 90–91; Don Hunter, with René Dahinden, *Sasquatch* (New York: Signet, 1975), 119–122.

17 Hunter and Dahinden, *Sasquatch*, 134.

18 Quoted in Regal, *Searching for Sasquatch*, 115–117.

19 Napier, *Bigfoot*, 107.

20 Quoted in Linda Witt, "Monster Hunter Roy Mackal Treks to Africa in Search of the Last of the Dinosaurs," *People*, December 7, 1981, http://www.people.com/people/archive/article/0,,20080868,00.html (accessed November 25, 2011).

21 Roy P. Mackal, *A Living Dinosaur? In Search of Mokele-Mbembe* (Leiden: Brill, 1987), 18.

22 Roy P. Mackal, *The Monsters of Loch Ness* (Chicago: Swallow Press, 1976).

23 Mackal, *Living Dinosaur?* 18.

24 Bernard Heuvelmans, review of *The Monsters of Loch Ness*, by Roy P. Mackal, *Skeptical Inquirer* 2, no. 1 (1977): 110–120.

25 Darren Naish, "Aquatic Proto-People and the ~~Theory~~ Hypothesis of Initial Bipedalism," March 17, 2008, Tetrapod Zoology, http://scienceblogs.com/tetrapodzoology/2008/03/initial-bipedalism/ (accessed November 25, 2011).

26 Ivan T. Sanderson, "There Could Be Dinosaurs," *Saturday Evening Post*, January 3, 1948.

27 Bernard Heuvelmans, *Sur la piste des bêtes ignorées* (Paris: Librarie Plon, 1955); *On the Track of Unknown Animals*.

28 See, for example, Ivan T. Sanderson, *Abominable Snowmen: Legend Come to Life: The Story of Sub-Humans on Five Continents from the Early Ice Age Until Today* (Philadelphia: Chilton, 1961); *Uninvited Visitors: A Biologist Looks at UFOs* (New York: Cowles, 1967); *Invisible Residents: A Disquisition upon Certain Matters Maritime, and the Possibility of Intelligent Life Under the Waters of This Earth* (New York: World, 1970); "The Twelve Devil's Graveyards Around the World," *Saga*, 1972; and *"Things" and More "Things": Myths, Mysteries, and Marvels!* (Kempton, Ill.: Adventures Unlimited Press, 2007).

29 Regal, *Searching for Sasquatch*.

30 Buhs, *Bigfoot*, 227–228.

31 Ibid.; Regal, *Searching for Sasquatch*.

32 Krantz was an enthusiastic early proponent of "dermatoglyphic" evidence for Bigfoot—the claim that some plaster casts of alleged Bigfoot footprints accurately record fine, fingerprint-like "dermal ridges" and even sweat pores. "When I first realized the potential significance of dermal ridges showing in sasquatch footprints," Krantz wrote, "it seemed that scientific acceptance of the existence of this species might be achieved without having to bring in a specimen of the animal itself. It was this hope that drove me to expend so much of my resources on it, and of my scientific reputation as well" (*Bigfoot Sasquatch Evidence* [Surrey, B.C.: Hancock House, 1999], 86). Dermal-ridge evidence still enjoys support from proponents of Bigfoot, including Jeff Meldrum and Texas forensics investigator Jimmy Chilcutt. However, skeptical investigator Matt Crowley has shown the argument for dermal ridges to be critically flawed, demonstrating through a series of experiments that ridged, fingerprint-like textures exactly replicating the patterns promoted by Bigfoot proponents are a predictable and common artifact of the footprint-casting process itself. Crowley explains that these ridges form spontaneously and frequently in "test casts in multiple substrates, with multiple casting cements, and at multiple slurry temperatures and thicknesses." So crushing is Crowley's analysis that we did not bother to discuss dermal-ridge evidence further in this book. Those who wish to learn more are referred to the dozens of relevant articles on Crowley's Web site, including "Dermal Ridges and Casting Artifacts," October 19, 2009, Orgone Research, http://orgoneresearch.com/2009/10/19/dermal-ridges-and-casting-artifacts/; "Ridges and Furrows," October 19, 2009, Orgone Research, http://orgoneresearch.com/2009/10/19/ridges-and-furrows-2/; "Arched Furrows," October 19, 2009, Orgone Research, http://orgoneresearch.com/2009/10/19/arched-furrows/; "Ridge Flow Pattern," October 19, 2009, Orgone Research, http://orgoneresearch.com/2009/10/19/ridge-flow-pattern/; and "Conclusion," October 19, 2009, Orgone Research, http://orgoneresearch.com/2009/10/19/conclusion/ (all accessed April 2, 2013).

33 Regal, *Searching for Sasquatch.*

34 Paul McCarthy, "Cryptozoologists: An Endangered Species," January 11, 1993, Bigfoot Encounters, http://www.bigfootencounters.com/articles/cryptozoologists.htm (accessed November 25, 2011).

35 CryptoMundo, http://www.cryptomundo.com/ (accessed November 25, 2011).

36 Buhs, *Bigfoot*; Regal, *Searching for Sasquatch.*

37 Both Mayor and Naish are on the peer-review panel of a recently announced new cryptozoology journal to be published by the Center for Fortean Zoology. See Karl Shuker, "Welcome to the *Journal of Cryptozoology*—A New, Peer-Reviewed Scientific Journal Devoted to Mystery Animals," February 27, 2012, ShukerNature, http://karlshuker.blogspot.com/2012/02/welcome-to-journal-of-cryptozoology-new.html (accessed February 27, 2012).

38 David Orr, "*Scaphognathus crassirostris*: A Pterosaur in the Historical Record?" June 14, 2011, Love in the Time of Chasmosaurs, http://chasmosaurs.blogspot.com/2011/06/scaphognathus-crassirostris-pterosaur.html (accessed November 25, 2011).

39 Cryptozoology Biographies, Cryptozoological Realms, http://www.cryptozoology.net/english/biographies/bios_a.html (accessed November 25, 2011).

40 William J. Gibbons, *Mokele-Mbembe: Mystery Beast of the Congo Basin* (Landisville, Pa.: Coachwhip, 2010), 9.

41 Niles Eldredge and Steven M. Stanley, eds., *Living Fossils* (Berlin: Springer, 1984).

42 Donald R. Prothero, *Evolution: What the Fossils Say and Why It Matters* (New York: Columbia University Press, 2007), 24–49.

43 According to Patriot Bible University's "Accreditation" page, the school is "*authorized by the State of Colorado Commission for Higher Education* to grant *religious degrees*. . . . As we do not exist to train secular career professionals, grant secular degrees, or educate those who see ministry as a profession . . . we have chosen not to spend our student's [*sic*] money on secular regional accreditation that costs tens of thousands of dollars—annually [emphasis in original]," Patriot Bible University, http://www.patriotuniversity.org/index.php?mod=Articles&menuid=67 (accessed February 26, 2012). They further specify,

> 1. Patriot Bible University is not eligible to participate in the Federal Student Loan/Financial Aid program.
> 2. Patriot Bible University is not authorized to accept the GI Bill.
> 3. Patriot Bible University is unable to guarantee acceptance of its degrees in other postsecondary institutions. . . .
> 4. Corporations are not required to recognize degrees from Patriot Bible University. ("Who Accredits Patriot Bible University?" Patriot Bible University, http://www.patriotuniversity.com/Secure/infoAccreditation.htm [accessed February 26, 2012])

44 Karen Bartelt, "The Dissertation Kent Hovind Doesn't Want You to Read," http://www.noanswersingenesis.org.au/bartelt_dissertation_on_hovind_thesis.htm (accessed January 26, 2012).

45 Michael Stewart, "A Decade for 'Dr. Dino,'" *Pensacola* [*Fla.*] *News Journal*, January 20, 2007, A1.

46 Bader, Mencken, and Baker, *Paranormal America*, 24, 218nn.9–11.

47 Susan A. Clancy, *Abducted: How People Come to Believe They Were Kidnapped by Aliens* (Cambridge, Mass.: Harvard University Press, 2005).

48 David W. Moore, "Three in Four Americans Believe in Paranormal," June 16, 2005, Gallup News Service, http://www.gallup.com/poll/16915/Three-Four-Americans-Believe-Paranormal.aspx (accessed March 1, 2012).

49 Bader, Mencken, and Baker, *Paranormal America*, 129.

50 Ibid., 94, fig. 4.3.

51 Ibid., 147–148, figs. 6.2 and 6.3.

52 Ibid., 56–57.

53 Quoted in David Briggs, "Paranormal Is the New Normal in America," January 31, 2011, Association of Religion Data Archives, http://thearda.com/trend/featured/paranormal-is-the-new-normal-in-america/ (accessed January 26, 2012).

54 Bader, Mencken, and Baker, *Paranormal America*, 147–148, figs. 6.2 and 6.3.

55 David Briggs, "Paranormal Is the New Normal in America."

56 Bader, Mencken, and Baker, *Paranormal America*, 155–156.

57 Quoted in John W. Morehead, "Paranormal America," January 19, 2011, *Sacred Tribes Journal*, http://www.sacredtribesjournal.org/stj/index.php?option=com_content&view=article&id=67:paranormal-america&catid=1:latest-news (accessed January 26, 2012).

58 Bader, Mencken, and Baker, *Paranormal America*, 141–156.

59 Ibid., 97–99, figure 4.6.

60 Ibid., 120.

61 Ibid., 130; Moore, "Three in Four Americans Believe in Paranormal."

62 Christopher D. Bader, "Supernatural Support Groups: Who Are the UFO Abductees and Ritual-Abuse Survivors?" *Journal for the Scientific Study of Religion* 42, no. 4 (2003): 669–678.

63 Bader, Mencken, and Baker, *Paranormal America*, 69–70.

64 Clancy, *Abducted*, 129.

65 Ibid., 129–136.

66 Question 76F: Have you ever read a book, consulted a Web site, or researched the following topics: Mysterious animals, such as Bigfoot or the Loch Ness Monster? Baylor Religion Survey, 2005, Association of Religion Data Archives, http://www.thearda.com/Archive/Files/Analysis/BRS2005/BRS2005_VAR390_1.asp (accessed March 1, 2012).

67 Matthew J. Sharps, Justin Matthews, and Janet Asten, "Cognition and Belief in Paranormal Phenomena: Gestalt/Feature-Intensive Processing Theory and Tendencies Toward ADHD, Depression, and Dissociation," *Journal of Psychology* 140, no. 6 (2006): 579–590.

68 Question 74K: To what extent do you agree or disagree with the following statements: Creatures such as Bigfoot and the Loch Ness Monster will one day be discovered by

science; Baylor Religion Survey, 2005, Association of Religion Data Archives, http://www.thearda.com/Archive/Files/Analysis/BRS2005/BRS2005_Var377_1.asp (accessed January 26, 2012).

69 "Belief in Bigfoot: In your opinion, does each of the following exist? Bigfoot (Baylor Religion Survey, Wave 2, 2007)," Association of Religion Data Archives, http://www.thearda.com/quickstats/qs_43.asp (accessed February 27, 2012).

70 "Americans More Likely to Believe in Bigfoot than Canadians," March 4, 2012, Angus Reid Public Opinion, http://www.angus-reid.com/wp-content/uploads/2012/03/2012.03.04_Myths.pdf (accessed March 4, 2012).

71 "TBRC Investigators Hear a Number of Close Knocks and Discover Fresh Tracks the Next Day," January 5, 2008, Texas Bigfoot Research Conservancy, http://www.texasbigfoot.com/reports/report/detail/455 (accessed March 1, 2012); Bader, Mencken, and Baker, *Paranormal America*, 122–126.

72 Bader, Mencken, and Baker, *Paranormal America*, 121.

73 Christopher Bader, F. Carson Mencken, and Joseph O. Baker, "Survey of Attenders: Texas Bigfoot Research Conservancy Annual Meeting 2009," Waco, Tex.: Baylor University, Department of Sociology, http://www.thearda.com/Archive/Files/Descriptions/BIGFOOT.asp (accessed March 1, 2012).

74 Bader, Mencken, and Baker, *Paranormal America*, 113–121.

75 Ibid.; "Survey of Attenders: Texas Bigfoot Research Conservancy Annual Meeting 2009: Gender," Association of Religion Data Archives, http://www.thearda.com/Archive/Files/Analysis/BIGFOOT/BIGFOOT_Var45_1.asp (accessed March 1, 2012).

76 Bader, Mencken, and Baker, *Paranormal America*, 120; "Survey of Attenders: Texas Bigfoot Research Conservancy Annual Meeting 2009: Education," Association of Religion Data Archives, http://www.thearda.com/Archive/Files/Analysis/BIGFOOT/BIGFOOT_Var105_1.asp (accessed March 1, 2012).

77 Mike Leggett, "Texas Bigfoot Conference More Boring Than You Would Think," *Austin American-Statesman*, October 4, 2009, http://www.mcclean.org/bigfoottimes/blog/2009_10_01_archives.html (accessed November 25, 2011).

78 Bader, Mencken, and Baker, *Paranormal America*, 120.

79 Chad Arment, *Cryptozoology: Science and Speculation* (Landisville, Pa.: Coachwhip, 2004), 11.

80 Loren Coleman, *Mothman and Other Curious Encounters* (New York: Paraview Press, 2002), 112.

81 Janet Bord and Colin Bord, *Bigfoot Casebook Updated: Sightings and Encounters from 1818 to 2004* (Enumclaw, Wash.: Pine Winds Press, 2006), 130–132.

82 Bader, Mencken, and Baker, *Paranormal America*, 137–138.

83 Bord and Bord, *Bigfoot Casebook Updated*, 137.

84 Ibid., 146–147.

85 Bader, Mencken, and Baker, *Paranormal America*, 138–139.

86 Bord and Bord, *Bigfoot Casebook Updated*, 136–137.

87 Michael McLeod, *Anatomy of a Beast: Obsession and Myth on the Trail of Bigfoot* (Berkeley: University of California Press, 2009), 145.

88 Jon-Erik Beckjord, "A Real Life Daily Application for Wormholes in Space-Time," February 2007, http://web.archive.org/web/20080414223831/; http://www.beckjord.com/wormholesinuse/ (accessed November 25, 2011).

89 Daniel Loxton, "New Video from Loch Ness," *e*Skeptic, June 13, 2007, Skeptic, http://www.skeptic.com/eskeptic/07-06-13/ (accessed March 3, 2012).

90 Tony Shiels, *Monstrum! A Wizard's Tale* (London: Fortean Times, 1990), 69.

91 Rikki Razdan and Alan Kielar, "Sonar and Photographic Searches for the Loch Ness Monster: A Reassessment," *Skeptical Inquirer* 9, no. 2 (1984–1985): 153–154.

92 Daniel Loxton, "An Argument that Should Never Be Made Again," February 2, 2010, Skepticblog, http://www.skepticblog.org/2010/02/02/an-argument-that-should-never-be-made-again/ (accessed March 3, 2012).

93 Michel Meurger, with Claude Gagnon, *Lake Monster Traditions: A Cross-Cultural Analysis* (London: Fortean Times, 1988), 23–24.

94 British Columbia Scientific Cryptozoology Club, "About," http://bcscc.ca/blog/About (accessed March 3, 2012).

95 Aaron M. Bauer and Anthony P. Russell, "A Living Plesiosaur? A Critical Assessment of the Description of *Cadborosaurus willsi*," *Cryptozoology* 12 (1996): 1–18.

96 John Kirk, "What on Earth Is Going On in the World of Cryptozoology?" *British Columbia Scientific Cryptozoology Club Quarterly*, January 2012, 1.

97 Benjamin Radford and Joe Nickell, *Lake Monster Mysteries: Investigating the World's Most Elusive Creatures* (Lexington: University Press of Kentucky, 2006), 50.

98 Richard J. Greenwell, "The Big Black Cat of Puerto Penasco," *Cryptozoology* 13 (1998): 38–46.

99 Richard J. Greenwell, "A Classificatory System for Cryptozoology," *Cryptozoology* 4 (1985): 1–14.

100 Elizabeth F. Loftus, *Memory: Surprising New Insights into How We Remember and Why We Forget* (New York: Addison-Wesley, 1980).

101 Sharon Hill, "Want to Shed the Pseudoscience Label? Try Harder," May 23, 2011, Doubtful, http://idoubtit.wordpress.com/2011/05/23/want-to-shed-the-pseudoscience-label-try-harder/ (accessed January 26, 2012).

102 Marcello Truzzi, "On the Extraordinary: An Attempt at Clarification," *Zetetic Scholar* 1 (1978): 11; "Carl Sagan on Alien Abduction," *Nova*, February 27, 1996, http://www.pbs.org/wgbh/nova/space/sagan-alien-abduction.html (accessed December 22, 2011).

103 Benjamin Radford, *Tracking the Chupacabra: The Vampire Beast in Fact, Fiction, and Folklore* (Albuquerque: University of New Mexico Press, 2011); Radford and Nickell, *Lake Monster Mysteries.*

104 Michael Shermer, *The Believing Brain: From Ghosts and Gods to Politics and Conspiracies: How We Construct Beliefs and Reinforce Them as Truths* (New York: Holt, 2011).

105 Daniel Cohen, *Myths of the Space Age* (New York: Dodd, Mead, 1967), 5.

106 Ray Hyman, *The Elusive Quarry: A Scientific Appraisal of Psychical Research* (New York: Prometheus Books, 1989), 446–447. A well-known critic of research into psychic phenomena, Hyman was a co-founder of CSICOP, the first major skeptical organization in the United States, in 1976.

107 Ibid.

108 Daniel Loxton, "Skeptics as Model Train Lovers," September 17, 2010, Skepticblog, http://www.skepticblog.org/2010/09/17/skeptics-as-model-train-lovers/ (accessed February 20, 2012).

109 Daniel Loxton, "LogiCon 2011 Keynote Available Now," July 1, 2011, Skepticblog, http://www.skepticblog.org/2011/07/01/logicon-2011-keynote-available-now/ (accessed February 20, 2012).

110 Durrell, introduction to Heuvelmans, *On the Track of Unknown Animals*, 20.

111 Celia Dugger, "Study Cites Toll of AIDS Policy in South Africa," *New York Times*, November 25, 2008, http://www.nytimes.com/2008/11/26/world/africa/26aids.html (accessed February 21, 2012).

112 Prothero, *Evolution*, 24–49.

113 California Academy of Sciences, "American Adults Flunk Basic Science: National Survey Shows Only One-in-Five Adults Can Answer Three Science Questions Correctly" [press release], February 25, 2009, http://www.calacademy.org/newsroom/releases/2009/scientific_literacy.php (accessed February 23, 2012).

114 National Science Foundation, Science and Engineering Indicators 2012, Appendix table 7-9: Correct Answers to Factual Knowledge Questions in Physical and Biological Sciences, 1985–2010," http://www.nsf.gov/statistics/seind12/append/c7/at07-09.pdf (accessed February 23, 2012).

115 Jon D. Miller, Eugenie C. Scott, and Shinji Okamoto, "Public Acceptance of Evolution," *Science* 313 (2006): 765–766.

116 National Science Foundation, Science and Engineering Indicators 2012, Appendix table 7-9.

117 "Galileo Was Wrong; the Church Was Right," www.galileowaswrong.blogspot.com/p/summary.html (accessed January 26, 2012).

118 Steve Crabtree, "New Poll Gauges Americans' General Knowledge Levels," July 6, 1999, Gallup News Service, http://www.gallup.com/poll/3742/New-Poll-Gauges-Americans-General-Knowledge-Levels.aspx (accessed February 23, 2012).

119 Eugenie Scott, "The Creation/Evolution Continuum" December 7, 2000, National Center for Science Education, http://ncse.com/creationism/general/creationevolution-continuum (accessed February 27, 2012).

120 Miller, Scott, and Okamoto, "Public Acceptance of Evolution."

121 "Canadians Believe Human Beings Evolved over Millions of Years," August 5, 2008, Angus Reid Public Opinion, http://www.angus-reid.com/wp-content/uploads/archived-pdf/2008.08.05_Origin.pdf (accessed February 22, 2012); Brian Alters, Anila Asghar, and Jason R. Wiles, "Evolution Education Research Centre," in "Darwin and the Evangelicals," *Humanist Perspectives*, no. 154 (2005), http://www.humanistperspectives.org/issue154/EERC.html (accessed January 26, 2012).

122 Organisation for Economic Co-operation and Development, "What Students Know and Can Do: Student Performance in Reading, Mathematics and Science," PISA 2009 Results: Executive Summary, http://www.oecd.org/dataoecd/54/12/46643496.pdf (accessed February 22, 2012).

123 Amanda Paulson, "New Report Ranks U.S. Teens 29th in Science Worldwide," December 5, 2007, *Christian Science Monitor*, http://www.csmonitor.com/2007/1205/p02s01-usgn.html (accessed January 26, 2012).

124 Richard Allen Green, "China Shoots Up Rankings as Science Power, Study Finds," March 29, 2011, CNN World, http://articles.cnn.com/2011-03-29/world/china.world.science_1_china-output-papers?_s=PM:WORLD (accessed January 26, 2012).

125 "Knowledge, Networks and Nations: Final Report," March 28, 2011, Royal Society, http://royalsociety.org/policy/projects/knowledge-networks-nations/report/ (accessed January 26, 2012).

126 Quoted in "New Countries Emerge as Major Players in Scientific World," March 28, 2011, Royal Society, http://royalsociety.org/news/new-science-countries/ (accessed January 26, 2012).

127 Richard Matthews, "China's Green Innovation and the Challenge for America," February 23, 2011, Global Warming Is Real, http://globalwarmingisreal.com/2011/02/23/chinas-green-innovation-and-the-challenge-for-america/ (accessed January 26, 2012).

128 Fiona Harvey, "UK Slips Down Global Green Investment Rankings," March 28, 2011, *Guardian*, http://www.guardian.co.uk/environment/2011/mar/29/uk-global-green-investment-rankings (accessed January 26, 2012).

129 "Germany Seventh in G-20 Members in Clean Energy Investments," March 25, 2010, Pew Environment, http://www.pewenvironment.org/uploadedFiles/PEG/Newsroom/Press_Release/Germany%20Seventh%20Among%20G-20%20Members%20in%20Clean%20Energy%20Investments.pdf (accessed January 26, 2012).

130 Jürgen Schmidhuber, "Evolution of National Nobel Prize Shares in the 20th Century," September 14, 2010, Dalle Molle Institute for Artificial Intelligence, idsia.ch/~juergen/nobelshare.html (accessed January 26, 2012).

131 Neil deGrasse Tyson, "Scientifically Literate See a Different World," September 22, 2009, Curious Cat Science and Engineering Blog, http://engineering.curiouscatblog.net/2009/09/22/neil-degrasse-tyson-scientifically-literate-see-a-different-world/ (accessed January 26, 2012).

132 Carl Sagan, *The Demon-Haunted World: Science as a Candle in the Dark* (New York: Random House, 1996), 26.

NUMBERS IN ITALICS REFER TO PAGES ON WHICH FIGURES APPEAR.